# Modern Time Series Analysis with R



Practical forecasting and impact estimation with tidy, reproducible workflows

**Dr. Yeasmin Khandakar**
**Dr. Roman Ahmed**

<packt>

# Modern Time Series Analysis with R



**Portfolio Director:** Sunith Shetty
**Relationship Lead:** Nilesh Kowadkar
**Project Manager:** Shashank Desai
**Content Engineer:** Gowri Rekha
**Technical Editor:** Seemanjay Ameriya
**Copy Editor:** Safis Editing
**Indexer:** Pratik Shirodkar
**Proofreader:** Gowri Rekha
**Production Designer:** Salma Patel
**Growth Lead**: Merlyn M Shelley

First published: February 2026

Production reference: 050226

Published by Packt Publishing Ltd.
Grosvenor House
11 St Paul's Square
Birmingham
B3 1RB, UK.

ISBN 978-1-80512-401-6
`www.packtpub.com`

*We would like to dedicate this book to our parents and Mahdiya and Safiyya.*

# Foreword

I have had the privilege of knowing **Dr. Yeasmin Khandakar** and **Dr. Roman Ahmed** for more than 20 years, during which time I have witnessed their remarkable contributions in implementing time series and forecasting methods across many organizations. They have also both substantially influenced the field of forecasting; Yeasmin as coauthor of the Hyndman-Khandakar algorithm, which underpins almost all automatic ARIMA modeling software; Roman as coauthor of the first forecast reconciliation method, now a standard approach in hierarchical and grouped time series forecasting.

What distinguishes Yeasmin and Roman from many academics and practitioners in this field is their wealth of hands-on experience across diverse business contexts. Between them, they have navigated the challenges of forecasting and time series analysis in retail, finance, telecommunications, technology, and transport sectors. This breadth of experience has given them unique insights into the practical realities that forecasting practitioners face daily: messy data, tight deadlines, stakeholder expectations, and the need to balance statistical rigor with business pragmatism.

*Modern Time Series Analysis with R* is a synthesis of their combined expertise. They understand that practitioners often struggle with knowing which approach to apply to a given business problem, how to prepare data appropriately, and how to communicate results effectively to non-technical stakeholders.

The book's structure reflects this practical orientation. It begins with accessible foundations in R programming before progressing through data wrangling, visualization, and a wide spectrum of forecasting techniques, from simple methods to more advanced machine learning approaches. Importantly, the authors don't just explain *how* to implement these methods; they provide crucial guidance on *when* and *why* to use each approach, drawing on their extensive experience to help readers avoid common pitfalls.

Novice forecasters often underappreciate the importance of data preparation, data visualization, and exploratory data analysis before launching into fitting the latest model. This book addresses these critical steps thoroughly, ensuring readers develop a solid understanding of their data before moving on to model building.

For business professionals who find themselves needing to deal with time series analysis and forecasting problems, this book will prove a valuable resource. Whether you're an analyst preparing monthly forecasts for board meetings, a data scientist building production forecasting pipelines, or a manager seeking to understand the capabilities and limitations of different approaches, you will find practical guidance tailored to your needs. The emphasis on reproducible workflows, clear code examples, and real-world context makes this book a great reference for anyone applying time series analysis in business settings.

*Rob J Hyndman*

*Professor of Statistics*

*Department of Econometrics and Business Statistics*

*Monash University*

# Contributors

## About the authors

**Dr. Yeasmin Khandakar** is a data scientist with over 15 years of experience across diverse sectors, including FinTech (Portland House Group), MedTech (Optalert), retail (Coles, Officeworks) and transport (Transurban). She has a PhD from Monash University and is the co-author of the paper Automatic time series forecasting: the forecast package for R, which has generated over 5,900+ citations. Dr. Khandakar specializes in solving strategic business challenges by integrating advanced statistical methods with machine learning and deep learning, and robust time-series techniques.

**Dr. Roman Ahmed** is an experienced statistician with a PhD specializing in time-series forecasting. He has more than two decades of experience across the corporate and academic sectors. With a career including prominent technical leadership at Optus, Xero, and ANZ Bank, he excels at applying high-impact forecasting, econometric, and machine learning solutions to business strategy. Roman has published methodological and applied research in top-tier journals and has presented work at prestigious conferences. His expertise lies in translating sophisticated methodological research into scalable, real-world tools, particularly within the R ecosystem.

*The authors would like to thank the technical reviewer, whose constructive suggestions and diligent oversight were instrumental in refining the manuscript into its final form. We extend our acknowledgement to the editorial team at Packt for coordinating such a thorough review process. The authors extend deep gratitude to everyone from Packt who was involved in various stages of the book. We would like to mention a few names: Tejashwini R (for highlighting the potential of this book), Farheen Fathima (for supporting the initial phase in the manuscript development), Hemangi Lotlikar (for being patient while the authors were taking time for content development), and Gowri Rekha (for serving as the diligent catalyst needed to reach the finish line).*

*The authors wish to acknowledge those colleagues and peers who have sought to integrate time-series analysis into their workflow but found themselves hindered by the continuous learning and implementation challenges. We will consider this book a success if it helps them navigate these obstacles and appreciate the elegance and utility of time-series analysis in their professional lives.*

*Yeasmin wishes to express gratitude to her employer, Transurban, for their support and flexibility that were instrumental during the initial development of the manuscript.*

*Finally, both authors extend their deepest gratitude to their respective families for their unwavering support. This journey was long and often required the suspension of cherished routines, disrupted weekends, and sacrificed holiday plans. Without their patience and encouragement, this work would not have come to fruition.*

## About the reviewers

**Rohit Choudhary** holds a master's degree in management science from the University of Texas at Dallas and has over a decade of experience in statistics, forecasting, and data analytics. He has worked across diverse domains, translating complex data into actionable business insights and driving data-informed decision-making. Rohit specializes in predictive modeling, performance measurement, and analytical strategy, with a strong passion for applying advanced analytics to solve real-world business problems. He continues to explore innovative approaches in data science and business intelligence.

**Rahul Singh** is a data science manager at Adobe, where he works at the intersection of product data, customer behavior, and data science to solve complex business problems. His experience spans customer value measurement, in-product personalization, applied machine learning, LLM-enabled systems, and understanding how users and AI agents interact through conversational and agentic traffic data. He regularly speaks and writes about how the data science field is evolving with the rise of AI agents and how people and technology can work together to support better decision-making. He also serves as an AI advisor to Zindi, supporting efforts to grow the data science ecosystem across Africa.

# Table of Contents

## Chapter 11: Exponential Smoothing 305



## Chapter 12: ARIMA Forecasting Models 339

## Chapter 14: Forecasting Models for Multiple Time Series 425



## Part 5: Plot Twists and Turning Points: Measuring Impact and Detecting Anomalies 471

## Chapter 15: Causal Impact Estimation 473

## Chapter 17: Anomaly Detection and Imputation 559



## Chapter 18: Unlock Your Exclusive Benefits 583



## Other Books You May Enjoy 588

## Index 591

# Preface

Welcome to the book *Modern Time Series Analysis with R*.

In today's world, we are overwhelmed by choice. An abundance of data and rapid technological advancements driven by both academic research and industry innovation are available at our fingertips. While collecting data has become remarkably inexpensive, performing the precise analysis needed to solve complex business problems has never been more challenging. With a wide array of methodologies and limited time to master them, practitioners often lack the critical skills to select the right approach. This book bridges the gap in a time series context by showing you how to assess business research problems, evaluate data characteristics, and determine the analytical roadmap required to extract value from time series data.

Time series analysis is a mature field of scientific literature, with the availability of resources serving the needs of practitioners in both academia and industry. Methods and models for time series analysis have been implemented in both open source and proprietary software for decades, and some of the best-known libraries are in the open-source space. Despite these facts, the adaptation of time series analysis is lagging in the industry.

The rise of **machine learning** (**ML**) in the early 2000s gave businesses an opportunity to harness the power of computing in the business decision-making process. Major enterprise platforms, such as Google BigQuery (BigQuery ML to ARIMA_PLUS), and Databricks (MLflow to AutoML for Time Series), had a couple of years' gap in adding dedicated time series forecasting to their respective automated ML suites. This again highlights the need for specialized treatment for time series using ML workflows and perhaps also sheds light on the benefits and value of using time-series-based techniques in the business domain.

From our experience across retail, finance and banking, telco, technology, and consulting, we realized that the challenge is not the availability of resources or tools. To us, the challenges are two-fold: (i) a gap in translating business problems into a time-series context, (ii) awareness of time-series-based techniques that could solve business problems with less complexity and cost.

To navigate this complexity, we leverage R, a language built by and for data professionals, which offers an unparalleled ecosystem for temporal data. The journey begins with the foundational building blocks of R, ensuring you have a stable environment to execute analysis and run examples presented throughout the book. We cover essential setup – from installation to mastering basic R functions and the mechanics of data input/output.

From this solid base, we move on to time series. We describe their characteristics and explain why they deserve special analytical treatment by highlighting their differences from other data types. We follow a "tidy" workflow of time-series data wrangling, transforming raw timestamps into structured formats. Through a visual-first approach, you will learn about specialized plotting techniques for time series to reveal hidden patterns. You will also learn about the wide array of time-series-based techniques to solve practical business problems.

The next step is the essential, often under-appreciated, stage of data preparation and pre-processing. We discuss techniques to adjust, transform, and decompose time-series data. You will learn the art of distilling raw data into meaningful descriptors by performing time-series feature extraction, alongside advanced filtering and smoothing methods to isolate the true underlying signal from the noise.

Building on these skills, the book explores the full evolution of modern forecasting, from classical statistical methods, such as ETS and ARIMA, to ML and deep learning. We tackle high-dimensional challenges such as hierarchical forecasting and multiple time series methods to ensure consistency across complex business structures.

Beyond forecasting, we investigate the "why" and "when" of data shifts by estimating causal impact. We discuss time-series-based methodologies to analyze business intervention data where traditional A/B testing is not applicable. You will learn how to identify critical turning points through changepoint analysis and anomaly detection. Ultimately, this book will equip you with the rigorous methodology and technical command necessary to transform volatile temporal data into a cornerstone of evidence-based decision-making.

# Who this book is for

This book is uniquely written by developers for developers. It moves beyond abstract theory to provide a practical manual for those tasked with building, deploying, and interpreting time-series models in high-stakes environments. We focus on the critical "missing link" in data science: the ability to connect complex business problems with the correct analytical solutions.

We cater to three distinct levels of professional expertise:

- **Early career practitioners**: For those just beginning their journey, this book provides a rigorous introduction to time-series analysis. It serves as a foundational roadmap, transforming you from someone who writes code into someone who designs robust, reproducible analytical workflows.

- **Mid-career professionals**: For analysts and engineers, we focus on methodological precision. You will learn how to navigate the nuances of time series data pre-processing, the selection of appropriate forecasting models, and conducting proper analysis of business interventions – ensuring your outputs are not just technically sound, but strategically relevant to your organization's objectives.
- **Expert leaders and researchers**: For senior leads, architects, and other data professionals, this book serves as a high-level strategic guide for building time-series-based solutions. A defining feature of this book is that for every methodology discussed, we explicitly highlight the rationale (why a specific model is the superior choice for given data characteristics) and the application (the precise business problems the model is designed to solve). It will empower business leaders by providing the methodological rigor required to lead teams toward evidence-based decision-making.

# What this book covers

*Chapter 1, R, Rstudio, and R Packages,* details the ways of installing R, RStudio IDE, and R packages on different operating platforms. It also includes guidelines for setting up the R ecosystem for seamless development.

*Chapter 2, Objects and Functions in R,* provides a foundational guide to R programming for beginners, detailing the properties of base and S3 objects and core data types, and demonstrating how to build custom functions using control flow and iterative techniques.

*Chapter 3, Data Input/Output in R,* provides a comprehensive guide to data input and output in R, detailing methods for importing and exporting files from different sources while demonstrating how to securely connect to relational databases, execute SQL queries, and maintain rigorous data governance standards.

*Chapter 4, Time Series Characteristics,* provides a formal introduction to the unique characteristics of time series data, distinguishing it from other time-indexed formats using the N-T decision space while detailing its fundamental components—trend, seasonality, and cycles—and explaining the diagnostic importance of stationarity and autocorrelation.

*Chapter 5, Time Series Data Wrangling and Visualization,* demonstrates how to parse date-time variables, transform data into tsibble objects, and apply specialized plotting techniques—such as seasonal, subseries, and lag plots—to identify underlying patterns.

*Chapter 6, Business Applications of Time Series Analysis,* presents an overview of business applications for time series analysis, categorizing problem domains into the study of inherent characteristics (including trend, seasonality, and anomaly detection), inference and attribution modeling, causal impact estimation for evaluating interventions, and predictive forecasting to support strategic decision-making.

*Chapter 7, Time Series Adjustments, Transformations, and Decomposition*, presents a comprehensive guide to time series pre-processing, detailing how to apply adjustments to correct for external biases, mathematical transformations such as Box-Cox to stabilize variance, and decomposition frameworks such as STL and X-13ARIMA to isolate inherent trend and seasonal components from random noise.

*Chapter 8, Time Series Features*, explores the extraction of diverse time series features—including descriptive statistics, autocorrelation profiles, STL-based strength measures, and windowed dynamics—to quantify temporal characteristics and facilitate advanced analytical tasks such as dimensionality reduction and the unsupervised clustering of large datasets.

*Chapter 9, Time Series Smoothing and Filtering*, explores techniques to reduce random noise and isolate meaningful signals through time series smoothing and filtering, detailing the application of **Exponentially Weighted Moving Averages** (**EWMA**) for quality control, linear filtering to manage error-term dependencies, the Hodrick-Prescott filter for separating long-term trends from business cycles, and the Kalman filter for recursively estimating latent states within complex dynamic systems

*Chapter 10, Basics of Forecasting*, establishes the foundational tidy forecasting workflow with the fable and fabletools R packages, detailing the implementation of baseline methods such as naïve, seasonal naïve, and average forecasting alongside **Time Series Linear Models** (**TSLM**) for scenario analysis, while explaining rigorous point forecast accuracy metrics and time series cross-validation techniques to ensure model generalizability.

*Chapter 11, Exponential Smoothing*, provides a comprehensive guide to the exponential smoothing family of forecasting, meticulously distinguishing between procedural methods and state-space models while detailing specific applications for simple exponential smoothing, Holt's linear trend, and Holt-Winters patterns alongside automated model selection leveraging information criteria.

*Chapter 12, ARIMA Forecasting Models*, explores the ARIMA forecasting framework, detailing its core components of **autoregressive** (**AR**) dependencies, **integration** (**I**) through differencing to achieve stationarity, and **moving average** (**MA**) error modeling while demonstrating the implementation of **seasonal variations** (**SARIMA**) and the automated Hyndman-Khandakar algorithm for efficient model selection.

*Chapter 13, Advanced Computational Methods for Forecasting*, explores advanced computational methods for forecasting by reframing time series as supervised learning problems, detailing the implementation of **Neural Network Autoregressive** (**NNETAR**) models, ML workflows using the forecastML package with regularized regression, the Prophet curve-fitting method, and the strategic use of forecast ensembles to enhance predictive accuracy.

*Chapter 14, Forecasting Models for Multiple Time Series*, explores methodologies for forecasting multiple time series simultaneously, specifically detailing the application of **Vector Autoregression** (**VAR**) to capture bidirectional feedback loops and various reconciliation frameworks—such as top-down, bottom-up, and optimal trace minimization—to ensure consistency across hierarchical and grouped data structures.

*Chapter 15, Causal Impact Estimation*, provides a comprehensive framework for causal impact estimation, detailing the equivalence between hypothesis testing and regression while demonstrating the application of **Interrupted Time Series** (**ITS**) analysis to isolate the effects of business interventions within stationary, trended, and autocorrelated data using specialized methods such as ARIMAX and robust statistical adjustments.

*Chapter 16, Changepoint Detection*, provides a guide to changepoint detection for identifying sudden, persistent shifts in time series behavior, detailing the recursive BFAST framework for trend and seasonal breaks, the optimization-based PELT algorithm for exact segmentation, and the BEAST probabilistic approach for quantifying uncertainty in the location and magnitude of structural breaks.

*Chapter 17, Anomaly Detection and Imputation*, discusses a framework for identifying point and collective anomalies using STL-based and Isolation Forest methodologies while detailing the application of linear interpolation for imputing missing values to preserve the integrity and temporal structure of time series data.

# To get the most out of this book

This book is structured to serve as both a step-by-step tutorial and a long-term technical reference. Each chapter follows a logical progression, starting with theoretical underpinnings before moving on to hands-on implementation using R. For the most effective learning experience, we recommend the following:

- **A sequential path**: Beginners should sequentially progress through the chapters to establish a solid foundation in computing and knowledge.
- **An applied path**: Experienced practitioners may choose to dive directly into specific chapters, such as the ones about forecasting or causal impact estimation.

- **Interactive learning**: All code snippets are designed for reproducibility and follow a standardized structure. We encourage you to execute the code as you read and benefit from the detailed explanation of the results.
- **Leveraging the modern framework**: The code throughout the book adheres to the tidyverse style guide, ensuring that your scripts are readable, maintainable, and scalable. You will work extensively with the R packages, such as tidyverse, fable, and so on, which provide a unified and consistent interface for time-series analysis, making it easier to transition between modeling techniques.

# Download the example code files

The code bundle for the book is hosted on GitHub at `https://github.com/PacktPublishing/Modern-Time-Series-Analysis-with-R.git`. We also have other code bundles from our rich catalog of books and videos available at https://github.com/PacktPublishing. Check them out!

# Download the color images

Your purchase includes a color, DRM-free PDF copy of this book, ideal for viewing color images, screenshots, and diagrams. Refer to the *Free benefits with your book* section at the end of the *Preface* to unlock your PDF copy.

# Conventions used

There are a number of text conventions used throughout this book.

`CodeInText`: Indicates code words in text, database table names, folder names, filenames, file extensions, pathnames, dummy URLs, user input, and Twitter handles. For example: "You have already used functions from this package, such as `ACF()`, `gg_season()`, `gg_subseries()`, and so on, in previous chapters."

A block of code is set as follows:

```
dimnames(mat_b) <- list(
paste0("row_",1:3), paste0("col_",1:3))
mat_b
```

Any command-line input or output is written as follows:

```
      col_1 col_2 col_3
row_1  TRUE  TRUE FALSE
row_2  TRUE  TRUE FALSE
row_3  TRUE  TRUE FALSE
```

**Bold**: Indicates a new term, an important word, or words that you see on the screen. For instance, words in menus or dialog boxes appear in the text like this. For example: "The **exponentially weighted moving average** (**EWMA**) method is applicable in such a scenario."

> **Note**
>
> Warnings or important notes appear like this.

> **Tip**
>
> Tips and tricks appear like this.

# Get in touch

Feedback from our readers is always welcome.

**General feedback**: If you have questions about any aspect of this book or have any general feedback, please email us at `customercare@packt.com` and mention the book's title in the subject of your message.

**Errata**: Although we have taken every care to ensure the accuracy of our content, mistakes do happen. If you have found a mistake in this book, we would be grateful if you reported this to us. Please visit , click **Submit Errata**, and fill in the form.

**Piracy**: If you come across any illegal copies of our works in any form on the internet, we would be grateful if you would provide us with the location address or website name. Please contact us at `copyright@packt.com` with a link to the material.

**If you are interested in becoming an author**: If there is a topic that you have expertise in and you are interested in either writing or contributing to a book, please visit `http://authors.packt.com/`.

# Free benefits with your book

This book comes with free benefits to support your learning. Activate them now for instant access (see the "*How to Unlock*" section for instructions).

Here's a quick overview of what you can instantly unlock with your purchase:

**DRM-Free PDF Version**

Download DRM-free PDF and ePub copies of this book.

**7-Day Packt Library Access**

Get 7-day unlimited access to 8,000+ books and videos. No credit card required.

Available for first-time Packt+ trial users only.

**Next-Gen Reader Access**

Read this book on Packt Reader with progress sync, dark mode and note-taking.

## How to Unlock

Scan the QR code (or go to `packtpub.com/unlock`). Search for this book by name, confirm the edition, and then follow the steps on the page.

UNLOCK NOW

*Note: Keep your invoice handy. Purchases made directly from Packt don't require one*

## Share your thoughts

Once you've read *Modern Time Series Analysis with R*, we'd love to hear your thoughts! Scan the QR code below to go straight to the Amazon review page for this book and share your feedback.

`https://packt.link/r/1805124013`

Your review is important to us and the tech community and will help us make sure we're delivering excellent quality content.

# Part 1

# Setting the Scene: R

*The most challenging thing in the world is not to learn fancy technologies but control your own wild heart.*

- Yuhui Xie

The contents presented in this part are essential to build a solid foundation on the tools required for time series analysis in R. This part of the book is more suitable for readers who are new to R. We have tried to present minimal R requirements that will support developers to perform time series analysis in a business setting.

*Part 1* comprises the following chapters:

- *Chapter 1, R, Rstudio, and R Packages*
- *Chapter 2, Objects and Functions in R*
- *Chapter 3, Data Input/Output in R*

# 1

# R, Rstudio, and R packages

This chapter focuses on the tools required to setup the R programming language. Before diving into time series analysis using the R language, the initial steps involve installing and configuring the software and its associated components. The R computational engine comes with companions such as RStudio and a vast collection of packages. Each component needs to be installed separately before it can be used coherently.

To ensure clarity and ease of use, the chapter is organized into the following sections:

- Section 1.1: Introduction to the R ecosystem
- Section 1.2: Downloading R and RStudio
- Section 1.3: Installing R and RStudio with admin access
- Section 1.4: Installation in a custom location
- Section 1.5: Installing R packages
- Section 1.6: Loading installed packages
- Section 1.7: Updating software and packages
- Section 1.8: How to choose R packages
- Section 1.9: RStudio for productive programming
- Section 1.10: The tidyverse package collection

Note

**Your purchase includes a free PDF copy + exclusive extras**

Your purchase includes a DRM-free PDF copy of this book, 7-day trial to the Packt+ library (no credit card required), and additional exclusive extras. See the *Free benefits with your book* section in the *Preface* to unlock them instantly and maximize your learning.

## 1.1 Introduction to the R ecosystem

Before going into the technical part, let's start with understanding the basics of the R programming ecosystem by answering three initial questions:

- What is R?
- What is RStudio?
- What are R packages?

### 1.1.1 What is R?

**R** is simultaneously used to indicate a programming language and an environment for statistical computing and graphics. It was developed as a coherent system for statistical analysis with a diverse collection of integrated intermediate tools for data analysis. The first release of R came in 2000 from the R Core Group as an open source project. The creators of R embedded the absolute concept of 'language' and 'environment' in a software suite. From the beginning until now one motto has been maintained—whatever is implemented in R, the users retain full control. This is highly appreciated by the R users. However, the *anything can be modified after it is created* (`Wickham, 2014`) attitude can be daunting to new users or users coming from a different programming language. It does make R somewhat a quirk as a programming language. Though what is a language without a bit of quirk in it anyway?

### 1.1.2 What is RStudio?

**RStudio** is an **Integrated Development Environment** (**IDE**) that traditionally used R as a computation engine. It provides a user-friendly interface with a collection of tools for data analysis, visualization, automated report writing, debugging, and workspace management. Over time, RStudio expanded its capabilities to support other programming languages such as Python, SQL, and more. You can also create a Jupyter notebook or use code editing software such as Visual Studio Code (VSCode). This not only enhances the capabilities of RStudio itself but also provides users with options to use multiple languages in one data science project to take collaboration to the next level.

In recent developments, the Positron IDE has emerged as a modern alternative, designed to offer a more flexible and language-agnostic environment for data science workflows. It supports seamless integration with R, Python, and other tools, and is built to enhance productivity in collaborative and reproducible research. Throughout this book, we will primarily utilize RStudio; however, the fundamental concepts and workflows presented are readily available and transferable to Positron.

RStudio offers multiple products. The open source RStudio Desktop can run locally and is free. RStudio Desktop Pro is commercially licensed for organizations. There are enterprise-level products such as Posit Team, Posit Workbench and Posit Connect, which organizations could use depending on their requirements and the maturity of their data science journey. You can find the product description on Posit site `https://posit.co/products/`.

This book references the free version of RStudio Desktop which allows you to run code locally without incurring any costs. Unless otherwise specified, the term RStudio in this book refers specifically to RStudio Desktop.

### 1.1.3 What are R packages?

**R packages** are extensions of the base R programming language and usually contain a collection of functions, methods, and algorithms to solve problems. R packages are written by a wide community of R users around the globe that comes from all avenues of expertise including both academia and industry. Usually, an R package is focused on solving a particular problem. For example, `dplyr` (`Wickham et al., 2023`) provides functions for data manipulation, `ggplot2` (`Wickham, 2016`) implements the grammar of graphics for statistics and computing, and the `forecast` package (`Hyndman et al., 2025`) is developed to visualize and analyze univariate time series with particular focus on forecasting. The vast collection of R packages will help you solve problems at hand without reinventing the research and development wheel.

Now that you have a foundational understanding of R, RStudio, and R packages, the remainder of this chapter will guide you through accessing and installing the required software across different operating systems.

## 1.2 Downloading R and RStudio

Before delving into using R for problem-solving, we must install and configure it. First, download the most up-to-date version of R and RStudio from the specified website that is compatible with your operating system. RStudio requires a 64-bit architecture for both software. If your machine is 32-bit you can download an older version of R and RStudio using the links provided on the respective sites:

- Download base R from `https://cloud.r-project.org/` and RStudio Desktop from `https://posit.co/download/rstudio-desktop/`.
- Alternatively, download both base R and RStudio desktop from `https://posit.co/download/rstudio-desktop/`.

R can be downloaded from other **Comprehensive R Archive Network** (CRAN) mirrors hosted closer to your location. The list is available at `https://cran.r-project.org/mirrors.html`.

> **Note**
>
> Always install R first, then RStudio.

After you have downloaded the software, let's move ahead to install it.

## 1.3 Installing R and RStudio with admin access

If you have admin access to your machine, the installation method is pretty much a download and click method on Windows and macOS. For Linux-based systems, there are three slightly different ways based on the flavors of the system: Debian, Ubuntu, or Fedora/Red hat. It is important to note that you need to read and follow the instructions that are relevant to your operating system.

### 1.3.1 Windows

By default, R and RStudio install under `C:\Program` Files. This creates relevant folders for it to get installed in. For package installation, R uses `thelibrary` folder inside the R installation. Follow these steps if you have admin access to the root `C:\` folder:

1. Double-click the downloaded `.`exe file for R.
2. Follow the instructions as they appear in the prompt to install R.
3. Double-click the downloaded `.`exe file for RStudio and follow the instructions to install.

4. Open the **RStudio** icon. Depending on the version of RStudio, it may ask you to choose a version of R.
5. In the RStudio console run (type and hit Enter) following commands:

```
R.home()
Sys.getenv("R_HOME")

Sys.getenv("HOME")
path.expand("~")
getwd()

.libPaths()
```

These commands point to the folder structure used during installation, such as the following:

- `R.home()` and `Sys.getenv("R_HOME")` return the location where R is installed; this will be `C:/Program Files/R/R-4.2.3`.
- `Sys.getenv("HOME")`, `path.expand("~")`, `getwd()` returns the user's home directory, typically `C:/Users//Documents`. It is also the default working directory for R. You can use the shorthand ~ character instead of writing the full path while coding.
- `.libPaths()` returns the default location for package installation, which will be `C:/Program Files/R/R-4.2.3/library`.

In the next section, you will learn how to perform the installation process on macOS.

## 1.3.2 macOS

When downloading R and RStudio for macOS, make sure to choose the correct architecture that matches your system: Apple Silicon or Intel. The installer for R and RStudio has the .pkg and .dmg extensions, respectively:

1. Click the downloaded `.pkg` file.
2. Follow the instructions and prompts (selects Next multiple times) to install R in the default location.
3. For RStudio, double-click the `.dmg` file. A folder will open with the **Application** and **RStudio** icons next to each other. Drag the **RStudio** icon to the **Application** folder to install it.

4. Navigate to the **Application** folder to check that both R and RStudio are available to use.
5. Open **RStudio** and type the following commands in the console to see the **installation** folder, **home** folder, and package installation path.

```
R.home()
Sys.getenv("HOME")
.libPaths()
```

The following section documents the installation process for Linux users.

### 1.3.3 Linux

On Linux systems, the steps are as follows:

1. Install a platform-specific system package manager.
2. Add the CRAN repository in the `/etc/apt/sources.list` file along with the Linux distribution name.
3. Install R via the linux system package manager and `apt` tool.

If you have writing permission, then you can omit `sudo` from the lines of codes provided in the following system-specific scenarios.

#### 1.3.3.1 Ubuntu

These steps are explained for Ubuntu Jammy Jellyfish (22.04, AMD64). The steps are applicable for any Ubuntu builds by replacing `jammy` or the build name where used:

1. Install Ubuntu's helper packages to install R.

```
sudo apt update -qq
sudo apt install --no-install-recommends software-properties-common dirmngr
```

2. Add CRAN repo in the `/etc/apt/sources.list` by typing below in the terminal.

```
sudo add-apt-repository "deb https://cloud.r-project.org/bin/linux/ubuntu $
(lsb_release -cs)-cran40/"
```

The `lsb\_release -cs` replaces the name of the Ubuntu build, such as `jammy` or `focal`.

Alternatively, open the `/etc/apt/sources.list` file in a text editor (such as `vim` or `nano`) and manually add `deb \ https://cloud.r-project.org/bin/linux/ubuntu jammy-cran40/` into it.

3. Install R and its dependencies by running the following code in the terminal:

```
sudo apt install --no-install-recommends r-base r-base-core r-recommended
r-base-dev
```

4. Test the installation by typing R in the terminal. Check the `R.home()`, `Sys.getenv()` and `.libPaths()` within R. Are you surprised there are three different folders in `.libPaths()`? We will learn more about this in the R package pathways subsection.
5. To install RStudio, first download it from the Posit site (previously `rstudio.org`) via the terminal:

```
wget https://download1.rstudio.org/electron/jammy/amd64/
rstudio-2023.03.0-386-amd64.deb
```

6. Then, install it after the download is complete:

```
sudo apt install -f ./rstudio-2023.03.0-386-amd64.deb
```

7. Launch installed RStudio either from `Application -> RStudio` or via the terminal:

```
rstudio
```

#### 1.3.3.2 Debian

At the time of writing, Debian Bookworm is the stable version; however, you can replace to other Debian version in the terminal commands when a new one is available in future. The following commands is run in the terminal:

1. Search `R add-ons` in the Debian package manager:

```
apt search "^r-.*" | sort
```

2. Add the CRAN entry to the `/etc/apt/sources.list.d` file directory and import the GPG key securely:

```
sudo apt update

sudo apt install --no-install-recommends software-properties-common dirmngr
gnupg ca-certificates
```

```
curl -sSL "keyserver.ubuntu.com" | sudo gpg --dearmor -o /usr/share/
keyrings/cran-archive-keyring.gpg

echo "deb [signed-by=/usr/share/keyrings/cran-archive-keyring.gpg] https://
cloud.r-project.org/bin/linux/debian bookworm-cran40/" | sudo tee /etc/apt/
sources.list.d/cran.list
```

3. Now, install R by running the following command:

```
sudo apt update
sudo apt install r-base r-base-core r-base-dev
```

   The `r-base-dev` is required when you compile R packages from sources that are not available as precompiled Debian or Ubuntu packages.

4. The RStudio installation is the same in Debian for Ubuntu, except you need to download the appropriate version of RStudio for Debian (e.g., Debian 12+):

```
wget https://rstudio.org/download/latest/stable/desktop/jammy/rstudio-
latest-amd64.deb -O rstudio-latest-amd64.deb

sudo apt install -f ./rstudio-latest-amd64.deb
rstudio
```

The `cran.r-project.org/bin/linux` site describes installing R on other Linux distributions, like Fedora, Red Hat, and SUSE. It is also a good resource if you have any issues during the installation process, such as a missing public key, and how to fix it.

While installing RStudio, Linux users may be restricted by the security policy of their operating system and need to import Posit's public code-signing key. If you have any issue in step 2, then navigate to `https://posit.co/code-signing/` and follow the instructions on importing the public key.

So far, you have learned how to install R suites into the default installation folder. Next you see how to install R and RStudio in a custom folder location.

# 1.4 Installation in a custom location

On a corporate-managed computer, you may not have admin privileges to install applications in system-level directories such as the root of the drive. While a system administrator can install R and RStudio in the default locations (e.g., `C:\Program` Files on Windows or `/Applications` on macOS), package installation may still fail if you don't have write access to R's default library folder. Raising a ticket to get support from an IT admin every time you install a package is not practical. Installing R and RStudio in a custom folder can solve the issue. You could also use a custom folder only for packages where the base R software can be installed in the default location.

Suppose that you have writing permission on the `D:\` drive in a Windows machine (or `/Users/<YourName>/` for macOS) and you can install R and RStudio there. First, you need to create a folder named R at this location (`D:\<Username>\R` for Windows or `/Users/<YourName>/R/` for macOS), and you need to use this path as a destination folder during the R and RStudio standard installation process. The same location can be used for installing R packages. Create a folder called `D:\<Username>\R\RPackages` (or `/Users/<YourName>/R/RPackages/` for macOS) for installing the personal package library.

The folder tree will look as follow on a Windows:

```
Username/R/
├── R-4.2.3/
├── RPackages/
└── RStudio/
```

Now, you can configure the destination folder for your package installation. Irrespective of whether you are using Windows or macOS, open RStudio and type `.libPaths()` in the console. It will print the default package installation folder path. A custom package installation location can be provided in the `.Rprofile` file. Type the following command in the console:

```
file.edit("~/.Rprofile")
```

The `.Rprofile` file from the default working directory will open, which is an empty file. Write the following snippet of code in it, save it, and close it. Replace the `<Username>` portion accordingly:

```
.First <- function() {
  path_to_package = "D:/<Username>/R/RPackages"
```

```
    .libPaths(c(path_to_package, .libPaths()))
}
```

Restart RStudio. You can check the change by typing `.libPaths()` in the console (the supplied folder name should appear as the first item) or opening `Packages->Install` from RStudio.

Now that we have learned how to install R and RStudio for various operating systems and situations, let's see how to install our packages.

# 1.5 Installing R packages

R packages are collections of functions, compiled code, and data bundled nicely together, hosted on CRAN (or GitHub) for you to download and use. At the time of writing this book, there are more than 19 thousand packages available on CRAN. There is no definite list of packages that are in the development stage hosted on GitHub. Looking at this staggering number of packages makes you realize how broadly and deeply the R community has grown over time. Each package is open source and free to use. Born as a statistical computing software, R has come a long way with the help of R packages.

The base R is shipped with about eight packages. You can get other packages from CRAN (the most stable version) or from GitHub (development version).

You need to install and load a package to use its capabilities in your R workspace:

- First, download and install the desired R package. Installing a package creates a package-specific folder inside the R installation directory, or in the custom directory (as described in *Section 1.4*) on your machine. This folder contains all binaries, code, and documentation for the specific package.
- Second, you need to call the package in the R working space to use it actively. There are two methods to install R packages: the RStudio menu and the R console. You will learn how to do both in the next two subsections.

## 1.5.1 Package installation via the RStudio menu

You can install any number of packages using RStudio's **Packages** menu. You can set the repository, see the package list as you type (a way to learn what other packages are available), and choose a location to install.

## 1.5.2 Package installation via R console

To install a package from CRAN, type `install.packages()` in the R console with the package name inside the parentheses enclosed in double quotes. The number of packages installed can vary from one to multiple in one execution:

```
# install a single package
install.packages("dplyr")

# install multiple packages
install.packages(c("DBI", "tibble", "dplyr",
                   "ggplot2", "fst", "knitr"))
```

You can specify a custom location on the fly by providing a path to the `lib` option within the `install.packages()` function if a custom package location has not already been set via `.Rprofile`.

Over time, you will have a list of R packages that you will always need. You will need to install those packages when you move to a new R project, update R, change machines, or move to another organization, and it is hard to remember which package is installed where. A way to install missing packages on the fly is using the `p_install()` or `p_load()` function from the `pacman` package which stands for "pac"kage "man"agement. The `p_install()` function installs missing packages, and `p_load()` functions installs and simultaneously loads the packages in the current workspace. Following code shows how to install `pacman` first (if not installed) and then use it to install other packages:

```
if (!requireNamespace("pacman", quietly = TRUE)) {
  install.packages("pacman")
}

pacman::p_load(DBI, tibble, dplyr, ggplot2, fst, knitr)
```

This code should be included at the beginning of any standalone R script or as a script in every R project. In this way you do not need to worry about code breaking due to missing packages when shared with collaborators.

On Linux, you can call the `install.packages()` function from within R. This works assuming that you have installed `r-base-dev` while installing R on the Linux system. First, call R from the terminal:

```
R
```

Then type the following within R:

```
install.packages("dplyr")
```

For the first time, it asks to create a personal library if the default installation path is not writable:

```
Installing packages into '/usr/local/lib/R/site-library'
(as 'lib' is unspecified)
Warning in install.packages("dplyr") :
'lib = "/usr/local/lib/R/site-library"' is not writable
Would like to use a personal library instead? (yes/No/cancel)
```

When you type yes, a folder of the home/<user>/R/<platform name>-library/<R version> type is created and asks for your permission to install into it. The <platform-name> is the Linux system name, such as x86-64-pc-linux-gnu, and the R-version is the version of R installed. This method compiles packages from source using the power of `r-base-dev`.

On Ubuntu you could also use `apt\ install` in the terminal after adding Ubuntu's R package repository `c2d4u`:

```
sudo add-apt-repository ppa:c2d4u.team/c2d4u4.0+
sudo apt install --no-install-recommends r-cran-dplyr
```

It may ask to create a user specific location for package installation when you don't have write permission in the default location. There are more than 5000+ R packages available in the `c2d4u` repo. Even though you can use this repo for any Ubuntu builds, the developers strongly suggest using the **Long Time Support** (**LTS**) release to receive the full benefit of the c2d4u repository.

You have seen that there are multiple methods to install R packages with their own folder hierarchy when installed. Is one method better than the others? Find the answer in the next section.

### 1.5.3 R package pathways

We already know that when an R package is installed, it creates a package-specific folder under a directory listed in `.libPaths()`. On Windows and macOS, all packages are installed in the library folder inside the R installation by default. A custom installation path can be used by assigning a folder as the first item in `.libPaths()`. All R packages are installed inside that folder can be found as `.libpaths()`[1].

On Windows and macOS there is no dominance in folder structure by package installation methods. Irrespective of the package installation methods, (either RStudio's Packages menu or using `install.packages()` from R console), all packages are installed under one folder.

On Linux systems, the R package pathways work slightly differently.

- `/usr/lib/R/library` contains binaries and libraries of R build, e.g., `r-base` and `r-recommends`
- `/usr/lib/R/site-library` contains binaries and libraries of precompiled Debian (or Ubuntu) R packages installed from `r-cran-*` and `r-bioc-*`
- `/usr/local/lib/R/site-library` contains binaries and libraries installed via `install.packages()` function inside R when admin/user has write permission in that folder
- `home/<user>/R/<platform-name>-library/<R-version>` contains binaries and libraries via `install.packages()` function within R when admin/user do not have write permission in `/usr/local/lib/R/site-library` folder.

The R packages installed from within R take precedence over the packages installed via the Linux platform's package management system, as shown in `the.libPaths()` hierarchy:

```
[1] "home/<user>/R/x86_64-pc-linux-gnu-library/4.1"
[2] "/usr/local/lib/R/site-library"
[3] "/usr/lib/R/site-library"
[4] "/usr/lib/R/library"
```

The folder structure indicates that you may need to look at different folders depending on the source of packages. One downside is that there may be multiple versions of one R package installed on your machine if you installed from different sources. In such cases, make sure to delete the version no longer in use to reduce complexity while coding.

To explore R packages that are in the development stage, use any of the methods listed in the next section to install them.

### 1.5.4 Installing R packages in development

There are several ways to install packages that are in development stage (not in CRAN yet) or a newer development version of a CRAN package. You can install from GitHub (`install_github()`), Bitbucket (`install_bitbucket()`), CRAN (`install_cran()`), Subversion (`install_svn()`) and many other sources. For example, the development version of the `purrr` package can be installed by using any of the following methods:

```
# using remotes package
remotes::install_github(repo = "tidyverse/purrr")

# using devtools package
devtools::install_github(repo = "tidyverse/purrr")

# using xfun package
xfun::install_github(repo = "tidyverse/purrr")

# using pacman package
pacman::p_load_gh(repo = "tidyverse/purrr")
```

The `remotes` package is the parent of the `install_*()` class of functions. The same functions in the `devtools` package allow users to pass additional arguments via the ... along with other base arguments, providing flexibility to users. The function in the `xfun` package provides faster auto-completion, allowing further flexibility to users. Just type `xfun::ig` in your RStudio console and see what happens. The fourth one, from pacman installs and loads simultaneously. Technically, you just need one of these functions that suits your purpose.

You might have noticed `::` between package name and the function name. This is used for calling a function via a namespace, that is, using the function from a package without loading the package. This is handy when you use only a limited number of functions from a package. It is also useful to avoid conflicts of function names coming from two different packages. For example, when `tidyverse` is loaded, you will notice that the `filter()` function from `dplyr` packages has a conflict with the `filter()` function from the `stats` package. The first one is used for data manipulation and the latter one is used for linear filtering of a time series. To avoid conflict and confusion you can call them via namespace; `dplyr::filter()` and `stats::filter()` when using both of them in one program/script.

After learning about several methods to install R packages, now it is time to load them into the R workspace.

# 1.6 Loading installed packages

After installation, you need to load/attach a package in your R workspace. The `library()` function is used for that. The function throws an error and stops working if the package is not installed, which is a good way to identify what is missing. You can also specify the custom location via the `lib.loc` argument if have not done via `.libPaths()` already:

```
# Load an installed package
library(dplyr)
```

You may have noticed that the `require()` function is also used to load packages. The main distinction is that the `require()` does not throw an error if a package is missing. In the worst case, you may have to spend hours figuring out why scripts that were working on your machine are not running on your colleague's machine. This was one of the reasons behind the development of the `pacman` package. When the `p_load()` function is used to simultaneously installs and loads packages in the current workspace, the use of the `library()` function for attaching packages becomes redundant.

After loading a package in the R workspace, you can use its' functionality while programming. However, it is also important to keep track of when an R package gets updated. In the next section you will find some best practices on how to update R suites without much hassle.

# 1.7 Updating software and packages

R, RStudio and R packages are continuously evolving due to updates, fixing bugs, adding new functionality and so on. It is good to be up to date with the new technologies. However, it will take a fair bit of time to install the full suite again if you decide to update every time a minor release is out, or a newer version of a single package is available. As each component is installed/updated separately, you can update only R, RStudio, or R packages without touching the other two. A commonly fair update time frame may look like this:

- Update R packages once a month
- Update R once or twice a year, only when a major release is out. Some advanced packages are built using the most recent R version. You have to update R to enjoy the full functionality of these packages in case your machine's version is older than what the packages are built on
- Update RStudio only when a major release is available

Usually, RStudio notifies you when a new release is available and the general dependency when a package is installed. You can get more information from the package-specific CRAN sites too.

> **Note**
>
> Remember to uninstall the previous version of R and RStudio before replacing with the newer version, and update the full list of packages when you update R.

## 1.7.1 Checking and updating R packages

The easiest way to check and update R packages is via RStudio Packages menu. Click **Packages-> Update -> Select All -> Install Updates**. All packages in the list have updated versions available. Hitting the **Install Update** button updates all packages wherever they are installed in your machine, either in the default or in a custom location.

Packages can be updated from the R console as well. First, make a list of the already installed packages and then re-install them using two lines of code given here:

```
# list of installed packages
pkg_list = as.data.frame(installed.packages(.libPaths()[1]),
                         stringsAsFactors = F)

# re-install packages
install.packages(pkg_list$Package)
```

## 1.7.2 Checking and updating R and RStudio

The availability of a recent RStudio update can be checked from the Help menu by clicking on **Help-> Check for Updates**.

For both R and RStudio, the simplest way is to uninstall the old version, download the newer one and follow steps for installing as you did for the first time. This method reduces the chance of conflicting setups when multiple R versions are present in one machine. You could save the content of an old `.RProfile` file, and copy-paste it into the new `.RProfile` file after re-installation is completed.

You could use a Windows specific package to update R from within R. Follow the instructions that appear after writing the following code snippet in the current R console:

```
install.packages("installr")
installr::updateR()
```

If a custom folder is used for packages, which is the standard case in Linux systems, make sure to update/re-install the full package list after R is updated/re-installed.

After installing the full R suites and some R packages of your choice, you are now ready to embark on the journey of learning about and using R packages. You may have noticed already that several R packages are available to do a single task. Which package is the best one? How are you going to choose an R package to learn from the pool? In the next sections, some key points are highlighted to make your choice of R packages (hence, the R learning) a smooth progression.

# 1.8 How to choose R packages

It is not simple to pick an R package (packages) given the availability of alternatives to choose from. If you are an experienced R user, there is a high chance you already have a list of packages that you always use. Keeping yourself updated with the existing packages and any new packages in your area of expertise is the key to your success.

For the newcomers, CRAN Task Views (`https://cran.r-project.org/web/views/`) is a great place to start. It provides guidance on package usage by grouping them under certain topics. For example, **Time Series Analysis Task Views** lists packages according to data type, modeling, functionality and applicability.

The short package description provides important information about the methods and algorithms used in the package. The function level descriptions and help file provide more detailed information about function usage and application to data.

You can use the **SMART approach** like any standard goal setting activity.

- **S**pecific:
    - What is my purpose for using packages: is it data wrangling, statistical analysis, forecasting, predictive modeling, visualization?
    - What can the package do that cannot be done in base R?
- **M**easurable:
    - Is it available on CRAN? For how long?
    - What algorithms and methods are used?
    - Do they appear in peer-reviewed journals, books, R-blogs, or other publications?
- **A**chievable:
    - Is there enough documentation available for me to learn?
    - Do I understand the methodologies and algorithms?

- **R**elevant:
    - Is it applicable to my problem?
    - Am I focused on a certain topic when searching for packages?
- **T**ime bound:
    - Can I apply the package to my problem in a reasonably quick time?

By answering these questions and researching package information, you will soon find some packages that will help solve your specific problems.

R packages are developed to make users' lives easy, productive, and efficient. The best way to learn about a package is by using it. The time you spend researching and learning about a package thoroughly will ultimately give you an edge in long term. When you properly understand a package, you gain knowledge on how to apply the methods to your problem, which provides the wisdom to identify the right situation to apply the package functionality in general.

Now, we move our focus to RStudio. You can use R without any development environment. RStudio was specifically built to work interactively with R, which has now been extended to other programming languages. In the next section, you will discover how RStudio can boost your productivity and efficiency as a programmer and/or data professional.

## 1.9 RStudio for productive programming

RStudio is an IDE with greater emphasis on the word "integrated". It does a lot of work that you may not notice under the hood to make you an efficient programmer. You will find plenty of documentation on using the various features of RStudio to reap the maximum benefit. The web page at `https://docs.posit.co/ide/user/` is a good start. This section will not describe at length how much of a productivity boost RStudio can give you. Rather, it includes some do's and don'ts that we have found helpful and believe you will benefit from too.

### 1.9.1 Do configure global and project level options

You can access and configure these options by clicking **Tools->Global** Options and **Tools->Project** Options. Most of the options are self-explanatory, and you can consult the documentation if needed. Some options are selected by default to optimize the user's experience. However, you should set up the environment that best suits your needs using the wide range of options available. Be aware that the project-level `.Rprofile` file will overwrite the `.Rprofile` file in the `Sys.getenv("HOME")` directory.

## 1.9.2 Do set up a working directory

We have seen that the default working directory is typically a user's **Documents** folder. In an organization provided machine, the **Documents** folder is usually mapped to a network storage, such as OneDrive, and constantly updates when it senses the addition or deletion of files. When you save a file frequently (scripts will be saved automatically if code is run when set via **Global Options**), that action constantly syncs the networked working directory downstream. You may experience that R is becoming slow as more and more projects are piling up in your network **Documents** folder, along with raw and intermediate data. To save yourself from agony, use a different location when creating projects rather than under the Documents folder, and change the default working directory via the **Tools->Global Options->R** menu.

Integration with Git is a better solution for keeping project backup than using network drive anyway.

## 1.9.3 Do start with an RStudio project

Always begin analysis by setting up a project, even when you will be performing a small data analysis that requires writing a single script. The small amount of time it takes to set up an RStudio project will save you many hours down the track. In business, it is common for stakeholders to come back with a request to re-run an analysis that was done months ago. They also have an eye for detail. You may need to give a good explanation if the numbers in the summary table you supplied last quarter do not match when calculated again from the same data source. Such situations become trivial when code/analysis is developed within a project. In a project, you should have your code, libraries, data sources, SQL, graphs, tables, and, most importantly, your thought process behind the analysis all in one place. It will be easy to perform an analysis again or extend it in various directions when required.

## 1.9.4 Do not preserve the workspace between sessions

Keep records of everything: used packages, data, data sources, SQL, code, analysis, plots, results, documentation, and presentation, but not the workspace between sessions. The previous session's objects may jeopardize a new block of code within the same project or cause conflicts when you run scripts from other projects. In both situations, debugging will be harder. You should be able to recreate a computational environment using written scripts, not carrying the environment from session to session. You can achieve this by not restoring `.RData` at start up and never saving `.RData` on exit (available under the **Global Options** menu).

There are many other ways RStudio can boost your productivity and increase efficiency in programming. It was already a great tool when used with R alone. As RStudio (or Positron) has extended its capability to include other programming languages, it has now become a *supercalifragilisticexpialidocious* IDE. Period.

After installing various R components and briefly learning how each of them brings value in your data analysis journey, it is time to take a sneak peek in the tidyverse package collection. The tidyverse packages share the same design principle: they are modular and highly cohesive. Some of these packages are used throughout the book.

## 1.10 The tidyverse package collection

The life of R can be separated into two parts: before tidyverse and after tidyverse. The tidyverse collection has changed the way we think, design and write programs in R. All the packages in tidyverse are built upon a shared design philosophy, grammar, and data structures.

There are 30 packages included in the `tidyverse` collection. You can install all of them at once by running the following:

```
install.packages("tidyverse")
```

You can see the names of all packages included in `tidyverse`:

```
tidyverse::tidyverse_packages(include_self = FALSE)
```

The following list will print in R console:

```
 [1] "broom"         "conflicted"    "cli"
 [4] "dbplyr"        "dplyr"         "dtplyr"
 [7] "forcats"       "ggplot2"       "googledrive"
[10] "googlesheets4" "haven"         "hms"
[13] "httr"          "jsonlite"      "lubridate"
[16] "magrittr"      "modelr"        "pillar"
[19] "purrr"         "ragg"          "readr"
[22] "readxl"        "reprex"        "rlang"
[25] "rstudioapi"    "rvest"         "stringr"
[28] "tibble"        "tidyr"         "xml2"
```

Not all `tidyverse` packages are needed in every R project. The four most common `tidyverse` packages that we will use in this book are the following:

- `dplyr`: A grammar of data manipulation
- `ggplot2`: Creates elegant data visualization using the grammar of graphics
- `tibble`: Simple data frames
- `lubridate`: Makes dealing with dates easier

We can see their monthly downloads from CRAN by fetching the stats using `dlstats` package and making a line plot using `ggplot2`. The code for generating the plot is intentionally omitted as data manipulation and visualization will be covered in detail in later chapters.

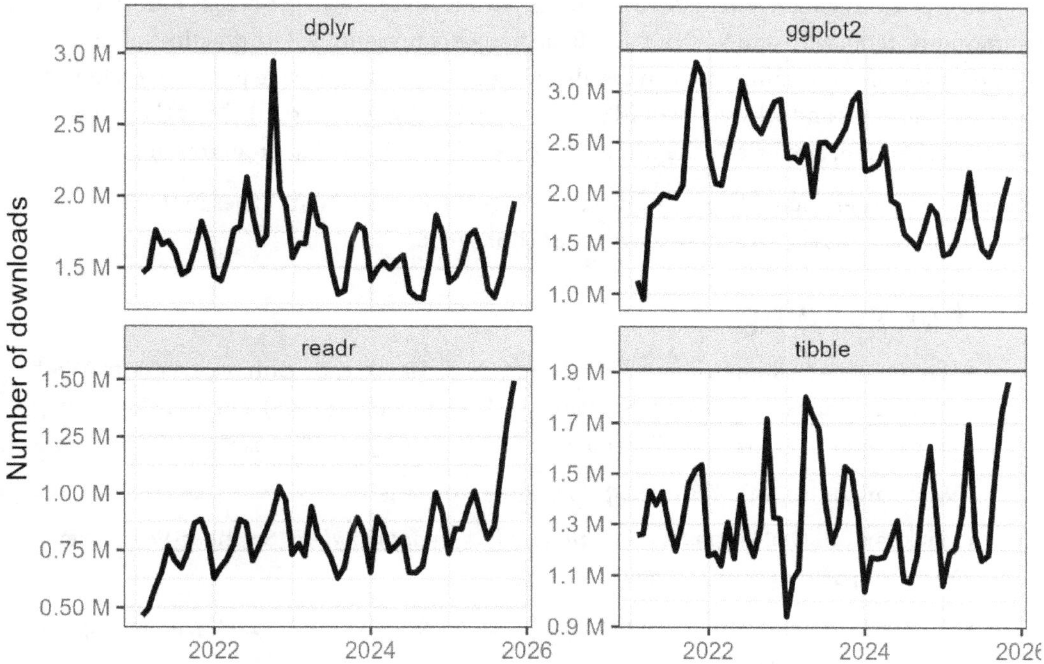


*Figure 1.1: Monthly downloads of a few tidyverse packages used in this book*

The figure illustrates the total monthly downloads for four widely used `tidyverse` packages over the period from 2021 to 2026. When aggregated across this five-year span, the cumulative downloads for these packages reach into the tens to hundreds of millions, underscoring their widespread adoption and popularity within the R community.

Throughout this book, you will learn how to apply the `tidyverse` approach to data analysis, including its principles of tidy data and consistent syntax. In addition to tidyverse tools, we will introduce other specialized R packages designed for time series forecasting and modeling temporal data. These packages will be presented in relevant chapters, along with practical examples demonstrating their usage and application to real-world datasets.

## Summary

We begin this chapter with a brief introduction to several companions of the R programming language. The download and installation of R, RStudio and R packages across multiple operating systems with and without admin access were presented. Best practices around updating any of these components were also presented. Some tips were provided on how to choose R packages and why you would use RStudio. Finally, you learned about the tidyverse package collection, and briefly the tidyverse approach that is used in programming.

In the next chapter you will be introduced to various objects and data types in the R programming language, and how to write functions in R.

## Further reading


- Hyndman, R., Athanasopoulos, G., Bergmeir, C., Caceres, G., Chhay, L., O'Hara-Wild, M., Petropoulos, F., Razbash, S., Wang, E., & Yasmeen, F. (2025). *forecast: Forecasting functions for time series and linear models.* `https://pkg.robjhyndman.com/forecast/`
- Wickham, H. (2014). *Advanced R.* CRC Press.
- Wickham, H. (2016). *ggplot2: Elegant graphics for data analysis.* Springer-Verlag New York. `https://ggplot2.tidyverse.org`
- Wickham, H., François, R., Henry, L., Müller, K., & Vaughan, D. (2023). *dplyr: A grammar of data manipulation* (Version 1.1.4).

## Get this book's PDF version and more

Scan the QR code (or go to `packtpub.com/unlock`). Search for this book by name, confirm the edition, and then follow the steps on the page.

UNLOCK NOW

*Note: Keep your invoice handy. Purchases made directly from Packt don't require an invoice.*

# 2

# Objects and Functions in R

In R, functions are designed to work with specific object types. They may have strict input/output object requirements. Functions from various packages are not necessarily optimized for your specific use cases. You may need to perform additional steps to prepare a dataset before passing it as input to a function. Similarly, the result from a function may require post-processing before you can analyze/report it. This is why basic knowledge about R objects, data types, and how to write functions in R is required. The materials discussed in this chapter are suitable for a beginner R user.

This chapter is organized into the following sections:

- Section 2.1: Objects and their names
- Section 2.2: Object types
- Section 2.3: Data types
- Section 2.4: Atomic vectors
- Section 2.5: Lists
- Section 2.6: Coercion
- Section 2.7: S3 atomic vectors
- Section 2.8: S3 lists
- Section 2.9: Missing values
- Section 2.10: Time series specific objects
- Section 2.11: Functions in R
- Section 2.12: Comparisons
- Section 2.13: Conditions

- Section 2.14: Conditional execution
- Section 2.15: Iterations

Note

**Your purchase includes a free PDF copy + exclusive extras**

Your purchase includes a DRM-free PDF copy of this book, 7-day trial to the Packt+ library (no credit card required), and additional exclusive extras. See the *Free benefits with your book* section in the *Preface* to unlock them instantly and maximize your learning.

# Technical requirements

We will use the following R packages in this chapter. Let's load them in your R workspace.

```
#install.packages("pacman")
library(pacman)
p_load("conflicted"
       ,"datasets"
       ,"gapminder"
       ,"purrr"
       ,"sloop"
       ,"tidyverse"
       ,"tsibble"
       ,"tsibbledata")

conflicts_prefer(
  dplyr::filter,
  purrr::compose)
```

Let's begin learning R programming by first understanding how R treats different objects and familiarizing ourselves with standard naming conventions.

# 2.1 Objects and their names

In R, first we create an object and then assign a name to that object. For example, a vector of numbers 1 to 5 is created using either of the following functions:

- `c()` is short for *combination* to create a vector
- `:` is a compact memory-efficient way of **alternative representation** that applies to certain vector types

R uses assignment operators to bind an object to its name. There are three types of assignment operators: `<-`, `<<-`, `=`. The most commonly used operator in R programming is `<-`. The assignment operators `<-` and `<<-` are bi-directional, meaning the object and its name can change sides as long as the arrow is pointing towards the name of the object. Unlike other programming languages, in R, the name `x1` has the value c(1,2,3,4,5), not the other way around.

```
x1 <- c(1, 2, 3, 4, 5)
x1
# [1] 1 2 3 4 5

x2 <- 1:5
x2
# [1] 1 2 3 4 5
```

When naming an object, you need to follow R's naming convention:

- A name consists of letters, digits, ., or _

```
.x <- 1:5
.x
# [1] 1 2 3 4 5

x_y <- 1:5
x_y
# [1] 1 2 3 4 5
```

- A name should not start with _ or a digit

```
_y <- 1:5
# Error: unexpected symbol in "_y"
```

```
1x <- 2:6
# Error: unexpected symbol in "1x"
```

- Do not use any reserved keywords, such as `if`, `TRUE`, `function`, or `break`. Type `?Reserved` for the full list of restricted keywords that cannot be used in an object name:

```
if <- 2:6
# Error: unexpected assignment in "if <-"
```

- There should be no white space between the combinations of letters, digits, . and _

```
a b <- 2:6
# Error: unexpected symbol in "a b"
```

- Some of these naming conventions can be relaxed by using back ticks around the object name; it's not encouraged though, due to the added overhead in typing as shown in the following code:

```
`_y` <- 1:5
`_y`
# [1] 1 2 3 4 5

`a b` <- 2:6
`a b`
# [1] 2 3 4 5 6
```

- Do not use double quotes in names and identifiers. If you do, a completely different method is required to retrieve values from that object as shown in the following case:

```
"a b" <- 1:5
"a b"
# [1] "a b"
```

After the initial introduction to R objects, let's delve into various objects and data types that are used in R programming.

## 2.2 Object types

Objects in R can be loosely defined into two groups: base objects and objects used for **Object-Oriented Programming** (**OOP**). There are various **Object-Oriented** (**OO**) systems in R, such as S3, R6, and S4, with the first one being the most common.

The metadata of an object is stored in **attributes**, which can be considered as name-value pairs for an object. Among attributes, names and dimensions are the common ones, which are preserved for an object. Different types of objects and data structures are created by adding various attributes to a base object. In *Section 2.7*, you will see that some S3 objects are created by adding attributes to the corresponding base objects.

## 2.3 Data types

**Vectors** are foundational building blocks of nearly all data structures in R. The simplest form of a vector is a scalar or a one-element vector, which represents a single value. Vector is an umbrella data type that has two different families of base objects underneath it: **atomic vectors** and **lists**. The two broad types of base vectors are further subdivided based on data types and their structure which are as follows:

- **Atomic vectors**: An atomic vector is a fundamental data structure that contains elements of only one data type. Common data types are:
    - Numeric
        - Double
        - Integer
    - Logical
    - Character
- **Lists**: A list can contain elements of different types and structures. It serves as the foundation for more complex objects, such as data frames.

By applying specific attributes to a base object, you can construct more complex data structures, such as **matrices** (two-dimensional rectangular structures) and **arrays** (multi-dimensional generalizations of matrices).

OO objects behave differently from regular base objects when passed to a generic function. An S3 object is built on top of base objects by assigning class attributes:

- S3 atomic vectors
    - Factors

    - Dates
    - Date-times (Posit / POSIXlt)
- S3 lists
    - Data frames
    - Tibbles

In the following sections, you will explore these data types and their attributes in detail, gaining the foundational knowledge needed for the rest of the book.

## 2.4 Atomic vectors

In an atomic vector, all elements must be of the same data type. Atomic vectors have four types: **integer**, **double**, **logical**, and **character**. The first two are commonly known as numeric vectors.

The functions:

- `typeof()` returns the type of an object. For atomic vectors, the type applies to the entire vector.
- `is.double()`, `is.integer()`, `is.logical()` and `is.character()` test whether a vector is of a particular type.
- `attr()` sets an individual attribute to an object. It can also be used to retrieve an individual attribute from an object.
- `structure()` sets multiple attributes to an object in one function call.
- `attributes()` retrieves attributes of an object that are already set.

**Note**

You can access R's help file for any function by using ? or `help()` functions in R. For example, `?typeof` or `help(typeof)` will take you to the relevant documentation of the `typeof` function in R.

## 2.4.1 Double

Double covers the decimals, scientific and hexadecimal numbers. The special values `Inf` (`infinity`), `-Inf` (`negative infinity`), and `NaN` (`Not a Number`, for example, zero divided by zero) can be added in double atomic vectors. The following code shows an example of handling a double vector:

```
dbl_vec <- c(7, 10.0, 19.025)
dbl_vec
# [1] 7.000 10.000 19.025

typeof(dbl_vec)
# [1] "double"

is.double(dbl_vec)
# [1] TRUE

is.integer(dbl_vec)
# [1] FALSE
```

You can set the names attributes to the object as shown in the following code:

```
attr(dbl_vec, "names") <- c('a', 'b', 'c')
dbl_vec
#     a      b      c
# 7.000  10.000  19.025
```

And also extract the names attributes from the `dbl_vec` object using the `attr()` function:

```
attr(dbl_vec, "names")
# [1] "a" "b" "c"
```

Or equivalently, using the `attributes()` function as shown in the following code:

```
attributes(dbl_vec)
# $names
# [1] "a" "b" "c"
```

## 2.4.2 Integer

Integer vectors only represent whole numbers. The trailing L is a must; otherwise, the numbers are considered as doubles. Let's see the differences in the following code snippets:

```
int_vec <- c(0L,7L,11L)
int_vec
# [1] 0 7 11

typeof(int_vec)
# [1] "integer"

int_vec2 <- c(0,7,11)
typeof(int_vec2)
# [1] "double"
```

## 2.4.3 Logical

Logical vectors have `TRUE` or `FALSE` entries. The abbreviated forms `T` and `F` are accepted too. Most mathematical functions work on logical vectors coercing `TRUE` to `1` and `FALSE` to `0` before being applied. Let's create a logical vector, `lg_vec` in the following code:

```
lg_vec <- c(TRUE, FALSE, TRUE)
typeof(lg_vec)
# [1] "logical"
```

The `is.logical()` function checks whether any vector is logical or not:

```
is.logical(lg_vec)
# [1] TRUE
```

Using the `which()` function you can extract the index of `TRUE` as follows:

```
which(lg_vec)
# [1] 1 3
```

The `abs()` function replaces the `TRUE` and `FALSE` by 1 and 0, respectively:

```
abs(lg_vec)
# [1] 1 0 1
```

In R, the logical constants `TRUE` and `FALSE` are reserved keywords, meaning they always represent logical values and cannot be redefined. However, their abbreviated forms, `T` and `F`, are not reserved and can be reassigned to other values.

This distinction is important because it can lead to unexpected behavior in your code. For example, when `T` and `F` are not explicitly defined in the workspace, R interprets them as logical values equivalent to `TRUE` and `FALSE`. However, if you assign any value to T or F, they become a vector:

```
T <- 5
T
# [1] 5
```

This demonstrates that while `TRUE` and `FALSE` are immutable logical constants, `T` and `F` are simply variables that can be overwritten. For clarity and reliability, it is recommended to use `TRUE` and `FALSE` explicitly in your code.

## 2.4.4 Character

Strings are a combination of letters and numbers when surrounded by either double quotes or single quotes. When a vector has multiple strings, it is called a character vector. Let's create a character vector and examine its characteristics as shown in the following code snippet:

```
char_vec <- c("this is", "a", "section of a book")
typeof(char_vec)
# [1] "character"

is.character(char_vec)
# [1] TRUE
```

When special characters, such as, `\t` (`tab`) and `\n` (`newlines`) are used then the `writeLines()` function processes those more appropriately than the `print()` function. Backslash-escaping is required when double quotes are used inside a double-quoted string. You will identify the differences as shown in the following code blocks:

```
char_vec2 <- c("This book\t discussed \n\"Time Series Forecasting\".")
print(char_vec2)
# [1] "This book\t discussed \n\"Time Series Forecasting\"."

writeLines(char_vec2)
```

```
# This book discussed
# "Time Series Forecasting".
```

The `nchar()` function counts the characters in each element of a character vector. Note that white spaces are also counted as characters as shown in the following code:

```
nchar(char_vec)
# [1] 7 1 17
```

This ends a brief introduction to the atomic vectors and their variants. Next, you will learn about lists.

## 2.5 Lists

Lists are generic vectors in R. In lists, each element can be any data type. This is the main distinction compared to atomic vector in which all elements must be of the same type. The `list()` function creates a list and the `is.list()` function checks whether an object is a list or not as shown in the following code:

```
lst1 <- list(
  dbl_vec
  ,int_vec
  ,lg_vec
  ,char_vec
)
lst1
# [[1]]
#      a      b      c
#  7.000 10.000 19.025

# [[2]]
# [1]  0  7 11

# [[3]]
# [1]  TRUE FALSE  TRUE

# [[4]]
# [1] "this is"           "a"
# [3] "section of a book"
```

```
is.list(lst1)
# [1] TRUE
```

A common practice is to first create an empty list and then add items sequentially to it. This is a useful method to store outputs from a function, especially when types of output object vary, as shown in the following code:

```
emp_lst <- list()
emp_lst[[1]] <- 1:5
emp_lst[[2]] <- c(letters[1:4])
emp_lst
# [[1]] [1] 1 2 3 4 5
# [[2]] [1] "a" "b" "c" "d"
```

The `str()` function displays the internal structure of an `R` object, and is used to inspect the structure of a list, especially if the list is a nested one. The `glimpse()` function (originally from the `pillar` package and re-exported in the `dplyr` package) also shows the internal data as much as possible:

```
str(emp_lst)
# List of 2
#  $ : int [1:5] 1 2 3 4 5
#  $ : chr [1:4] "a" "b" "c" "d"
```

As an example, let's create a nested list, `lst2`. It includes `lst1` (a four-element list), and a new list object with an integer vector created using the `seq()` function and a three-element logical vector. The `str(lst2)` or `glimpse(lst2)` displays the nested list in more informative way than just typing `lst2` in the console:

```
# create a nested list
lst2 <- list(
  lst1
  ,list(
    seq(0,20,5)
    ,c(TRUE, TRUE, FALSE)
  ))

glimpse(lst2)
```

This returns the following output:

```
List of 2
 $ :List of 4
  ..$ : Named num [1:3] 7 10 19
  .. ..- attr(*, "names")= chr [1:3] "a" "b" "c"
  ..$ : int [1:3] 0 7 11
  ..$ : logi [1:3] TRUE FALSE TRUE
  ..$ : chr [1:3] "this is" "a" "section of a book"
 $ :List of 2
  ..$ : num [1:5] 0 5 10 15 20
  ..$ : logi [1:3] TRUE TRUE FALSE
```

There is only one attribute attached, with the first element in the first list object inside `lst2`. Unlike previous scenarios, the `attributes()` function on a nested list returns NULL:

```
attributes(lst2)
```

`NULL` to retrieve the attribute of the list items, the corresponding sub-setting needs to be passed properly as shown in the following code:

```
attributes(lst2[[1]][[1]])
# $names
# [1] "a" "b" "c"
```

Next, we will look at two important data types, matrix and array, which are indeed base objects with additional attributes.

## 2.5.1 Matrix

A matrix represents vectors in a two-dimensional data structure. The dimension (`dim`) attribute transforms a vector to a matrix by passing a vector of size 2, specifying the number of rows and columns, respectively. By default, the `dim` function fills the matrix column-wise using the elements from a vector:

```
mat_a <- 1:9
dim(mat_a) <- c(3,3)
mat_a
```

The code returns the following matrix:

```
     [,1] [,2] [,3]
[1,]    1    4    7
[2,]    2    5    8
[3,]    3    6    9
```

The `is.matrix()` function will check whether an object is a matrix or not. The attributes on a matrix returns the dimension of it:

```
attributes(mat_a)
# $dim
# [1] 3 3
```

The easiest way to create a matrix is using the `matrix()` function along with specifying `nrow` and `ncol` options:

```
mat_b <- matrix(
  c(rep(TRUE,3), rep(TRUE,3), rep(FALSE,3))
   ,nrow=3
   ,ncol=3)
mat_b
```

The code returns the following matrix:

```
     [,1] [,2]  [,3]
[1,] TRUE TRUE FALSE
[2,] TRUE TRUE FALSE
[3,] TRUE TRUE FALSE
```

The row names and column names can be specified either using `rownames()` and `colnames()` as shown in the following code example:

```
rownames(mat_a) <- c("a", "b","c")
colnames(mat_a) <- paste0("col_",1:3)

mat_a
```

We will get the following matrix:

```
  col_1 col_2 col_3
a     1     4     7
b     2     5     8
c     3     6     9
```

The same operation can be performed by passing a list in the `dimnames()` function:

```
dimnames(mat_b) <- list(
  paste0("row_",1:3), paste0("col_",1:3))
mat_b
```

The resulting matrix will look like the following:

```
      col_1 col_2 col_3
row_1  TRUE  TRUE FALSE
row_2  TRUE  TRUE FALSE
row_3  TRUE  TRUE FALSE
```

To create an identity matrix, you can use the shorthand `diag()`:

```
diag(3)
```

Following is the identity matrix:

```
     [,1] [,2] [,3]
[1,]    1    0    0
[2,]    0    1    0
[3,]    0    0    1
```

## 2.5.2 Array

Array is a multidimensional data structure and a more generalized version of matrix. A 3-dimensional array can be created by passing a vector specifying dimensions inside the `array()` function:

```
b_arry <- array(1:18, c(3,3,2))
b_arry
```

Following is the output:

```
, , 1
     [,1] [,2] [,3]
[1,]    1    4    7
[2,]    2    5    8
[3,]    3    6    9

, , 2
     [,1] [,2] [,3]
[1,]   10   13   16
[2,]   11   14   17
[3,]   12   15   18
```

If you have an array object and want to know its dimension, just use the following:

```
dim(b_arry)
# [1] 3 3 2
```

Next, we will discuss the concept of data coercion, which is crucial in functional programming. You will see what happens when different data types are assigned in one vector or list. Does data of one type overpower other data types?

# 2.6 Coercion

The rule for a vector object in R is that all elements must be of the same type. You have already used `typeof()` and various `is.*()` functions to check the type of specific atomic vectors. **Coercion** occurs when two different data types are mixed into one vector. When different types of atomic vectors are combined deliberately, then they follow a fixed order of implicit coercion: `character -> double -> integer -> logical`.

In the following example, the integer type coerced logical entries:

```
# integer overwrite logical
cr1 <- c(1L, TRUE, FALSE, 5L)
typeof(cr1)
# [1] "integer"
```

The single character entry coerces both double and logical entries as shown in the following code:

```
# character coerce all other data types
cr2<- c("Pen", 5.30, TRUE )
typeof(cr2)
# [1] "character"
```

Often in programming, we need to coerce one type of vector into another type. This is called an explicit coercion. The `as.character()`,`as.double()`,`as.integer()`, and `as.logical()` functions are used for explicit coercion. Let's try to coerce a logical vector as double:

```
# character coerce all other data types
as.double(c(TRUE, FALSE, FALSE))
# [1] 1 0 0
```

A vector with mixed elements can be coerced to the highest precedent one. For example, you can coerce a vector with logical and double element to a double vector:

```
# character coerce all other data types
as.double(c(TRUE, FALSE, FALSE, 1.5))
# [1] 1.0 0.0 0.0 1.5
```

However, a mixed vector with a character element cannot be coerced as double. It will produce NA and warns when it fails to coerce:

```
as.double(c(TRUE, FALSE, "Paper", 1.5))
# [1] NA NA NA 1.5
```

So far you have learned the basics of R's base objects and data types: atomic vectors and lists. Next, you will see how they are extended in OOP by allowing them to become an S3 object.

## 2.7 S3 atomic vectors

An S3 object allows for flexible and extensible OOP in R. S3 is used to create custom data structures that are tailored to specific needs. In S3, an object must have a class attribute that describes the type of data stored in that object. The class attribute is set and retrieved by the `class()` function.

Along with class, there are other attributes/metadata information that make S3 objects a powerful tool for data analysis and modeling in R. In time series data analysis, we need some commonly used S3 objects that are based on base atomic vectors.

Factors, dates, and date-times are three common S3 atomic vectors that we will discuss here.

## 2.7.1 Factors

A factor is internally an integer atomic vector with two attributes: class and levels. It stores categorical data and contains predefined values. Created using the `factor()` function, it does a few things:

- Evaluate how many categories there are
- Remembers which category a data point belongs to
- Produces a vector of strings associated with the names of each of the categories, aka, the levels.

Let's create a character vector of three elements and then make it a factor:

```
ch_vec <- rep(c("a", "b","c"), times=2)
fct_vec <- factor(ch_vec)
fct_vec
# [1] a b c a b c
# Levels: a b c
```

You can examine its attributes too:

```
attributes(fct_vec)
# $levels [1] "a" "b" "c"
# $class [1] "factor"
```

An ordered factor has an internal hierarchy in the factor levels created using either the `ordered()` or `factor()` function. First, see how you can use the `ordered()` function in the following code example:

```
cust_resp1 <- ordered(
  c("bad", "good", "good", "fair")
  ,levels = c("bad", "fair", "good"))
cust_resp1
# [1] bad good good fair
# Levels: bad < fair < good
```

You can specify the labels argument in either function to make them pretty while plotting and reporting. Let's see an example using the `factor()` function:

```
cust_resp2 <- factor(
  c("bad", "good", "good", "fair")
  ,levels = c("bad", "fair", "good")
  ,labels = c("Bad", "Fair", "Good"))
cust_resp2
# [1] Bad Good Good Fair
# Levels: Bad Fair Good
```

You could use the `cut()` function to convert a numeric vector to a factor:

```
score <- c(0, 10, 2, 6, 8, 8, 1, 9)
nps_score <- cut(
  score
  ,breaks=c(-Inf, 7, 9, Inf)
  ,labels = c("Detractors", "Neutral", "Promoters")
  ,include_lowest = TRUE)

nps_score
# [1] Detractors Promoters Detractors Detractors Neutral
# [6] Neutral Detractors Neutral
# Levels: Detractors Neutral Promoters
```

The `cut()` function is a useful function for data preparation while modeling or presentation, such as generating a table from the `nps_score` vector:

```
table(nps_score)
# nps_score
# Detractors  Neutral  Promoters
#          4        3          1
```

To validate that a factor is actually an S3 object, you can use the `otype()` function from the `sloop` package. The `otype()` on a factor vector returns S3:

```
otype(fct_vec)
# [1] "S3"
```

Whereas the `otype()` function on a character vector returns base type:

```
otype(ch_vec)
# [1] "base"
```

In the next section we will discuss the date data types, crucial data types that you need to understand while handling time series data and performing analysis on them.

## 2.7.2 Dates

In R, the date is counted as the number of days from the epoch, which is January 1, 1970 at 00:00:00 UTC (Coordinated Universal Time). Using this double vector as a base and assigning the class attributes on top we get the S3 Dates vectors.

The value 2023-01-01 is passed as a character while the `as.Date()` function coerces it to an S3 date object. Let's examine the outputs from the following code snippets to understand the differences:

```
date_vec <- as.Date("2023-01-01")
typeof(date_vec)
# [1] "double"

attributes(date_vec)
# $class [1] "Date"

otype(date_vec)
# [1] "S3"
```

Epoch is a point in time and serves as a reference for measuring time in a computer software system. Microsoft Excel uses a different epoch (Jan 01, 1900) than R. Although rare, loading date variables from Excel may cause trouble until properly formatted in R's date convention.

You can easily calculate how many days have passed since R's epoch using `unclass()` on `Sys.Date()`, or in general by stripping the class attribute from any date object:

```
unclass(Sys.Date())
# [1] 20465
```

## 2.7.3 Date-times

**POSIXt** is a class of objects that store the date and time information with a higher degree of precision, including year, month, day, hour, minute, second, and time zone. The POSIXt class is based on the **POSIX** (**Portable Operating System Interface**) standard for representing date and time values, which is used by many operating systems and programming languages. POSIXt objects in R are stored as the number of seconds since the epoch (January 1, 1970, at 00:00:00 UTC), with fractional seconds stored as a decimal fraction of a second.

You can create an S3 date-time object using the `as.POSIXct()` or `as.POSIXlt()` functions, which convert dates supplied as character strings or numeric values to `POSIXt` objects. The internal `POSIXt` object is a double atomic vector. The `ct` represents the calendar time and `lt` represents local time. The following code shows how to convert a character date and time to a specified time zone transfer to a date-time object:

```
dtm_vec <- as.POSIXct("2023-01-01 22:41:00",
                      tz="Australia/Melbourne")
dtm_vec
# [1] "2023-01-01 22:41:00 AEDT"
```

The `tz` argument in the `as.POSIXct()` function specifies the time zone to use for the object. You can see almost 600 time zones by typing `OlsonNames(tzdir = NULL)` in the console. The attributes store the following information:

```
attributes(dtm_vec)
# $class [1] "POSIXct" "POSIXt"
# $tzone [1] "Australia/Melbourne"
```

When converting a numeric value to date, you need to provide the origin as shown in the following code:

```
as.POSIXct(8^10, origin="1970-01-01")
# [1] "2004-01-11 00:37:04 AEDT"
```

## 2.7.4 Duration

You can calculate the difference between two dates or date-time objects, which becomes the duration data type. The difference between two dates or dates-time is stored in the `difftime` object:

```
dur1 <- as.Date("2025-12-07") - as.Date("2023-03-28")
dur1
# Time difference of 985 days
```

You can extract the numeric difference by applying `as.numeric()` on a `difftime` object as shown in the following code:

```
as.numeric(dur1)
# [1] 985
```

The base object type of a time duration vector is double:

```
typeof(dur1)
# [1] "double"
```

A difftime object has two attributes, class and units. The unit attributes determines the unit of the date/time difference as shown in the following example:

```
attributes(dur1)
# $class [1] "difftime"
# $units [1] "days"
```

The `as.difftime()` function is used for creating a `difftime` object from other existing objects. You can specify the units argument in which the results are desired. In the following example, a numeric value is coerced to a `difftime` object with weeks as units:

```
dur2 <- as.difftime(14, units="weeks")
dur2
# Time difference of 14 weeks
```

Different software systems store date-time information differently, including having their own epoch. Date, time, and duration variables that come from outside of R may not follow R's logic and manipulating them becomes frustrating from time to time. A strategy is to spend time understanding how a date-time variable will behave after loading data into R and then transforming it into R's date object. You could do this using functions from base R or from the `lubridate` package, which is dedicated to working with date-time variables.

In the next section, you will delve into S3 lists, which encompass two essential data storage concepts in the R language, widely employed in modern programming.

# 2.8 S3 lists

S3 lists are built on top of regular R lists, but they have specific structures. Their behaviors are defined by the set of methods or functions that can be applied to them. We will discuss the two most important S3 lists, data frame and tibble.

## 2.8.1 Data frames

A data frame represents data in a tabular manner, where each column represents a variable, and each row represents an observation. In a data frame, a column must be of the same data type, and different columns could have different data types. The data frame is created using the `data.frame()` function.

Let's create a data frame with three columns; a character (`country`), a date (`date`), and a double (`population_mil`) as shown in the following code:

```
pop_df <- data.frame(
  country = c(rep("USA", 2), rep("AUS", 2))
  , date = as.Date(rep(c("2000-01-01", "2023-01-01"), times = 2))
  , population_mil = c(282.2, 334.3, 19.03, 26.6))
pop_df
```

The generated `pop_df` data frame looks like as follows:

```
  country       date population_mil
1     USA 2000-01-01         282.20
2     USA 2023-01-01         334.30
3     AUS 2000-01-01          19.03
4     AUS 2023-01-01          26.60
```

Interestingly, the internal object type of a data frame is a regular list:

```
typeof(pop_df)
# [1] "list"
```

Using a regular list as a base object, a data frame is different because of three attributes:

- `names`: The column names
- `class`: `data.frame`
- `row.names`: The name of the rows. Uses a numeric sequence if not provided explicitly.

Let's examine the attributes of the `pop_df` data frame:

```
attributes(pop_df)
# $names [1] "country" "date" "population_mil"
# $class [1] "data.frame"
# $row.names [1] 1 2 3 4
```

A data frame does not require you to specify the dimension structure explicitly. However, each column vector should be of equal length. In the above code, the `rep()` and `c()` are used repeatedly to maintain this constraint of data frame structure. Otherwise, it throws an error as shown in the following output:

```
data.frame(id = paste0("00", c(1:6)), score = c(10:14))
# Error in data.frame(id = paste0("00", c(1:6)), score = c(10:14)) :
# arguments imply differing number of rows: 6, 5
```

The `data.frame()` function can recycle shorter column vectors only when they are an integer multiple of the longest column. The following code output provides an example:

```
df_ok <- data.frame(id = paste0("00", c(1:6)), score = c(10:12))
```

By default, data frames transform non-syntactic names automatically when created in R or while loading data externally. The following code output shows an example of column name sanitization:

```
df_nm <- data.frame(
  `1` = c("rock", "paper", "scissors"),
  `2` = c(5, 3, 2))

names(df_nm)
# [1] "X1" "X2"
```

This behavior could be overwritten using `check.names=FALSE` if you are happy to type back-ticks around column names while coding:

```
df_wnm <- data.frame(
  `1` = c("rock", "paper", "scissors"),
  `2` = c(5, 3, 2), check.names = FALSE)

names(df_wnm)
# [1] "1" "2"
```

The `stringAsFactors=TRUE` argument controls the coercion of character vectors to factors. For example:

```
char_become_factor <- data.frame(
  option = c("rock", "paper", "scissors"),
  n = c(5, 3, 2),
  stringsAsFactors = TRUE)
glimpse(char_become_factor)

# Rows: 3
# Columns: 2
# $ option <fct> rock, paper, scissors
# $ n      <dbl> 5, 3, 2
```

By default, it is `FALSE` for `R 4.0.0` and higher. You can set this argument to `FALSE` when the coercion was not desirable:

```
char_remain_char <- data.frame(
  option = c("rock", "paper", "scissors"),
  n = c(5, 3, 2),
  stringsAsFactors = FALSE)
glimpse(char_remain_char)

# Rows: 3
# Columns: 2
# $ option <chr> "rock", "paper", "scissors"
# $ n      <dbl> 5, 3, 2
```

The `is.data.frame()` function checks whether an object is a data frame, and `as.data.frame()` coerces other classes of `objects`, `character`, `list`, `matrix`, or `array` into a data frame. Custom row names can be allocated to data frames using `row.names()` function:

```
row.names(char_remain_char) <- paste0("game_", 1:3)
char_remain_char
#          option n
# game_1     rock 5
# game_2    paper 3
# game_3 scissors 2
```

However, it is not easy to manipulate row names of a data frame, especially when the information is required in analysis or modeling. That is why tibble is here to help!

## 2.8.2 Tibbles

A tibble is a modernized S3 object for storing rectangular data. It provides with stricter checking and better formatting than the regular data frames and makes data manipulation a simple yet fast process. The `tibble` object was introduced in the `tibble` package (`Müller et al., 2023`). Depending on the context, the word tibble refers to the package, the data structure, the specific function (`tibble()`), or the method for printing and data manipulation.

Some of the benefits of tibble include, but are not limited to, the following:

- **Improved printing method**: A `tibble` object is optimized for printing only a few columns and rows that fit on an active screen. The dimensions and data types of each column are printed along with the summary of the remaining rows and columns. This makes it easier to inspect and understand the data without using any functions at all. Colors for positive, negative, and missing values highlight the important features of data at first glance. This improved printing method is beneficial for large rectangular data with many rows and many columns.
  For example, you may try printing the gapminder data from the `gapminder` package in your R console as shown in the following code:

  ```
  gapminder::gapminder
  ```

The output may look like the following (depending on the size of your console). The default print method of a `tibble` object provides useful information, such as number of rows and columns in the data, along with column types:

```
# A tibble: 1,704 × 6
   country     continent  year lifeExp      pop gdpPercap
   <fct>       <fct>     <int>   <dbl>    <int>     <dbl>
 1 Afghanistan Asia       1952    28.8  8425333      779.
 2 Afghanistan Asia       1957    30.3  9240934      821.
 3 Afghanistan Asia       1962    32.0 10267083      853.
 4 Afghanistan Asia       1967    34.0 11537966      836.
 5 Afghanistan Asia       1972    36.1 13079460      740.
 6 Afghanistan Asia       1977    38.4 14880372      786.
 7 Afghanistan Asia       1982    39.9 12881816      978.
 8 Afghanistan Asia       1987    40.8 13867957      852.
 9 Afghanistan Asia       1992    41.7 16317921      649.
10 Afghanistan Asia       1997    41.8 22227415      635.
# i 1,694 more rows
# i Use `print(n = ...)` to see more rows
```

- **Tibbles are lazy but effective**: They do not compute or evaluate anything until it is necessary. This is highly memory-efficient and makes data manipulation operations faster. In the following example, a tibble with two columns is created, each with one million rows. The tibble is not created immediately until the `head()` function is applied to it. In that instance, only the first 5 rows are computed and returned instead of creating the entire tibble:

```
set.seed(2)
large_tbl <- tibble(x = 1:1e+06, y = rnorm(1e+06))
large_tbl |>
  head(5)

# # A tibble: 5 × 2
#         x       y
#     <int>   <dbl>
# 1       1 -0.897
# 2       2  0.185
# 3       3  1.59
# 4       4 -1.13
# 5       5 -0.0803
```

You may have noticed the piping operators (|> or %>%) used for chaining functions and/or instructions in the previous code example. The operator |> is native to RStudio, and %>% is widely used.

- **Tibble maintains an inheritance hierarchy (the classes)**: A `tibble` object has three classes. In order of priority they are, `tbl_df` (specific class for a tibble), `tbl` (a more general table class), and `data.frame` (the standard, built-in R data frame class). The `str()` function could show the structure of a tibble:

```
str(large_tbl)

# tibble [1,000,000 × 2] (S3: tbl_df/tbl/data.frame)
#  $ x: int [1:1000000] 1 2 3 4 5 6 7 8 9 10 ...
#  $ y: num [1:1000000] -0.8969 0.1848 1.5878 -1.1304 -0.0803 ...
```

When you call a function on a `tibble` object, R first checks if there is a method for `tbl_df` and then for `tbl`. If R cannot find a specific method for either of the previous classes, then it falls back to the default `data.frame` methods. Being a modern data frame, tibble also knows how to behave like a standard R data frame when necessary.

- **Tibble never coerces input**: Unlike data frames, different types of columns are not coerced, there is no need for the `stringsAsFactors` argument when working with a tibble object.
- **Refer variables during construction**: A tibble lets you reference a newly created column when defining subsequent columns during its construction. In the following code output, the column y uses x, and z uses both x and y even before the `ref_tbl` being created.

```
ref_tbl <- tibble(x = 0:4, y = x^2, z = 2 * y + x)
ref_tbl

# # A tibble: 5 × 3
#         x     y     z
#     <int> <dbl> <dbl>
# 1       0     0     0
# 2       1     1     3
# 3       2     4    10
# 4       3     9    21
# 5       4    16    36
```

- **No messy row names manipulation**: Tibble considers row names as metadata that should be stored in the same way as the rest of the data is stored, such as a column. This design also avoids complicated string operations on row names during re-sampling or bootstrapping. The `rownames` argument in the `as_tibble()` function, or the `tibble::rownames_to_columns()` function adds row names as a column. The following code output shows an example:

```
rwnam_tbl <- as_tibble(ref_tbl, rownames = "counters")
rwnam_tbl

# # A tibble: 5 × 4
#   counters      x     y     z
#   <chr>     <int> <dbl> <dbl>
# 1 1             0     0     0
# 2 2             1     1     3
# 3 3             2     4    10
# 4 4             3     9    21
# 5 5             4    16    36
```

- **Easy operation with list columns**: Both data frames and tibbles can have a column that is actually a list. However, it is easy to manipulate list columns in a tibble. A list column can be anything from a simple vector to complex model output objects. The list column could itself be a tibble. The following is an example of list column manipulation in a tibble object using `mtcar` data. First, we will create a tibble object sub-setting the gear variable and running linear model (`lm()`) on each subset. The model is saved as a list-column, while each element of the column is a model object. This is an example of how tibbles can help in functional programming:

```
mtcars_tbl <- as_tibble(mtcars)

# create a list column tibble
mtcars_lm <- mtcars_tbl |>
    group_by(gear) |>
    nest() |>
    mutate(model = map(.x = data, ~lm(mpg ~
        ., data = .x)))
mtcars_lm
```

The code produced the following output where the data and the model columns are in list column format:

```
# A tibble: 3 × 3
# Groups:   gear [3]
   gear data              model
  <dbl> <list>            <list>
1     4 <tibble [12 × 10]> <lm>
2     3 <tibble [15 × 10]> <lm>
3     5 <tibble [5 × 10]>  <lm>
```

You will learn how to use other data objects that are built on tibble flavors and how to use them in time series analysis via functional programming throughout this book.

You will commonly need to handle missing values while working with data. In the next section, it is briefly touched upon so that you understand how R handles missing values in general.

## 2.9 Missing values

Dealing with missing values is a common phenomenon when you work with real data. In general, R represents missing values with NA (Not Applicable) although it has data type-specific variants: `NA_integer_` (for integer), `NA_real_` (for double), and `NA_character_` (for character). Let's create a vector with missing values as shown in the following code:

```
mv_vec1 <- c(2, NA, 1, NA, 0)
```

The `is.na()` checks the element wise existence of missing values:

```
is.na(mv_vec1)
# [1] FALSE TRUE FALSE TRUE FALSE
```

The `anyNA()` function checks for missing values over the entire object:

```
anyNA(mv_vec1)
# [1] TRUE
```

It is important to understand that the string "`NA`" is not the same as NA:

```
mv_vec2 <- c("NA", NA)
is.na(mv_vec2)
# [1] FALSE TRUE
```

When checking missing values in a list, the `is.na()` should be applied on each element using the proper member from the apply family of functions, as seen in the following code example:

```
mv_lst <- list(1:3, mv_vec1, mv_vec2)
sapply(mv_lst, is.na)
# [[1]] [1] FALSE FALSE FALSE
# [[2]] [1] FALSE TRUE FALSE TRUE FALSE
# [[3]] [1] FALSE TRUE
```

You may encounter NaN (Not a Number), which is a byproduct of mathematical operations rather than indicating missingness in the data. The `is.nan()` checks whether a numeric value is a number or `not.` `Inf` (divided 0) and `-Inf` are numbers, and truthfully `is.nan()` returns `FALSE` for them as shown the following code outputs:

```
is.nan(0/0)
# [1] TRUE

is.nan(5/0)
# [1] FALSE

is.nan(c(5, NA))
# [1] FALSE FALSE
```

Unlike relational databases, `NULL` does not indicate missing values in R. It has two special use cases,

- Assign an empty object of any type, such as `c()` or `my_list <- NULL`
- As a place holder in a function argument. When `NULL` is used in an argument, it suggests the function will compute the argument's value from previous steps, but takes the user supplied value if provided.

So far, you have learned about the generic objects and data types in R. Now, we look at the special objects and data structures that are specific to time series data.

# 2.10 Time Series Specific Objects

Having a date-time variable along with a numeric vector does not make observed data a time series. Specific information is required to treat a numeric vector as a time series, which can be derived from the date-time variable.

Time series data is indexed or ordered by time. A time series object is a specific data structure designed to store and analyze time series data. These time series objects provide specialized methods and functions for analyzing, manipulating, and visualizing time series data. They often come with built-in support for time-based indexing, sub-setting, and applying time-specific operations. To create a time series object, you typically need a vector or matrix of data values along with an associated time index.

R provides many time-series objects that are designed to perform specific tasks. The common objects are `ts`, `zoo`, `xts`. The `ts` object is from the `stats` package and considered native in R. The `ts` object is particularly useful for time-dependent analyses, forecasting, and modeling.

The `tsibble` object extends the principles of the tidyverse, making it easier to manipulate, analyze, and visualize time series data in a consistent and intuitive manner. tsibble represents temporal data in a rectangular form and provides a data-and-model-oriented framework.

Let's look at the time-series objects in more detail.

## 2.10.1 ts

The `ts` object is an important time series object in R. It is used to represent observations at regular intervals over time. The `ts` object stores data values along with information about the time index, such as the start time, frequency, and number of observations. A `ts` object is constructed using the `ts()` function. In the following code, a monthly time series (frequency=12) is created from a random sample of 36 observations:

```
set.seed(100)
month_ts <- ts(rnorm(36, 25, 7), start = "2020", frequency = 12)
month_ts
```

The following output will be generated:

```
          Jan      Feb      Mar      Apr      May      Jun
2020 21.48465 25.92072 24.44758 31.20749 25.81880 27.23041
2021 23.58856 30.17888 25.86366 24.79478 22.27802 28.57599
2022 19.29935 21.93085 19.95845 26.61661 16.89589 26.72953
          Jul      Aug      Sep      Oct      Nov      Dec
```

```
2020 20.92747  30.00173  19.22318  22.48097  25.62920  25.67392
2021 18.60330  41.17208  21.93337  30.34842  26.83373  30.41383
2022 24.36221  37.30163  24.03449  24.22165  20.16990  23.44744
```

Let's look at the arguments in the `ts()` function:

- `start`: the starting time of the series
- `frequency`: the number of observations per unit of time, such as year(1), monthly(12), daily(365.25), and so on.

`ts()` is an S3 object with two attributes, `class` and `tsp`. The `tsp` attributes return the start time, end time and the frequency for a series as shown in the following code output:

```
attributes(month_ts)
# $tsp [1] 2020.000 2022.917 12.000
# $class [1] "ts"
```

The `ts` object has built-in support to handle seasonal data, where a pattern repeats over a fixed interval of time. Almost all methods in the famous `forecast` package expect a `ts()` object as an input. Other packages, such as `stats` and `TSA`, offer functionality for advanced modeling and diagnostics using `ts`  object as well.

## 2.10.2 tsibble

The `tsibble` object was introduced by the package of the same name (Wang et al., 2022). The `tsibble` object emphasizes the organization of temporal data into rectangular data frames with variables as columns and observations as rows. The `tbl_ts` is the new class attributes introduced for tsibble and by default become an S3 object. It can handle both regular and irregular time series data, providing flexibility for handling various types of time series data under the same framework.

You can create a `tsibble` object by passing either a vector, data frame or a tibble along with a time index variable. The following code is an example of creating a `tsibble` object:

```
set.seed(101)
sales_tsb <- tsibble(
  year = 2000:2023,
  sales = rnorm(24, 500, 80),
  index = year)
sales_tsb
```

The following output will be generated:

```
# A tsibble: 24 x 2 [1Y]
    year sales
   <int> <dbl>
 1  2000  474.
 2  2001  544.
 3  2002  446.
 4  2003  517.
 5  2004  525.
 6  2005  594.
 7  2006  550.
 8  2007  491.
 9  2008  573.
10  2009  482.
# i 14 more rows
# i Use `print(n = ...)` to see more rows
```

The `as_tsibble()` functions coerces other objects, such as a `ts` object into a `tsibble` object. The benefit of using `tsibble` objects shines when it presents multiple time series in a single object. It maintains the hierarchy of uniquely identified time series via the key argument.

For example, transfer the gapminder data to a `tsibble` object, using the year as index (time index of observation) and continent and country as `key` (grouping variables):

```
gpm_tsb <- gapminder |>
  as_tsibble(index = year, key = c(continent, country))
gpm_tsb
```

The code generates the following output:

```
# A tsibble: 1,704 x 6 [5Y]
# Key:       continent, country [142]
   country continent  year lifeExp      pop gdpPercap
   <fct>   <fct>     <int>   <dbl>    <int>     <dbl>
 1 Algeria Africa     1952    43.1  9279525     2449.
 2 Algeria Africa     1957    45.7 10270856     3014.
 3 Algeria Africa     1962    48.3 11000948     2551.
 4 Algeria Africa     1967    51.4 12760499     3247.
 5 Algeria Africa     1972    54.5 14760787     4183.
 6 Algeria Africa     1977    58.0 17152804     4910.
```

```
 7 Algeria Africa     1982    61.4 20033753     5745.
 8 Algeria Africa     1987    65.8 23254956     5681.
 9 Algeria Africa     1992    67.7 26298373     5023.
10 Algeria Africa     1997    69.2 29072015     4797.
# i 1,694 more rows
# i Use `print(n = ...)` to see more rows
```

Errors occur when the key-index combination has duplicates and time series are not uniquely identified. If you run the following code that will generate a detailed error message with indication where the issue has arisen:

```
tibble(
  country = c(rep("USA", 4)),
  date = as.Date(
    rep(c("2000-01-01","2023-01-01"), times = 2)
  ),
  population_mil = c(282.2, 334.3, 19.03, 26.6)
) |>
  as_tsibble(index = date, key = country)
```

Some key benefits and features of `tsibble` include time-aware data manipulation and time-based indexing, and seamless integration with other `tidyverse` packages. `tsibble` integrates well with the `fable` package (`O'Hara-Wild et al., 2023`), which provides a unified framework for time series forecasting in R.

The R packages for time series forecasting evolved as R evolved as a language. The introduction of the tidy principle of data analysis and tibble for data representation changes how we interact with those packages. Almost all methods in the `forecast` package expect a `ts()` object as input and have their own way to specify various time series models, whereas the `fable` package uses the `tsibble` object as input. While the underlying modeling mechanism does not change, the face value interaction has changed significantly between these two packages. You will get exactly the same model outputs when you apply `forecast::ets()` and `fable::ETS()` in a series. However, the two functions expect two different input objects, different ways to specify models, and a different way to extract results from their respective output objects.

Forecasting projects with real-life data do not adhere to a single prescribed script. You may need to apply algorithms and functions from multiple R packages that require different input/output objects. Hence, you need to know how to create and interact with them with sufficient clarity.

After going through the main objects and data types in R, it is now time to use them in functions! The next section touches on the basics of writing functions. When applied to real data, these building blocks serve as the foundation for conducting intricate analyses.

# 2.11 Functions in R

Functions are an integral part of interacting with any programming language. The inherent flexibility of R offers users the ability to write their own functions in parallel with using functions from existing packages. Functions from R packages are optimized for generic situations. You may need to apply multiple potential forecasting models on one dataset, wrapping around a conditional if-else statement. The same if-else condition may need to extend further under a `for` loop in case you want to run the models using a rolling window.

As you progress through your data science journey with R, writing functions becomes your second habit. The first habit would be learning how to efficiently and effectively use functions from existing packages.

Writing functions in R is easy! You need to provide the following:

- As self-explanatory name
- The assignment symbol (<-)
- Follow that with the word `function`
- Various arguments in parentheses, along with the default value if available
- The beginning of a curly brace \{
- Fill up the body with code to instruct what the function is supposed to do
- Return output either explicitly or implicitly
- End the curly brace \}

It may look like the following:

```
# my_func <- function(arg1, arg2){ #
#  lines of code ....
return something
# }
```

Before writing a function, you should possess a clear understanding of the following at the very least:

- **Objectives**:
  - What the function will do. This fills up the body of the function.

    - You can use any base R code, and functions from other packages. Make sure to load the packages and call the functions appropriately inside your function to avoid breaking and getting errors.
    - Write modular functions by separating code into smaller, manageable chunks. It is better to have multiple unique functions rather than a single big, complex one if your function has a long to-do list.
- **Input**:
    - The data type, structure, and object type of input. From the lessons on R objects in previous sections, you know different objects behave differently when passing through the same function.
    - What is the input? Is it data to analyze or a model object to perform further analysis?
    - Does the code in the body require a specific data type: numeric, character, factor?
    - What is the data structure: vector, list, or tibble?
    - How will the code inside the body interact with the input data?
- **Output**:
    - What result do you want to keep?
    - What is the structure and object of the output; a list, a tibble, or a model?
    - Do you want to print all the outputs on the console as the function runs, or save them as objects?
- **Arguments**:
    - Controls the behavior of code in various scenarios.
    - Which part of the function will be fixed (default behavior)?
    - Which part of the function will change based on the user's input (argument with no default value)?
    - Which section of the code is optional (argument passed with a default value)

It is normal if you don't have all the information when writing your first function. You will learn as you go with testing and debugging, and you can always come back to improve your function at a later date.

In programming languages, a function is generally built to perform one of the four main tasks: comparison, evaluate a condition, conditional execution, or iteration. Functions can be simple when built to perform only one type of task or can be complex when multiple tasks of several types are combined. Let's learn about each of them and how to implement them in functions.

## 2.12 Comparisons

For comparison, a full set of operators includes >, >=, <, <=, != and ==. They are optimized for vector operations. They return either values or logical outputs depending on how they are used.

Let's create two vectors, a and b and see some examples of comparison using them:

```
a <- c(1, 3, 5)
b <- c(1, 6, 10)
```

From a vector, you can extract values that are smaller (or greater) than a specific value:

```
a[a < 3]
# [1] 1
```

When comparison is performed on entire vectors, they perform element wise comparison between the two vectors and return a logical vector:

```
a >= b
# [1] TRUE FALSE FALSE
```

The operators can be used for complex sub-setting. One simplest example is shown in the following code, where `a[a!=b]` returned only those values from `a` that were not in `b`:

```
a[a != b]
# [1] 3 5
```

On a `tibble` object, comparison operations are passed in the `filter()` function or its variations. Let's extract information of all countries in the year 1992 from the gapminder data as shown in the following code output:

```
gapminder |>
  filter(year == 1992)
```

This generates the following output:

```
# A tibble: 142 × 6
   country     continent  year lifeExp       pop gdpPercap
   <fct>       <fct>     <int>   <dbl>     <int>     <dbl>
```

```
 1 Afghanistan Asia       1992    41.7  16317921      649.
 2 Albania     Europe     1992    71.6   3326498     2497.
 3 Algeria     Africa     1992    67.7  26298373     5023.
 4 Angola      Africa     1992    40.6   8735988     2628.
 5 Argentina   Americas   1992    71.9  33958947     9308.
 6 Australia   Oceania    1992    77.6  17481977    23425.
 7 Austria     Europe     1992    76.0   7914969    27042.
 8 Bahrain     Asia       1992    72.6    529491    19036.
 9 Bangladesh  Asia       1992    56.0 113704579      838.
10 Belgium     Europe     1992    76.5  10045622    25576.
# i 132 more rows
# i Use `print(n = ...)` to see more rows
```

## 2.13 Conditions

**Conditions** evaluate logical expressions and return a value when `TRUE`.

- | and & are vectorized operations that we can apply on multiple values. Some examples are presented for you in the following code:

  ```
  b <- c(1, 6, 10)

  b[(b < 3) | (b >= 10)]

  gapminder |>
    filter((lifeExp < 40) | (lifeExp >= 80))

  gapminder |>
    filter((lifeExp < 40) & (lifeExp >= 80))

  gapminder |>
    filter((lifeExp >= 40) & (lifeExp < 80))
  ```

  You can examine their output when running in your R console.

- * || (`OR`) and && (AND) combine multiple logical expressions. However, they stop evaluation as soon as they get the first `TRUE` or first `FALSE`, respectively.

## 2.14 Conditional execution

In conditional execution, a part of the code is executed after a logical expression is evaluated. There are multiple ways to use conditional execution in R in various scenarios,

- `if` statements
- `ifelse` statements. These can be extended by chaining multiple `ifelse` statements
- `switch` functions
- The `dplyr::case_when()` function

Basically, these statements decide to run part of the code based on whether the conditions are `TRUE` or `FALSE`. A single `if` statement executes when the condition is `TRUE`. An example of such condition evaluation is presented in the following output:

```
x <- 10
if (x > 0) {
  x + 10
}
# [1] 20
```

In an `ifelse` statement, a code chunk executes the condition is `TRUE`, and the rest of the code executes when the condition is not met. An example of an `if-else` statement is shown in the following code output:

```
x <- -5
if (x > 0) {
  x + 10
} else {
  abs(x)
}
# [1] 5
```

An `ifelse` statement is extensible by adding multiple `else if` in between. Let's write a simple function, `check_zlogic()` with multiple if-else statements, as shown in the following code snippet:

```
check_zlogic <- function(z) {
  if (z <= 0) {
    abs(z)
```

```
  } else if (z > 0 & z <= 9) {
    z^2
  } else if (z > 10 & z <= 20) {
    z + 10
  } else {
    print("z is beyond limit")
  }
}
```

You can test the function for different values of z and compare what it returns in each case. The following output shows the results:

```
check_zlogic(z = -4)
# [1] 4

check_zlogic(z = 12)
# [1] 22

check_zlogic(z = 35)
# [1] "z is beyond limit"
```

The `switch()` function provides a way to choose from multiple options using a simple and concise syntax. Avoid the return of invisible `NULL` due to unmatched input; force `switch()` to throw an error using `stop()` as the last component. Even though `switch()` can be used for numeric input, it is recommended to use `switch()` with character input only. The following code presents an example of using `switch()` and the print method:

```
select_model <- function(x) {
  switch(
    x,
    lm = print("Selected model: Linear regression"),
    ets = print("Selected model: Exponential smoothing"),
    arima = print("Selected model: ARIMA"),
    stop(str_c("A ", str_to_title(x), " model is not available"))
  )
}
```

The `switch()` performs the respective operation when the chosen option belongs to the available expression list:

```
select_model("arima")
# [1] "Selected model: ARIMA"
```

An error is raised when the chosen option is not available in the expression list the `switch()` function is operating:

```
select_model("deep learning")
# select_model('deep learning') Error in
# select_model('deep learning') : A Deep Learning
# model is not available
```

The `case_when()` function is from the `dplyr` package and is used for conditional re-coding, or creating new variables based on multiple conditions. It provides a flexible and readable way to handle complex conditional logic.

The `case_when()` function takes multiple conditions separated by ~, where each condition is followed by the value to assign if the condition is `TRUE`. The conditions are evaluated in the order specified, and the first condition that is `TRUE` triggers the corresponding value assignment. Let's see an application of `case_when()` in the following output:

```
score_cat <- tibble(
  score = 1:6,
  category = case_when(
    score < 3 ~ "Low",
    score %in% c(3, 4) ~ "Medium",
    TRUE ~ "High"
  )
)
score_cat

# # A tibble: 6 × 2
#   score category
#   <int> <chr>
# 1     1 Low
# 2     2 Low
# 3     3 Medium
# 4     4 Medium
```

```
# 5     5 High
# 6     6 High
```

These examples demonstrate how conditional evaluation is performed in R. They provide a flexible and readable approach to deciding which block of code to execute after evaluating conditional logic. They can be used to handle multiple conditions, for code evaluation, or to create new variables based on some logic.

## 2.15 Iterations

Loops are used to perform repetitive tasks. You need to tell R which code section to repeat over and over, how many times to run the loop and when to stop repeating. There are two ways to incorporate looping in R functions:

- A `while` statement
- A `for` statement

A while loop includes any number of expressions. To avoid accidental *infinite* loop creation that runs forever, it is recommended to add a `stop()` function to terminate the loop when needed. An example of using the `while` loop to increment x by 1 until it reaches to 4, is presented in the following code:

```
x <- 1
while (x <= 5) {
  print(x)
  x <- x + 1
  if (x == 4) {
    stop("x reached 4. Loop terminated.")
  }
}
```

The `for` loop in R allows you to iterate over a sequence of values and execute a block of code for each iteration. Use a `for` loop when you know exactly how many times the loop should iterate. A `for` loop could have one to many expressions to execute. The output can be printed on console or saved in an object. The latter is the most common when a `for` loop is used inside a function.

```
x <- 10:13
for (i in 1:4) {
  print(x[i]^2)
```

```
}
# [1] 100 [1] 121 [1] 144 [1] 169
```

Executing iterations is a common task performed by a function. The way to write an iteration function depends on the data structure. For example, a function to iterate over an array item is supposed to be different from a function to iterate over the rows of a data frame. R offers two broad families of functions, apply and map to generalize the iteration task to many data and object types. The functions belonging to each family provide a flexible yet concise way to perform iterations. These two families of functions are discussed in the next sections.

## 2.15.1 The apply family

In base R, the `apply` family of functions provides a set of powerful tools to iterate a function over data structures such as vectors, lists, matrices, arrays, and data frames. These functions allow you to efficiently perform operations across dimensions of your data, avoiding the need for explicit loops.

The main functions in the `apply` family are as follows:

- `apply()`: This applies a function to margins (rows or columns) of an array, matrix, or data frame. The general syntax is `apply(X, MARGIN, FUN, ...)`:
    - `X` is the data structure,
    - `MARGIN` specifies the dimension to apply the function: 1 for rows, and 2 for columns
    - `FUN` is the function to be applied
    - `...` represents additional arguments of `FUN`
    - It returns a vector, array or list

  The following code shows an application of `apply()` on a 3x3 matrix to calculate column sums. We replace an observation by NA to show you how to pass additional argument (`na.rm=TRUE`) to the sum function inside `apply()`:

```
mat_dt <- matrix(1:9, nrow = 3, ncol = 3)
mat_dt[2, 3] <- NA
apply(mat_dt, 2, sum, na.rm = TRUE)
# [1] 6 15 16
```

- `lapply()`: This function applies a function to each element of a list or a vector. The general syntax is `lapply(X, FUN, ...)`:
    - `X` is the list or vector
    - `FUN` is the function to be applied
    - `...` represents additional arguments to the function
    - It returns a list of the same length as X

Let's create a list object adding different vector types as shown in the following code:

```
comb_lst <- list(
  dbl_vec = c(7, 10, 19.025),
  int_vec = c(0L, 7L, 11L),
  lg_vec = c(TRUE, FALSE, TRUE),
  char_vec = c("this is", "a", "section of a book"))
```

Let's extract summary information of the `comb_lst` object using `lapply()` which will provide the following output if you want to use the function to extract summary information for each list item:

```
lapply(comb_lst, summary)
```

This generates the following output:

```
$dbl_vec
   Min. 1st Qu.  Median    Mean 3rd Qu.    Max.
   7.00    8.50   10.00   12.01   14.51   19.02

$int_vec
   Min. 1st Qu.  Median    Mean 3rd Qu.    Max.
    0.0     3.5     7.0     6.0     9.0    11.0

$lg_vec
   Mode   FALSE    TRUE
logical       1       2

$char_vec
   Length     Class      Mode
        3 character character
```

- `sapply()`: This is similar to `lapply()` and simplifies the result if possible. Following the same syntax, `sapply(X, FUN, ...)`, the main distinction from `lapply()` is that it tries to `return` a vector, a matrix, or an array instead of a list if the output is homogeneous. In the following example, you will create a list of vectors of values that are randomly drawn from three statistical distributions:

```
set.seed(102)
dist_list <- list(
  normal = rnorm(n = 50, mean = 0, sd = 1),
  binomial = rbinom(n = 80, size = 20, prob = 0.3),
  poisson = rpois(30, lambda = 4)
)
```

  The `sapply()` function will compute mean from each list item and presents the results in a vector, as shown in the following output:

```
sapply(dist_list, mean)
#     normal   binomial     poisson
# 0.2986312  6.0125000  3.3666667
```

- `vapply()`: This is similar to `sapply()`, but it allows you to specify the output format explicitly via a custom function. This is useful when you know the expected type and length of the result. The general syntax is `vapply(X, FUN, FUN.VALUE, ...)`:
  - `X` is the list or vector
  - `FUN` is the function to be applied
  - `FUN.VALUE` is the format of the expected output
  - `...` represents additional arguments to the function

  Let's use `vapply()` on the `dist_list` list to round the computed mean to two digits:

```
vapply(dist_list, function(x) {
  round(mean(x), 2)
}, numeric(1))
# normal  binomial  poisson
#   0.30      6.01     3.37
```

- `mapply()`: The mapply function is useful when you want to apply a function to multiple vectors or lists element-wise. It iterates over corresponding elements of the given arguments and performs operations on them. This can be particularly useful when you have multiple vectors or lists of the same length and need to perform operations on them in a pairwise manner. The general syntax is `mapply(FUN, ..., MoreArgs = NULL)`:

- `FUN` is the function to be applied
- `...` represents the vectors or lists to iterate over
- `MoreArgs` is an optional list of additional arguments to the function

We will explain the functionality of `mapply()` with a simple example. Suppose you have two vectors, v1 and v2 and you want to multiply them element-wise. Write a custom function, `vec_multiply()` as shown in the following code:

```
v1 <- 1:3
v2 <- 5:7

vec_multiply <- function(x, y) {
  return(x * y)
}
```

Now, pass the custom function with the two vectors into the `mapply()` function. You will see the result in the following output:

```
mapply(vec_multiply, v1, v2)
# [1] 5 12 21
```

One of the things that makes R a functional programming tool is the concept of using a function inside another function without writing explicit for loops. The apply family of functions does exactly that using base R functionality. The function you are passing could be simple to complex. You can either define a custom function before passing it inside `apply` or define it on the fly within `apply`.

Once you understand how the apply family works in base R, you can leverage the knowledge to learn about the `map` family of functions from the `purrr` package.

## 2.15.2 The map family

The map family of functions makes iterations in R an elegant and succinct business. It offers a set of functions for applying operations (ranging from simple functions to models) to elements of lists or vectors. These functions are designed to provide a consistent and flexible interface for iteration and data manipulation. You saw an example in *Section 2.8.2* where a linear model was fit on multiple datasets that were saved in a list-column object.

There are many members of the map family in terms of increasing complexity in both depth and breadth. For example, some members apply one function on one piece of input data, preserving the input-output structure. Others apply multiple functions on one input, and some other members apply multiple functions on multiple inputs simultaneously. For illustration, we will just look at the simplest members of `map()` functions. The use of other functions will be explained appropriately when used in the book.

The `map_*()` branch includes functions that apply **one function** on each element of an atomic vector or list and return a vector of the same length preserving the names. The family of `map()` functions along with their input/output is presented in *Table 2.1*.

| Function name | Input | Output |
|---|---|---|
| `map()` | A list or vector | A list with the same length as the input list |
| `map_lgl()` | A list or vector | A logical vector |
| `map_int()` | A list or vector | An integer vector |
| `map_dbl()` | A list or vector | A double(numeric) vector |
| `map_chr()` | A list or vector | A character vector |

*Table 2.1: Variations of the `map()` function*

These functions take the syntax map(.x, .f, ...):

- `.x` is the input data
- `.f` is a function: it takes any one of the following forms:
    - An existing named function (for example, `mean`)

    - An anonymous custom function (`function(x){#do something}` or `\(x)#do something )`
    - A formula representation (`~.x^2`). The .x must be referred to indicate the first argument.
    - A string, integer or list
- `...`: supplies additional arguments

As an example, two different functions are applied through the `map_dbl()` function on a tibble called `wdt`. The `map_dbl()` function returns results in a scalar after applying the functions on each column. The application of map functions is flexible and intuitive without needing to write an explicit `for` loop to perform a repetitive task.

In the following example, a `tibble` object is created from random samples drawn from a Standard Normal distribution (mean 0 and standard deviation 1):

```
set.seed(103)
wtbl <- tibble(
  a = rnorm(10), b = rnorm(10),
  c = rnorm(10), d = rnorm(10)
)
```

There are multiple ways the map family functions can apply on an object. For example, you can pair them with a built-in function (for example, `mean`) to compute mean from each column in the `wtbl` tibble:

```
map_dbl(wtbl, mean)

#           a           b           c           d
# -0.35070052  0.17806119  0.07854021 -0.02179665
```

Any custom function can also be passed within any map family function. An example of rounding results to two digits after mean is computed is shown in the following code:

```
wtbl |>
  map_dbl(function(x) {
    round(min(x)^2, 2)
  })
```

```
#    a    b    c    d
# 3.48 0.87 2.18 4.60
```

Coupling a tibble data structure with the map family of functions allows you to have more time thinking about how you solve a problem than actual time taken to code it.

# Summary

In this chapter, you learned about base types and S3 objects in R. The crucial rectangular data representations, `data.frames` and tibble will help you to prepare and manipulate data in an appropriate format. You also learned about `ts` and `tsibble`, two data representations specific to time series analysis. Knowing different data structures and their behavior will help you to handle them properly in your own analysis. You learned the basics of writing functions. Also, you grasped the concept of performing comparisons, conditions, and iterations in multiple ways in R. The `apply` and `map` functions will help you to develop efficient code. Even though the discussion in this chapter mostly covered writing functions in base R, these are the building blocks of developing more complex functions. You can apply this learning to write your own functions or use them to understand how functions are developed in various packages.

In the next chapter, we will concentrate on how to handle data in R. You will learn how you can load data from multiple sources, how to connect various data bases and how to apply SQL directly from R.

# Further reading


- Müller, K., & Wickham, H. (2023). *tibble: Simple data frames* (Version 3.2.1). `https://CRAN.R-project.org/package=tibble`
- O'Hara-Wild, M., Hyndman, R., & Wang, E. (2023). *fable: Forecasting models for tidy time series* (Version 0.3.3). `https://CRAN.R-project.org/package=fable`
- Wang, E., Cook, D., Hyndman, R., & O'Hara-Wild, M. (2022). *tsibble: Tidy temporal data frames and tools* (Version 1.1.3). `https://tsibble.tidyverts.org`

# Join our community on Discord

Join our community's Discord space for discussions with the author and other readers:

`https://discord.gg/G4YmB6DQvf`

# 3

# Data Input/Output in R

In any data analysis and reporting project, the ability to seamlessly import and export data is essential. This chapter covers the foundational processes of data input and output (I/O) in R. The use of the term "data" is generic and covers a wide range of objects. You might need to load an external raw data file or results, or a complex model object from prior analysis into R. After loading, the objects can be used to perform an analysis, and save the results and artifacts, which could be a data frame or a model object. Whether you are conducting statistical analyses, building predictive models, or managing large-scale data workflows, understanding how to import and export data into R from various sources is crucial. Besides reading data from a local machine into an R workspace, this chapter shows you how to connect to a database from RStudio to run a SQL query in the database. Additionally, we touch on key principles of data governance, including secure handling of credentials when connecting to a database.

This chapter covers the following topics:

- Section 3.1: General structure of input/output functions
- Section 3.2: Saving and loading files generated by R
- Section 3.3: Data files from R packages
- Section 3.4: Importing external files
- Section 3.5: Importing data from relational databases
- Section 3.6: Using SQL from R
- Section 3.7: Data governance

Note

**Your purchase includes a free PDF copy + exclusive extras**

Your purchase includes a DRM-free PDF copy of this book, 7-day trial to the Packt+ library (no credit card required), and additional exclusive extras. See the *Free benefits with your book* section in the *Preface* to unlock them instantly and maximize your learning.

# Technical requirements

Let's load the following packages that you will need to run the examples included in this chapter.

```
#install.packages("pacman")
library(pacman)
p_load(
  "DBI",
  "dbplyr",
  "forecast",
  "fst",
  "glue",
  "keyring",
  "nycflights13",
  "readxl",
  "RSQLite",
  "ROracle",
  "tidyverse"
)
```

The code file is available in the GitHub repo `https://github.com/PacktPublishing/Modern-Time-Series-Analysis-with-R.git` for you to download and practice.

# 3.1 General structure of input/output functions

In R, many functions related to I/O are intuitively named. They often start with `read` for loading into R and `write` for saving from R. There are a variety of functions, and all of them include one of the following keywords,

- `read` or `load`, which specifies that these functions import objects into R
- `write` or `save`, which indicates that these functions export objects from R.

They include a mandatory `file` or `path` argument to specify the input or output location, which can often accept a file path string or a connection object. These functions provide flexibility via various optional arguments, such as `col_names` or `compression`.

The `path` or `file` argument usually assumes the project folder as the root directory. You can check your root directory by `getwd()` function. Before that, check what objects are available in your R global environment by typing `ls()` in the console and clear the session using `rm(list=ls())`.

```
# lists of objects in global environment
ls()

# clear R workspace
rm(list = ls())

# prints root directory
getwd()
```

Let's create a folder named data inside your project folder if it does not exist already.

```
# check existence of data folder
dir.exists("data")
dir.create("data")
```

The `dir.exists()` function checks for the existence of the data folder, and the `dir.create()` function creates a folder. For this chapter, we will use a folder called `data_chap3` that resides inside the data folder. The following code checks for the existence of the folder. If it is available, it sets the full path as `data_path`; if it is not available, it creates the folder and then saves it to `data_path`:

```
# Check if the folder exists and create if it doesn't

if (dir.exists(file.path(getwd(), "data", "data_chap3"))){
  data_path <- file.path(getwd(), "data", "data_chap3")
} else {
  dir.create(
    file.path(getwd(), "data", "data_chap3"),
    recursive = TRUE
  )
  data_path <- file.path(getwd(), "data", "data_chap3")
}
```

When creating folders after checking for their existence, wrapping the actions within an `if-else` block reduces the chance of accidentally deleting or overwriting folders. You can run this code block multiple times while keeping the existing folder and its component intact.

During the remainder of this chapter, we will use `data_path` as the location for write and read files. Saving the location in an object will save significant typing time in the following examples. You can use any other location of your choice by changing the path or file arguments inside a function.

Now you can programmatically check the existence of a folder and create one if needed. Next, you will see how to save and load R data files into an R workspace.

## 3.2 Saving and loading files generated by R

The `saveRDS()` and `readRDS()` functions save and load any R objects with extension `.rds` or `.RDS`. These functions are commonly used to save model objects and then load them later for further analysis. Windows and macOS use a case-insensitive file system, which means both the file extensions are treated equally. However, Linux/Unix file systems use a case-sensitive file system which treats `.RDS` differently than `.rds`.

The `saveRDS()` function predominantly saves a single object. Let's save the `mtcars` data locally using the following code:

```
saveRDS(mtcars,
        file = file.path(data_path,"mtcars.rds"))
```

You can save multiple objects by organizing them in a list as shown in the following code output:

```
saveRDS(list(mtcars, austres),
        file = file.path(data_path, "mtcars_austres.rds"))
```

It is always better to assign a name to a loaded object unless you just want to print it and do nothing afterwards:

```
singl_obj <- readRDS(file.path(data_path, "mtcars.rds"))
singl_obj
```

When multiple objects are loaded from the same RDS file, access them in the standard way that you extract elements from a list:

```
mult_obj <- readRDS(file.path(data_path,"/mtcars_austres.rds"))
df1 <- mult_obj[[1]]
df2 <- mult_obj[[2]]
```

In this section, you have learned how to read and write R data files. You have also learned how to load multiple RDS files with a single command.

Next, you will look at how to access datasets available in R packages.

# 3.3 Data files from R packages

Many R libraries are shipped with some data files in them. Those data files are available when the package is loaded in an R session. To learn about the data files of a package even before loading it, type `data(package="package name")`, which will pop up a tab with the data file names. For example, there are five data files included in the `forecast` package. You can see their name using the following code:

```
data(package="forecast")
```

You can use the `data()` function again to load a specific dataset into the R environment. Let's load the gas data from the forecast package as shown in the following code:

```
data(gas)
```

Now use the head() function to print the first 10 observations:

```
head(gas,10)

# Time Series:
# Start = c(2000, 1)
# End = c(2000, 10)
# Frequency = 52
#  [1] 70.636 71.040 68.490 65.137 67.918
#  [6] 75.117 72.970 76.106 78.158 82.272
```

You can use numerous data files that come with the various R packages to learn and practice R. Next, you will see how to load data files that are generated externally from R.

## 3.4 Importing external files

There are multiple convenient packages available that are dedicated to importing and exporting files into/from R. Your choice of packages may depend on the following:

- File formats
- How parsing happens while loading data into R
- The size of a file
- How fast the reading/writing is

This is just a guideline to consider when choosing packages to import external files.

The file import and export functionality of some key packages is discussed in the next sections. This is not an exhaustive list of packages that R offers for data I/O. However, the discussion and examples will help you to choose appropriate functions that are aligned with your R workflow.

### 3.4.1 The readr package

The readr package from tidyverse provides convenient functions to load and parse different types of rectangular data in fast and flexible ways. Parsing involves taking flat files of any format and outputting them as a tibble object while ensuring each column has the correct type.

The set of functions are read_*() and write_*() where the second part after the underscore indicates which file formats the functions can be applied to, such as comma-separated values (csv), tab-separated value (tsv), fixed-width text file (fwf). The readr versions of the base saveRDS() and readRDS() functions are write_rds() and read_rds(), respectively.

Let's add a character column to the `mtcars` data by taking the row names and save it as `mtcars2`. Use the `write_csv()` function to save the `mtcars2` object externally as a csv. Remove `mtcars2` from global environment. The following code block is showing you the operations:

```
mtcars2 <-
  rownames_to_column(mtcars, "make") |>
  as_tibble()

write_csv(
  mtcars2,
  file = file.path(data_path,"mtcars2.csv"))

rm("mtcars2")
```

Now, load the `mtcars2` object into the workspace. You will see the `read_csv()` function has parsed 1 character and 11 numeric columns accurately:

```
mtcars2 <- read_csv(
  file.path(data_path,"mtcars2.csv"))
```

You can check the column types of a data file before loading it into workspace applying `spec_csv()` function as shown in the following output:

```
> spec_csv(file.path(data_path,"mtcars2.csv"))

cols(
  make = col_character(),
  mpg = col_double(),
  cyl = col_double(),
  disp = col_double(),
  hp = col_double(),
  drat = col_double(),
  wt = col_double(),
  qsec = col_double(),
  vs = col_double(),
  am = col_double(),
  gear = col_double(),
  carb = col_double()
)
```

The `spec_*()` function checks the first 1,000 rows to guess the column specification.

`readr` provides convenient functions for parsing one column type to another. It is a common issue for external files to have date/time formats different from what R expects. The package provides three functions to parsing a date/times vector:

- `parse_date()`: Uses 1970-01-01 as the epoch and the format specified by `locale()`. The default uses an automatic date parser and provides output in the format of `Y-m-d` or `Y/m/d`, where `Y` is a four-digit year, `m` is a two digit month, and `d` is a two digit day. You need to provide the format of the input date(s) in the format argument inside the function, not the output date format. Some examples are presented in the following code snippets.

  The input date is in month/day/year format. The format argument should reflect that. The `%Y` format the year in a four-characters form:

```
parse_date("10/05/2023", format="%m/%d/%Y")
# [1] "2023-10-05"
```

  Here, the input date is in day/month/year format and the corresponding `%d/%m/%Y` format is used:

```
parse_date("20/05/2023", format="%d/%m/%Y")
# [1] "2023-05-20"
```

  The following format could not parse the input date due to the mismatch between the input date format and the requested parsing format:

```
parse_date("10/05/2023", format="%Y-%m-%d")
# [1] NA
```

  The delimiter, hyphen (-) or a front slash (/) plays no part in accurate date parsing as long as the requested format matches with the input date as shown below:

```
parse_date("2023-05-10", format = "%Y-%m-%d")
# [1] "2023-05-10"
```

- `parse_datetime()`: This function uses the number of seconds since midnight 1970-01-01 as the epoch.

An example of parsing a date-time input without the format argument:

```
parse_datetime("2023-05-14 16:42")
# [1] "2023-05-14 16:42:00 UTC"
```

You can pass any custom date-time input and corresponding matching format to parse properly as shown in the following code:

```
parse_datetime("14 May, 2023", "%d %B, %Y")
# [1] "2023-05-14 UTC"
```

- `parse_time()`: Uses the number of seconds since midnight as epoch. By default, the automatic time parser recognizes times of the form H:M. Let's see an example of the `parse_time()` function:

```
parse_time("4:42pm")
# 16:42:00
```

Next, you will see how to load specific part of an Excel file using the `readxl` package.

## 3.4.2 The readxl package

The `readxl` package is designed to import Excel files in R. You can use the extension specific `read_xls()` and `read_xlsx()` functions, or the generic `read_excel()`. The last one identifies the proper extension from the file itself. In these functions, you can specify the sheet argument to direct which sheets to read from. The cell ranges to read can be provided in multiple ways via the range argument:

- `read_excel("test.xlsx", range = "A1:D10")` reads all cells in the range A1 to D10, regardless of whether they are empty
- `read_excel("test.xlsx", range = cell_rows(1:10))` reads all non-empty rows from rows 1 to 10, across all columns
- `read_excel("test.xlsx", range = cell_cols("A:D"))` reads all non-empty rows from columns A to D
- `read_excel("test.xlsx", range = anchored("B4", c(3, 4)))`, starting from the cell B4, it reads 3 rows below and 4 columns in the right.

Like the `readr` package, `readxl` provides helper functions for column type guessing. Parsing the date column properly is important as a numeric cell is interpreted as the serial number of days from the epoch of Excel's two different date systems, 1900 or 1904. For illustration, we will use the `type-me.xlsx` data from the `readxl` package. First, load the data as shown in the following code:

```
typeme <- readxl_example("type-me.xlsx")
```

Let's load the data from a specific sheet of the excel file without specifying any column types:

```
read_excel(typeme ,sheet = "date_coercion")
```

This returns the following output:

```
# A tibble: 7 × 2
  `maybe a datetime?` explanation
  <chr>               <chr>
1 NA                  "empty"
2 41051               "date only format"
3 41026.479166666664  "date and time format"
4 TRUE                "boolean true"
5 cabbage             "\"cabbage\""
6 4.3                 "4.3 (numeric)"
7 39448               "another numeric"
```

Notice, the column name `maybe a datetime`? in the preceding output. This suggested that the first column in the file could be a datetime, even though there were mixed cell entries. The actual data types for each row are provided in the explanation column. Now, try loading the file again with defined column types as provided in the following code:

```
read_excel(typeme
           ,sheet = "date_coercion"
           ,col_types = c("date", "text"))
```

This returns the following output:

```
# A tibble: 7 × 2
  `maybe a datetime?` explanation
  <dttm>              <chr>
1 NA                  "empty"
2 2016-05-23 00:00:00 "date only format"
3 2016-04-28 11:30:00 "date and time format"
4 NA                  "boolean true"
5 NA                  "\"cabbage\""
```

```
6 1904-01-05 07:12:00 "4.3 (numeric)"
7 2012-01-02 00:00:00 "another numeric"
Warning messages:
1: Expecting date in A5 / R5C1: got boolean
2: Expecting date in A6 / R6C1: got 'cabbage'
3: Coercing numeric to date in A7 / R7C1
4: Coercing numeric to date in A8 / R8C1
```

The previous output shows that all the numeric entries have parsed as datetime, along with insightful information where parsing failed.

It is almost impossible to maintain a unified Excel file saving protocols throughout an organization. This particular functionality of the `read_excel()` function will save you a lot of time when you will handle many Excel files shared by different business units.

Next, we briefly discuss the fst package which is useful for handling large files.

## 3.4.3 The fst package

If you work with large files (100MB) and save intermediate data regularly during an analysis, then the `fst` package becomes helpful. Converting large data files into `fst` format allows random access to stored data. That means a subset of the complete table can be loaded from an fst file depending on the range of rows and by column name. The multithreaded serialization performs lightning-fast compression and decompression while writing on disk or importing in R using `write_fst()` or `read_fst()` function, respectively. By default, `read_fst()` outputs a data frame, which can be transformed into another data object (e.g., `tibble`).

For illustration, let's use the `write_fst()` function to compress the `mtcars` data in a `.fst` format. We first convert the row names in a column before saving it as an `mtcars3.fst` file format as shown in the following code:

```
write_fst(
  as_tibble(mtcars, rownames= "make"),
  path = file.path(data_path, "mtcars3.fst"),
  compress = 100)
```

Now, load a subset of the `mtcars3.fst` file by selecting specific columns and rows. The following code will show you how to perform the operations:

```
read_fst(
  path = file.path(data_path,"mtcars3.fst"),
  columns = c("make", "mpg", "cyl"),
```

```
    from = 5,
    to = 10) |>
    as_tibble()
```

This returns the following output:

```
# A tibble: 6 × 3
  make                 mpg   cyl
  <chr>              <dbl> <dbl>
1 Hornet Sportabout   18.7     8
2 Valiant             18.1     6
3 Duster 360          14.3     8
4 Merc 240D           24.4     4
5 Merc 230            22.8     4
6 Merc 280            19.2     6
```

You can now use three R packages, `readr`, `readxl`, and `fst` for data I/O depending on the file types and sizes. You have also learned how to parse date columns to the formatting required in your analysis.

Next, you will see a convenient option in RStudio for importing files.

### 3.4.4 Import data via RStudio's menu

RStudio's **Import Dataset** menu provides options to import data in R. You can use `readr` and `readxl` packages to load CSV/text and Excel files. Using the menu you can import data from other software, such as SPSS, SAS, and STATA via the `haven` package.

The menu pops up a window in which you can specify the file location/URL, choose the file format, and provide other arguments interactively. You can also preview the table before importing it into R. This is a convenient way to learn about data when you start working with a file for the first time.

Nowadays, you need to extract data from relational databases into R. This is quite a common practice in business. In the next section, you will learn how to set up a connection to databases and extract data straight into your R workspace.

# 3.5 Importing data from relational databases

The use of cloud-based databases to store relational data is a common aspect of modern business. As a data professional, you may need to import data into R from those databases for analysis. You can directly connect to databases from R, as well as extracting data and loading it into the current workspace. Conveniently, you can run SQL code from R or use `dplyr` to translate standard R code into SQL. Importing data from a relational database via the R connector is a broad topic. This section includes minimal examples and some tips that specifically apply when you extract time series data from relational databases into R.

The `nycflights13` `(Wickham, 2021)` package includes airline data for all flights departing NYC in 2013 including multiple relational tables, airlines, airports, planes, and weather. We will use the tables in our examples. The first step in extracting data from a database is to set up a successful connection using the **Open Database Connectivity** (**ODBC**) or **JDBC** driver. In the next subsection, you will learn how to setup this connection using the ODBC driver. The method is quite similar and easily reproducible for JDBC.

## 3.5.1 Connecting to databases

Using the DBI package as a core communicator, R offers direct connection to several open source databases via respective packages, for example, the `RMySQL` package for MYSQL database, and bigrquery for Google's BigQuery database etc. The commercial databases, such as **Snowflake** and **Oracle**, come with their own **ODBC** (**Open Database Connectivity**) drivers that are required to setup a connection along with R's `DBI` and `odbc` packages.

After installing the respective ODBC driver for a database, you can set up a connection from R using the following parameters. A **Database Administrator** (**DBA**) in your organization should have this information in case you don't know:

- `drv`: An object that facilitates connection to a database. For example, `drv=RSQLite::SQLite()` indicates to use the `SQLite()` connector object that comes with the `RSQLite` package. Similarly, in `drv=odbc::odbc()`, `odbc()` is the database connection object included in the package of the same name.
- `dsn`: DSN stands for Data Source Name. A user can set a custom name while installing an ODBC driver on a single machine. In the example "Snowflake" is set as the DSN during ODBC installation. There are file DSNs and system DSNs, which are applicable when multiple users share a single file for driver configuration or by all users under a single system.
- `uid`: The user identifier.

- pwd: Password to use. Do not under any circumstances include your password in R script. See *Section 3.5.2* for a brief discussion on how to use credentials securely.
- authenticator: Some sort of authentication method to establish the connection. Two common methods are using an external browser or **Single Sign On (SSO)**. It depends on the organization's set up and you need to comply with that.

Not all databases need every parameter. On the other hand, some databases provide extra parameters to control hierarchies, such as a **warehouse**, **database**, and **schema** in the connection setting.

It is common to have many schemas within a warehouse/database combination. If you need to import data from multiple schemas that reside within a warehouse simultaneously, you need to set up multiple connections, one for each schema, by changing schema names, with the remaining parameters staying the same.

Let's create an in-memory SQLite database connection using the RSQLite package:

```
sqlite_con <- DBI::dbConnect(RSQLite::SQLite(), ":memory:")
```

Write the weather data frame from the nycflights13 package into it, and use "WEATHER" as a database table name. The dplyr::copy_to() function copies a local data frame to a remote data source:

```
copy_to(sqlite_con,weather,"WEATHER")
```

The commercial databases require some form of authentication to connect. Usually, it is the user ID and password, or SSO method. You need to follow the organization's guidelines when connecting to databases. An example of an external browser authentication method using user ID and password to connect Snowflake database is shown next. It is worth noting that the Snowflake connection examples are intended to illustrate best practices. They require a properly configured Snowflake environment, DSN, and authentication method to work. You might not be able to execute this code unless you have access to a Snowflake account and ODBC setup:

```
# Connecting to Snowflake
snowflake_conn <- DBI::dbConnect(
  drv = odbc::odbc(),
  dsn = "Snowflake" ,
  warehouse = "<warehouse name>",
  database = "<database name>",
  schema = "<schemaname>",
```

```
  uid = "<user id>",
  pwd  = "<securely_retrieve_password_here>",
  authenticator="externalbrowser"
)
```

The ROracle package connects to Oracle by passing the database information as a string:

```
# Loading library prior to setup avoids unexpected errors
library(ROracle)

protocol <- "protocol"
host <- "<host of the database>"
port <- "<port of the database>"
sid <- "<database name>"

connect.string <- paste(
   "(DESCRIPTION=",
  "(ADDRESS=(PROTOCOL=" , protocol,
  ")(HOST=", host,
  ")(PORT=", port, "))",
   "(CONNECT_DATA=(SID=", sid, ")))",
  sep = ""
)

oracle_con <- DBI::dbConnect(
  dbDriver("oracle"),
  dbname = connect.string,
  uid,
  pwd
)
```

There is no message that appears in the R console when a connection is established successfully. Some error messages appear if there are issues, such as the username and password not being correct, or the connection parameters not being properly set. There could be many reasons for a database connection failing. However, broader discussion on them is out of the scope of the book.

It is vitally important to secure credentials when connecting to a database from R. There are multiple options available to the users of R when it comes to securing credentials in code, which you will learn about next.

## 3.5.2 Credential security

You need to use some form of authentication along with the username and password to connect with a database. However, it is extremely crucial to secure credentials and avoid every possibility of exposing them. A common mistake is to save the username and password in a script and accidentally store that in a version control system (e.g., Git). In the following sections, you will learn four different ways to secure your credentials. So, there is no excuse for exposing them even accidentally.

### 3.5.2.1 Pre-configuring credentials during ODBC set-up

The username and password can be added when ODBC is configured in your computer system. In such a case, you do not need to supply the ``uid'' and ``pwd'' explicitly while connecting to a database. Your R script is free from any sensitive information. This is the best way to secure credentials.

### 3.5.2.2 The keyring package

The keyring package encrypts the username/password combination for each service using the operating system's credential storage. Consider it as one key in a single keyring that opens a unique lock (service). A service-username-password combination is set by the `key_set()` function. The username and password are retrieved using the `key_list()` and `key_get()` functions, respectively:

```
key_set(service = "<database service name>",
        username = "<user id>")

# Retrieved credentials are used in the connection directly
snowflake_conn <- DBI::dbConnect(
  drv = odbc::odbc(),
  dsn = "<database service name>",
  warehouse = "<warehouse name>",
  database = "<database name>",
  schema = "<schema name>",
  uid = keyring::key_list("<database service name>")[1,2],
  pwd = keyring::key_get("<database service name>","<user id>"))
```

Using this method, you can safely secure your credentials for multiple database connectivity and use them simultaneously to extract data for your analysis.

### 3.5.2.3 The config package

You can make a `config.yml` file to store credentials along with all connection arguments and place them in the root directory of your R project. Required information for the arguments is retrieved via the `config::get()` function to pass in the database connection function.

### 3.5.2.4 The RStudio API

The RStudio API provides a prompt to supply credentials interactively. You can input both the username and the password by using the `askForPassword()` function twice.

Now you have learned several methods for securely connecting to databases to extract data.

Most relational databases use the SQL query language. You may need to apply SQL to extract data from a database once the connection is set up. In the next section you will see examples of using SQL directly from R for data extraction.

## 3.6 Using SQL from R

There are multiple ways to use SQL from R, and each of them has advantages and disadvantages. The choice of methods is also dictated by a database's expectation of SQL queries. As an example, parameterized queries on date/time columns behave differently in Oracle than it does in Snowflake. After a successful connection is set to a database, the connection name is used along with a query to extract data from a relational database table.

The packages that are usually used in SQL query executions are the following:

```
p_load( "DBI"
       ,"dplyr"
       ,"dbplyr"
       ,"glue"
       ,"tidyr")
```

We will use the **WEATHER** table copied to an in-memory SQLite database. The database connection is setup as follows:

```
sqlite_con <- DBI::dbConnect(RSQLite::SQLite(), ":memory:")

# Write weather table
copy_to(sqlite_con ,weather ,"WEATHER")
```

## 3.6.1 Inline SQL query

The query is passed as an inline string directly inside the `DBI::dbGetQuery()` function, which executes and extracts data into R:

```
weather_ewr <- dbGetQuery(
  sqlite_con,
  statement = "
    select *
    from weather
    where origin = 'EWR'
  "
) |>
  as_tibble()
weather_ewr
```

The full query should be enclosed by double quotes to avoid cancellation with the single quotes around EWR. The extracted data will produce the following table:

```
# A tibble: 8,703 × 15
   origin  year month   day  hour  temp
   <chr>  <int> <int> <int> <int> <dbl>
 1 EWR     2013     1     1     1  39.0
 2 EWR     2013     1     1     2  39.0
 3 EWR     2013     1     1     3  39.0
 4 EWR     2013     1     1     4  39.9
 5 EWR     2013     1     1     5  39.0
 6 EWR     2013     1     1     6  37.9
 7 EWR     2013     1     1     7  39.0
 8 EWR     2013     1     1     8  39.9
 9 EWR     2013     1     1     9  39.9
10 EWR     2013     1     1    10  41
# i 8,693 more rows
# i 9 more variables: dewp <dbl>,
#   humid <dbl>, wind_dir <dbl>,
#   wind_speed <dbl>, wind_gust <dbl>,
#   precip <dbl>, pressure <dbl>,
#   visib <dbl>, time_hour <dbl>
```

This is a convenient way to run a small number of queries. No change needs to be made in the SQL query itself.

## 3.6.2 SQL query from a file

This method works well for complicated and/or large queries. The query is saved in a `.sql` file and then passes through the `readr::read_file()` function. Open a new SQL Script from the **New File** drop down menu, paste the following query, and save it as `weather.sql` in the `data/data_chap3` folder. The folder location is saved as `data_path` in the following code:

```
select *
from weather
where temp >= 50
and wind_dir > 250
```

Now pass the file location in the `dbGetQuery()` function as shown in the following code example:

```
weather_temp <- dbGetQuery(
  sqlite_con,
  statement = readr::read_file(file.path(data_path,"weather.sql"))
) |>
  as_tibble()
```

You can examine the output in the R console after the query execution.

## 3.6.3 SQL translations via dplyr

You can apply `dplyr` ways of coding directly on the database table via the `tbl()` function and then extract the result in R console using the `collect()` function. This method requires minimal SQL knowledge. However, it may give error when R's data type handling clashes with a specific database's data type definition. Not all databases support the `dplyr-style` syntax, such as **Athena** and **SalesForce** at the time of writing. An example of such `dplyr-style` sql query is provided in the following code:

```
weather_summary <- tbl(sqlite_con, "WEATHER") |>
  group_by(month, origin) |>
  tally() |>
  ungroup() |>
  pivot_wider(id_cols = "month"
              ,names_from = "origin"
              ,values_from = "n") |>
  collect()
```

## 3.6.4 Parameterized query using glue_sql()

Often, a query needs to be parameterized based on user input. The `dbGetQuery()` function does not work with a parameterized query. Instead, a combination of `glue_sql() -> dbSendQuery() -> dbBind() -> dbFetch()` is used to apply different parameter values at every iteration. The following code is showing an example of parameterized `sql` while extracting weather information for three airports:

```
origin_par <- c("EWR", "JFK" ,"LGA")
origin_sql <- glue_sql(
  "select
    origin,
    year,
    temp
  from weather
  where origin = ?"
)

origin_data <- list()
for(i in seq_along(origin_par)){
  origin_query <- dbSendQuery(sqlite_con, origin_sql)
  dbBind(origin_query, list(origin_par[i]))
  origin_data[[i]] <- dbFetch(origin_query) |>
    as_tibble()
  dbClearResult(origin_query)
}
```

Not all database drivers process the parameterized SQL query through `dbBind()`. A common issue occurs when date/time columns are parameterized since most databases have a unique way of storing date-time data. In such a scenario, you need to use the `sqlInterpolate()` function. Supply the parameters twice, once inside the query and again inside the `sqlInterpolate()` function.

SQLite does not have a specific type to store date-time. In the following example, below, `time_hour` field is formatted to a character before applying filtering based on the user's input. After loading data to R, the date should be formatted as a date type. It should be noted that `sqlite_con` is used twice, first in the `sqlInterpolate()` function and second in the dbGetQuery() function. For example, the following sql query first formats time_hour column and then uses start_date and end_date to extract data:

```
start_date <- "2013-01-01"
end_date <- "2013-05-31"

query <- "
  select *
  from(
    select
      origin,
      strftime(
        '%Y-%m-%d',
        datetime(time_hour,'unixepoch', 'localtime')
      ) as date,
      temp
    from weather
  )
  where date >= ?start_ts
    and date <= ?end_ts
"
query_interpolate <- sqlInterpolate(
  sqlite_con,
  query,
  start_ts = start_date,
  end_ts = start_date )

query_extract <- dbGetQuery(
  sqlite_con, query_interpolate) |>
  as_tibble()
```

You have now learned multiple ways of using SQL programmatically to extract data from a database and bring it into R. You can keep the connection information, data extraction queries, and your analysis, in R scripts or notebooks. All of the information stays in one place, which helps to ensure reproducibility.

As a data professional, you should be aware of the governance around data. Some key data governance practices will be highlighted next for you.

## 3.7 Data governance

When you use R for analysis, it is essential to incorporate key principles of data governance to ensure the integrity, transparency, and accountability of the analytical process. You should consider at least three important aspects: data quality, data privacy and security, and documentation and metadata management. To support these practices, the R ecosystem offers a variety of packages. We will briefly introduce core concepts and highlight selected packages that can assist you in implementing sound data governance strategies within your workflows:

- **Data quality assurance**: Ensuring data quality is crucial for accurate and reliable analysis. It involves verifying the accuracy, consistency, completeness, and validity of the data being used. It includes data profiling, data cleaning, and data transformation. With the help of various R packages and functions, you can assess data quality, identify missing values, and use standardized data formats. Before performing any analysis, you should check the data quality so that the results obtained are actionable and reliable. The following set of packages might prove useful based on your data analytics needs:
    - `dataMaid`: Automates data quality checks by generating reports that identify missing values, outliers, and potential issues in variable types and distributions.
    - `janitor`: Provides simple functions for cleaning data, including removing empty rows/columns, cleaning column names, and identifying duplicate entries.
    - `validate`: Allows you to define and check data validation rules (e.g., ranges, formats, and logical conditions) to ensure data integrity.
    - `skimr`: Offers a compact summary of datasets, including missing values, data types, and basic statistics, which is great for quick data profiling.
    - `naniar`: Specializes in handling and visualizing missing data, making it easier to understand patterns of NA values and impute them if needed.
    - `Hmisc`: Includes tools for data analysis and cleaning, such as imputation, labeling, and descriptive statistics.
    - `summarytools`: Provides detailed summaries of datasets, including frequency tables, descriptive stats, and missing value counts.
- **Data privacy and security**: Protecting the privacy and security of data is paramount, especially when handling sensitive and personally identifiable information. In R you could apply appropriate measures to safeguard data from unauthorized access, breaches, or misuse. This includes practices such as anonymization, encryption, and access controls. The keyring package, introduced earlier in this chapter for securely managing passwords is an example of implementing privacy and security measures. It

enables the safe storage and retrieval of credentials and reduces the risk of exposing sensitive information in code. There are several other R packages that support data protection practices such as anonymization, encryption, and secure access control. Some notable examples are as follows:

- `digest`: Offers cryptographic hash functions (e.g., MD5, and SHA-256), which is useful for verifying data integrity and creating secure identifiers.
- `openssl`: Facilitates encryption, decryption, and secure communication using OpenSSL, making it suitable for protecting data during storage or transmission.
- `credentials`: Assists in managing Git credentials securely, particularly when working with private repositories or deploying code that accesses sensitive data.
- `httr`: Supports secure web communication and authentication protocols, including OAuth and token-based access, which are essential for accessing protected APIs.

Together, these packages provide a robust foundation for safeguarding sensitive data and ensuring compliance with privacy standards in analytical workflows.

- **Documentation and metadata management**: Proper documentation and metadata management contribute to data governance by promoting transparency, reproducibility, and understanding of the analysis process. It involves but is not limited to documenting data sources, data transformations, variable definitions, and any assumptions or decisions made during analysis. In R, several packages support these practices by enabling structured documentation, automated reporting, and metadata handling. For instance, you can use **rmarkdown** or **Quarto** to create reproducible reports that integrate R code, SQL queries, visualizations, and narrative text in a single document. Beside this:
  - `knitr`: Works in a streamlined way with rmarkdown to execute embedded code chunks and generate well-formatted outputs
  - `codebook`: This package can assist in generating comprehensive metadata documentation for datasets, including variable descriptions, value labels, and summary statistics.
  - `dataMeta`: This provides tools for managing and attaching metadata to data frames, ensuring that contextual information remains linked to the data throughout the analysis.

Through proper use of these tools, you can maintain clear and up-to-date documentation that fosters collaboration, facilitates knowledge transfer, and ensures the integrity of results from analysis.

## Summary

In this chapter you learned about a range of packages and functions to import data in R from flat files, such as CSV and TSV. You also learned how to connect and extract data from various databases that modern businesses frequently use. You also learned how to use SQL from R in different scenarios, from standard to complex parameterization. Finally, the importance of data governance while doing analysis in R was highlighted and actions to practice were presented.

You have just finished the last chapters of *Part 1* of the book. You have learned about R as a software suit and a programming language in *Chapter 1*. You learned about different objects and data types and how to write functions in *Chapter 2*. You also learned various techniques to import and export data in R in *Chapter 3*.

Now you are about to embark on the time series journey using R as a programming language. The chapters in *Part-2* of this book cover an introduction to time series. *Chapter 4* discusses time series and its components. *Chapter 5* discusses special techniques required when wrangling and visualizing timeseries data. *Chapter 6* presents some business applications of time series analysis and forecasting.

## Further reading

- Wickham, H. (2021). *nycflights13: Flights that departed NYC in 2013* (Version 1.0.2).

# Get this book's PDF version and more

Scan the QR code (or go to `packtpub.com/unlock`). Search for this book by name, confirm the edition, and then follow the steps on the page.

UNLOCK NOW

*Note: Keep your invoice handy. Purchases made directly from Packt don't require an invoice.*

# Part 2

# The Main Character: Time Series

This part is focused on introducing time series data. Time series components, characteristics, and types are presented. Some data wrangling and visualization techniques specialized for time series data are covered with code examples. We highlight some practical and real-life areas where data problems can be solved using time series methods and modeling.

*Part 2* comprises the following chapters:

# 4

# Time Series Characteristics

This chapter provides a formal introduction to time series data and distinguishes it from other types of time-indexed data. It lays out the conceptual groundwork for why specialized methods, models, and techniques are essential for analyzing time series. By exploring the unique characteristics of time series data, you will gain the intuition necessary for effective analysis.

This chapter is organized as follows:

- Section 4.1: What is a time series?
- Section 4.2: Distinction of time series from other time-indexed data
- Section 4.3: How to classify time-indexed data to apply suitable analytics
- Section 4.4: Components of a time series
- Section 4.5: Combining time series components
- Section 4.6: Stationarity
- Section 4.7: Autocorrelation
- Section 4.8: Types of time series

## 4.1 What is a time series?

A time series is a sequence of observations recorded over multiple time periods for a single unit of interest. These measurements are typically taken at regular intervals, such as hourly, daily, monthly, or annually, creating a natural temporal ordering in the data. This chronological ordering might introduce dependence between successive observations. This means observations at the current time point may be influenced by past observations. This dynamic behavior is a defining feature of time series data and allows it to be analyzed in ways that are distinct from other types of data.

This book primarily explores the analysis of univariate time series, which refers to the study of temporal data collected from a single variable or unit of interest. Univariate time series are foundational in time series analysis and serve as the starting point for understanding more complex temporal structures.

However, as we progress through the chapters, we will extend our focus beyond univariate series to include both multivariate and multiple time series. These broader frameworks allow us to capture richer dynamics and interdependencies that arise in real-world data.

Multiple time series are collections of univariate time series that might be interrelated or independent in terms of time series dynamics. Several interrelated univariate time series observed simultaneously are generally referred to as multivariate time series. This structure is commonly encountered in macroeconomic datasets, for instance, the **gross domestic product (GDP)** growth rate, the unemployment rate, and the official interest rate of a country. Even though each of these can be analyzed independently as a univariate time series, when studied together, they often exhibit dynamic interactions. Such interaction can help to explain both short-term fluctuations and long-term equilibrium relationships between the series; both of them must be accounted for in a comprehensive analysis. In short, multivariate time series analysis incorporates modelling interdependence among multiple series collected at the same period.

It is important to note that multivariate time series typically involve only a small number of time series. This limitation arises because the complexity of the analysis increases rapidly with the number of series involved. Specifically, the volume of data required to perform robust multivariate analysis grows rapidly with each additional series, often making it impractical to include a large number of time series without compromising analytical feasibility. To avoid these constraints multivariate time series with a large number of univariate series (say, 10 or more) are often analyzed as independent univariate time series in the first stage and, in a subsequent stage of analysis, between series time dynamics are overlaid. In this book, this is categorized as a univariate time series.

## 4.2 Distinction of time series from other time-indexed data

In real-life analysis, you might encounter situations where multiple units have data collected over time but are independent or need to be assumed independent due to an inadequate amount of data for enabling either a univariate or a multivariate time series analysis. This typically happens when univariate time series for tens to thousands of units are relatively short and/or have missing values for modeling the time dynamic adequately from individual series.

For example, suppose you have annual debt and turnover data for small-business customers (with turnover between $500K and $1M) for a bank for the last 10 years from the bank's customer database. While it may seem intuitive to apply time series methods to each business individually to analyze their debt-to-turnover ratio, practical challenges often arise. Small businesses may evolve, disappear, or emerge over time, resulting in datasets with inconsistent temporal histories and varying levels of noise. These complexities can limit the effectiveness of standard time series techniques. This results in three distinct types of data, all of which can have a time dimension but are not necessarily considered time series data.

An effective approach to analyzing the debt profile of small businesses over time is through panel data analysis. Panel data combines both cross-sectional and temporal dimensions by capturing repeated observations for a relatively large number of units, such as businesses across multiple time periods. This structure enables you to study how changes in turnover influence debt levels while accounting for characteristics unique to each business, such as management practices or industry-specific factors. Even when some businesses have missing data for certain years, forming an unbalanced panel, the method remains robust. By leveraging both within-unit and between-unit variation, panel analysis allows for stronger attribution and inference and is particularly well suited to answer questions such as: Do businesses with growing turnover tend to reduce their debt ratios over time?

An alternative analytical strategy is longitudinal data analysis, which is particularly well suited to studying within-unit changes over time. Longitudinal data typically consists of observations for a relatively small number of units tracked across a moderate to extended time span. The central aim of this approach is to model individual trajectories and examine how characteristics evolve within each unit.

For instance, you might focus on a cohort of businesses established in 2015 and still operating in 2025, allowing for an in-depth exploration of the debt evolution over the business life cycle. This method is ideal for identifying patterns such as early-stage borrowing, mid-phase expansion, or late-stage deleveraging. Because longitudinal analysis relies on consistent tracking of the same units, it requires relatively complete data histories, and missing observations can significantly constrain the scope of analysis. Nevertheless, when such data is available, longitudinal methods offer powerful insights into how financial behavior develops and transforms within stable business units over time.

Cross-sectional analysis offers a valuable approach for examining the financial characteristics of businesses at a specific point in time. Unlike longitudinal or panel data, cross-sectional data does not capture temporal variation within units, making it unsuitable for analyses that require tracking changes over time. Instead, it provides a snapshot of the population, allowing researchers to explore patterns such as the distribution of debt-to-turnover ratios, identify high-risk businesses, or compare financial metrics across industries and regions. This approach can also be extended across multiple years by treating each year as a separate cross-section, enabling comparisons over time without the need to follow individual businesses longitudinally. Cross-sectional analysis is particularly useful when historical data is incomplete or when the analytical focus is on population-level insights rather than individual trajectories.

## 4.3 How to classify time-indexed data to apply suitable analytics

There are formal statistical tests available to determine the most suitable analytical framework. When deciding whether to use panel, longitudinal, or time series methods, a simple heuristic may be helpful to guide the initial decision. One such approach is the use of the **N–T** decision space, where *N* represents the number of units (e.g., businesses, individuals, or households) in the dataset, and *T* denotes the number of time points covered. This framework provides a practical starting point for evaluating the structure of the data. *Figure 4.1* presents a convenient framework for classifying data structures commonly used in empirical analysis.

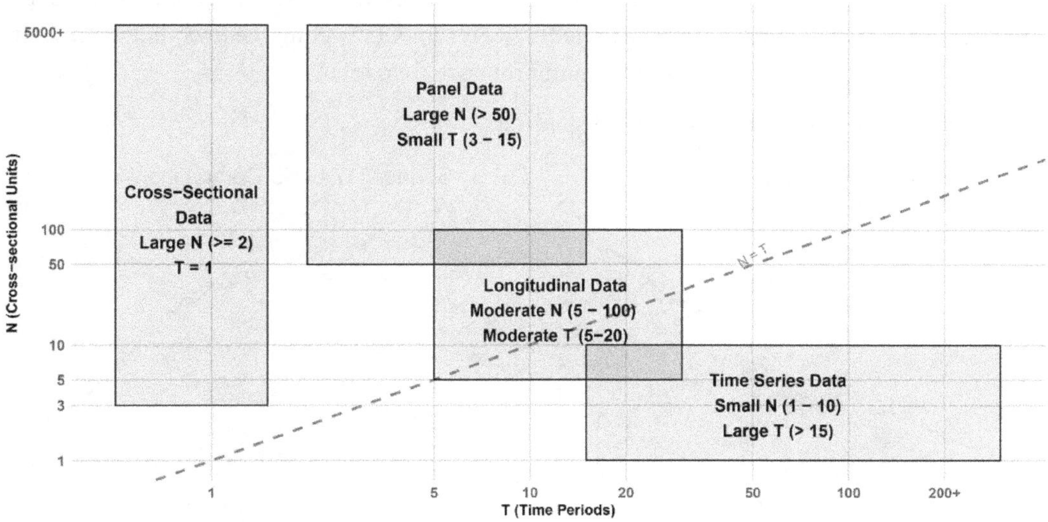


*Figure 4.1: Approximate data type classification in N-T decision space*

When $N$ is large and $T$ is small, panel data methods are typically appropriate, as they leverage cross sectional breadth with limited temporal depth. Conversely, when $T$ is large and $N$ is small, longitudinal analysis becomes more suitable, focusing on individual trajectories over time. If the dataset consists of a single unit observed over many time points, time series analysis is the natural choice.

This simple classification can help you align your analytical strategy with the inherent structure of the data, especially when dealing with incomplete records or limited metadata. However, it is important to recognize that this classification is approximate. The thresholds for the number of units (N) and time periods (T) are not rigidly defined; rather, they are context-dependent and often shaped by the specific analytical objectives at hand.

## 4.4 Components of a time series

Components, also known as **features**, **patterns** or **characteristics**, are the building blocks of an observed time series. There are three main components: trend ($T$), seasonal ($S$), and cyclical ($C$). These components, along with random error ($E$), generate the observed time series.

### 4.4.1 Trend

A trend is the long-term behavior in a time series depicted by an upward or downward movement over an extended period. This reflects a persistent increase or decrease in the measurements of an univariate time series and is often driven by factors, such as long term economic growth, technological progress, or both.

The main distinguishing factor to notice for trend is the persistent and consistent movement over a longer period. Identifying the trend is crucial for understanding the overall dynamics of a time series and analyzing what could account for such dynamics.

For illustration, two series are plotted in *Figure 4.2*:

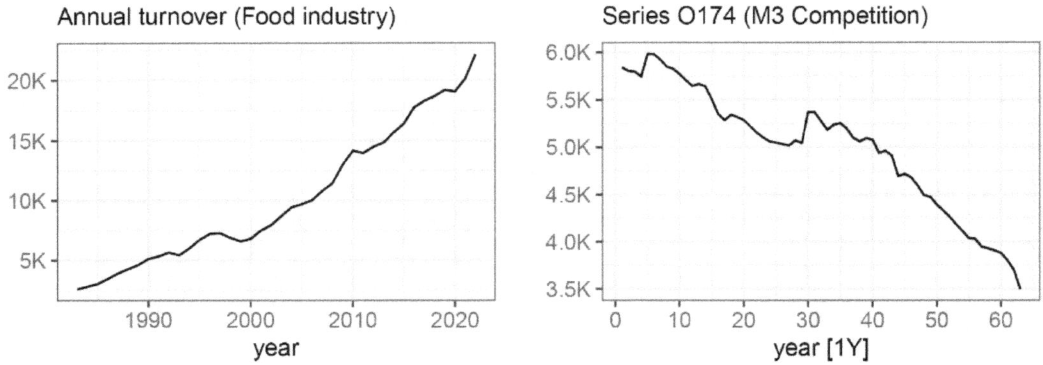


*Figure 4.2: Examples of upward and downward trends*

The left panel is the monthly takeaway food services turnover aggregated at a yearly level, showing a strong, positive trend. On the right is an annual series from the telecommunications industry, which is the 3003-th series in the M3 competition (Series no. O174), showing a downward trend.

## 4.4.2 Seasonality

Seasonality refers to the presence of repeating patterns in time series data that occur at regular intervals. These intervals, known as seasonal periods, are typically driven by systematic and recurring external factors such as the day of the week, the time of day, or the season of the year.

Each seasonal period is associated with a fixed number of observations, which is referred to as the frequency of the data. For example, in monthly time series data, the frequency is 12, corresponding to the 12 months in a year. This means that if a seasonal pattern exists, the data for a given month is expected to resemble the data from the same month in previous years.

Such patterns arise because similar conditions tend to recur at the same relative time each cycle. For instance, retail sales may consistently rise in December due to holiday shopping, or electricity usage may peak during summer months due to increased air conditioning. These recurring swings in the data reflect the influence of seasonal factors and are key to understanding and forecasting time series behavior.

Recognizing and modeling seasonality is essential for accurate predictions, as it allows you to separate short-term fluctuations from long-term trends and make more informed decisions based on historical patterns.

The monthly turnover of takeaway food services during 1983-2022 is plotted in *Figure 4.3*. Compared to the annual data presented in *Figure 4.2*, this series shows a distinct pattern that repeats every month. For better visibility, a five-year window is plotted on the right panel that shows the clear seasonal pattern by month.

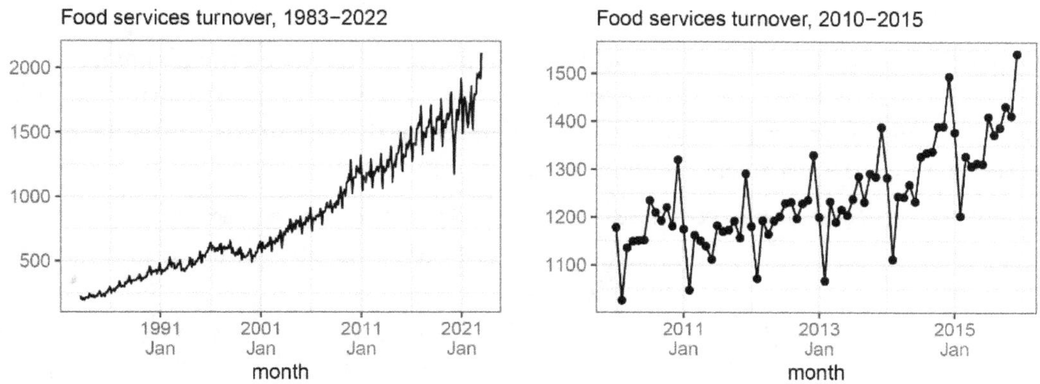


*Figure 4.3: Examples of seasonality and trend*

For yearly data, there are no options for seasonality as each seasonal period is shorter than a year. Seasonal periods are usually integer, such as monthly (12) and quarterly (4), or fractional, such as annual seasonality in weekly observations (52.18), and daily observations (365.25). Usually, we consider integer seasonal periods and approximate fractional values while modeling.

Data might show multiple seasonality depending on the granularity of data collection. Daily data might have two seasonal periods, one weekly (7) and one annual (365.25). For hourly data, the possible seasonal periods are daily (24), weekly ($24 \times 7 = 168$), and annual ($24 \times 365.25 = 8766$). To capture the seasonality while modelling it is important to identify the correct seasonal period and whether there are multiple seasonalities present in the observed data.

### 4.4.3 Cyclical

In time series, a cyclical pattern refers to a long term fluctuation that occurs over an extended period. Generally, a peak-trough pattern repeats over many years. The cyclical pattern has variable width and magnitude, and these characteristics differentiate the cyclical pattern from a seasonal pattern, which has a fixed width.

An example of a cyclical pattern in time series is business cycles. Expansion and recession are common business cycles that appear in a lot of economic data. During an expansion phase, economic activity increases, often driven by rising consumer confidence, investment, and production. Conversely, a recession marks a period of declining economic performance, reduced spending, and contraction in business activity. However, each cycle does not repeat at fixed intervals. When a business senses it is about to enter a recession, it does everything to delay it, which makes the width wider than the previous recession cycle. In the expansion stage, a business adopts steps that make the upward movement steeper than the previous expansion period.

Understanding cyclical behavior is crucial for your analysis involving strategic planning, particularly in fields such as economics, finance, and policy making. Along with historical data and external factors and market dynamics that influence the timing and intensity of these cycles, you can perform a much deeper analysis.

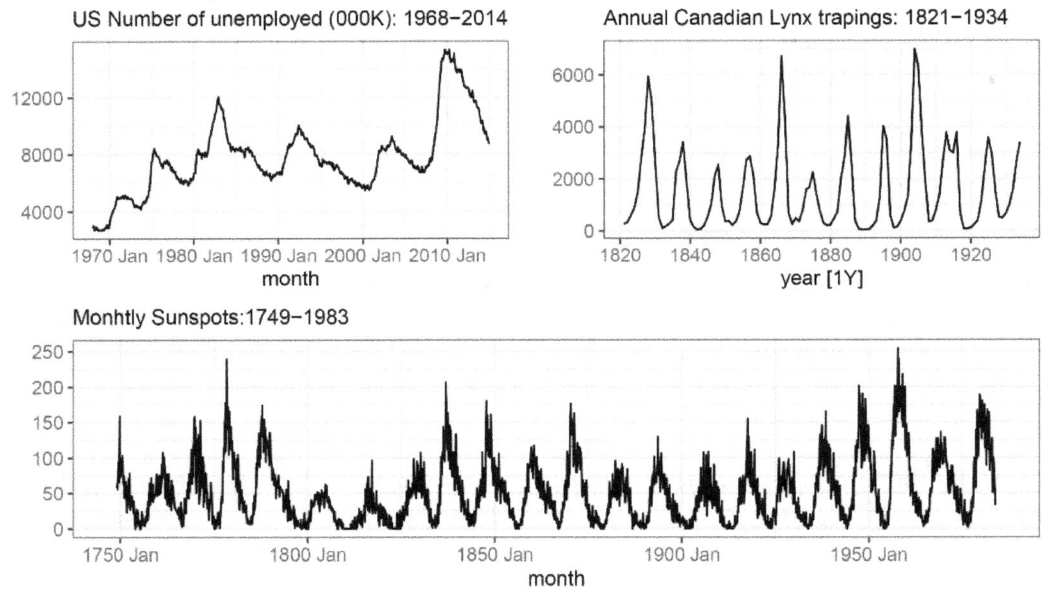


*Figure 4.4: Examples of cyclical series*

A cyclical pattern is a recurring pattern of rises and falls in data that does not have a fixed period. This means that the length of the cycle can vary from one occurrence to the next, making it difficult to predict and estimate. Having lengthy data, such as monthly sunspots or annual lynx trappings in *Figure 4.4*, might make cyclical series more predictable than, for example, the number of unemployed people in the US per month.

## 4.4.4 Random errors or remainders

Like any form of data, time series data is subject to random error or unpredictable variations that cannot be attributed to identifiable patterns such as trend, seasonality, or cyclical behavior. Once these structured components are accounted for, what remains is the residual or error component, which captures the influence of irregular, unforeseen events.

These errors often arise from disruptions in the underlying data generation process. For example, the COVID-19 pandemic had a major impact on sales, employment, and economic indicators across industries. Such events are not part of the regular dynamics of the data and cannot be anticipated by analyzing historical data.

While the random fluctuations are valuable for exploratory analysis often to understand the impact of past disruptions, they hold limited relevance for predictive modeling. Forecasting future values based on random error would require predicting future unexpected events such as pandemics, natural disasters, or geopolitical shocks, which is highly speculative.

Now that you know the different components of time series, let's discuss how they can be present in data. Does one component always dictate the data, or is it a combination of components?

# 4.5 Combining time series components

A time series is rarely composed of a single component. Instead, it typically reflects a combination of components: trend, seasonality, cyclical variation, and random error. The presence and strength of these components can vary significantly across datasets, and understanding their interplay is essential for effective analysis.

To exhibit temporal dynamics, a time series must contain at least one of the structured components: trend, seasonality, or cyclical patterns. In practice, it is common to encounter time series that exhibit all four components, including random error. The presence of different components can dictate how the data needs to be analyzed and modeled.

It is worth noting that the error component is always present in any time series. This is because no data is perfect, and there will always be some random fluctuations that cannot be explained. Conceptually, it is possible to imagine a time series that is made up of the error component only. However, such data will be better treated as cross sectional data and will benefit from using cross-sectional data analysis. This is because such a series lacks time dynamics.

The way the patterns are combined when they are present is also an important determinant for analysis. Two commonly assumed structures are the following:

- Additive: $Y_t = T_t + S_t + R_t$, where $t = 1,2,\ldots,T$ represent time
- Multiplicative: $Y_t = T_t \times S_t \times R_t$

In the preceding expressions, the notation $Y_t$ stands for an observed value of the time series at time $t$. $T_t$ represents combined impact of trend and cycle. $S_t$ is the seasonal component and $R_t$ is the random error.

Let's visually examine two series that are presented in *Figure 4.5*.

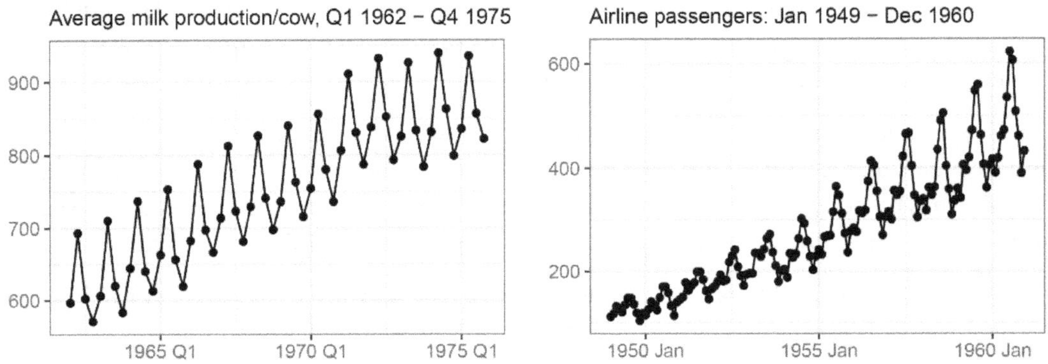


*Figure 4.5: Examples of time series with an additive effect of trend and seasonality*

The quarterly milk production series in *Figure 4.5* has an additive relationship between seasonality and trend, as the magnitude of seasonality stays constant with an increase in trend. In contrast, in the monthly international airline passengers plot, the seasonality increase indicates that the trend and seasonality are combined multiplicatively.

The first step in any successful time series analysis is to identify which components are present and how they are combined. In real world data, components may appear with varying intensities. A strong trend, for instance, might obscure a weaker seasonal pattern. Therefore, it is crucial to decompose the series and study each component individually before proceeding with modeling. You will learn systematic methods for decomposing time series data to isolate and analyze its underlying components in *Chapter 7*.

Next, we turn our attention to two fundamental properties of time series, stationarity and autocorrelation. Not only are these characteristics essential for distinguishing time series data from other types of data, but they also play a critical role in the proper analysis of data that exhibits time dynamics.

## 4.6 Stationarity

Stationarity is a fundamental property of a time series that indicates whether its statistical properties, such as mean, variance and the correlation structure, remain constant over time. In simple terms, a stationary series behaves consistently throughout its time duration. The way components are combined and repeated does not depend on the specific point in time at which the measurements are taken. This stability allows you to analyze the series with greater confidence as the underlying dynamics of the process are assumed to be time invariant.

In a stationary time series, the average level remains roughly constant as time progresses. Pure trend and trend-seasonal patterns are not stationary as the average level changes as time moves forward. Some pure cyclical patterns are stationary, such as the Canadian lynx in *Figure 4.4*. The number of unemployed in the US and sunspots series in the same figure are non-stationary due to the trend-cyclical pattern. All other example series you have seen in this chapter are non-stationary.

If a dataset shows departure from stationarity, such as the presence of trends or seasonality, it is a strong indicator that the data is a time series rather than cross-sectional. In this sense, departure from stationarity helps distinguish time series data from other types and thus considered a characteristic of a time series.

There are proper statistical methods to test stationarity in an observed series. If data is non-stationary, then suitable transformations such as differencing and seasonal differencing are performed to make it stationary before analysis.

Next, we will extend the well-known concept of correlation in the context of time series. Let's see what autocorrelation means in a time series context.

# 4.7 Autocorrelation

In cross-sectional data, correlation measures the strength of the linear relationship between two variables. A conceptually similar measure for time series data is known as autocorrelation, a defining feature of a time series. In fact, this characteristic is so important that its absence might indicate that data could be adequately analyzed without using a specialized time series analysis.

The autocorrelation is calculated between the current and lagged values of the same series. In general terms $k$-th autocorrelation is calculated between $y_t$ and $y_{t-k}$:

$$r_k = \frac{\sum_{t=k+1}^{n}(y_t - \bar{y})(y_{t-k} - \bar{y})}{\sum_{t=1}^{n}(y_t - \bar{y})^2} \qquad (4.1)$$

Typically $k$ ranges between $1$ to $\frac{n}{3}$. $k = 0$ is the correlation of the time series, $y_t$, with itself, and the autocorrelation is $1$. In an autocorrelation plot, the autocorrelation is generally plotted for the entire range of $k$, along with significance boundaries, which is also known as the **autocorrelation function** (**ACF**). Any spike outside the significance boundary implies a strong dependence of the current value on that lagged value.

The ACF plot is a great tool to identify various time series components, such as trend, cycle, and seasonality that you need to account for in the analysis.

*Figure 4.6* shows an autocorrelation plot of the original monthly takeaway food services turnover and the de-trended ACF of the same series.

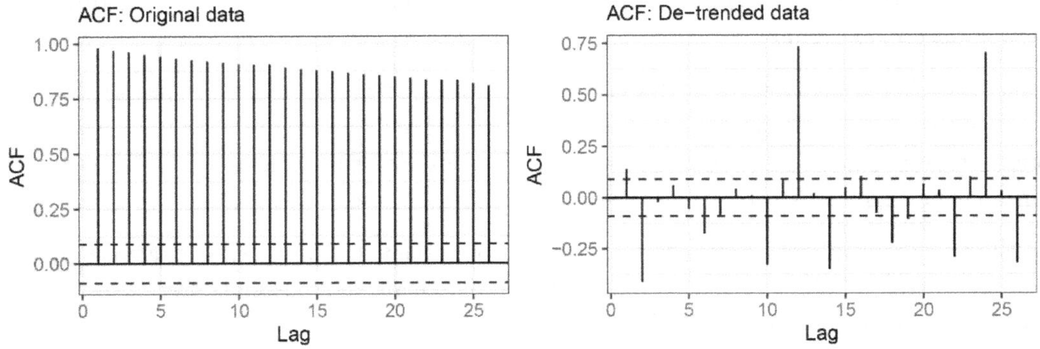


*Figure 4.6: Autocorrelation of monthly takeaway food turnover: original data and detrended data*

The ACF of the original series decaying slowly and shows significant spikes up to the 26-th lag. This indicates that there is a strong trend in the data. This does not mean that the current value depends on the previous 26 values. This shows the longer impact of serial dependency carried forward as time increases. The seasonality is also not visible as the strong trend component masks the seasonal pattern. The de-trend ACF of the same series gives significant seasonal spikes at lag 2, 6, 10, and 12 with various magnitudes.

These visual tools (time series plot and ACFs) are helpful for identifying time series patterns, how they are combined, and whether they are stationary or not. These tools also work as diagnostics to assess the quality of the analysis. Understanding time series data by visual inspection is an important skill to gain that comes with experience. It is not possible to inspect every series visually when you must analyze many series in a short time. Most modeling approaches are automated to identify patterns and their magnitude, such as presence of stationarity, by conducting various statistical tests. However, there might be situations where even a good model does not produce desirable results. In such a scenario, these visual tools provide information on what is missed by the model and an opportunity to improve when you know how to use them effectively.

Time series can also be grouped according to their shape, or how often they are collected. Let's learn more about this in the next section.

# 4.8 Types of time series

So far, we have discussed time series that are collected over equal spaces of time and have a linear relationship between the current and past values. This can be explained via a general equation:

$$y_t = \beta_0 + \beta_1 y_{t-1}, \ldots, +\beta_k y_{t-k}.$$

Here, both the parameters $(\beta_0, \ldots, \beta_k)$ and the past values $(y_{t-1}, \ldots, y_{t-k})$ lie in linear space.

They are known as regular time series with a linear relationship. All the examples you have seen from *Figure 4.2* to *4.4* are regular time series. Even though their frequencies are different, observations are collected at constant intervals; some are monthly and some are quarterly.

There are many other types of time series available in real scenarios:

- **Intermittent time series**: In an intermittent series, the observations between time intervals are not constant. In this type of series, there is a period of observations followed by periods of no data or missing or zero observations. This can be due to a variety of factors, such as the nature of the process being measured, the availability of data, or the limitations of the measuring instrument. Sales data of a specialized product that depends on cultural or religious festivals are examples of intermittent time series. For example, Father's Day cards are exclusively accessible a few weeks prior to the celebration, and the rest of the year the sale value would be 0.
- **Irregular time series**: In irregular time series, the intervals between the observations are inconsistent, but they still can form a continuous sequence without gaps. The irregularity can occur due to varying measurement schedules, data collection methods, or other factors. For example, sales at different times of day will become an irregular time series. There will be an observation for each day, but due to inconsistent intervals, it will be hard to depict any patterns from the observed data within a day.
- **Non-linear time series**: In non-linear time series data, relationships, patterns, or trends exhibit non-linear progression. This could be due to either parameters being non-linear $(\beta_i^2)$ or past values having a non-linear impact $(y_{t-i}^2)$ on the current value. The spreading of infectious disease is a perfect example of a non-linear series. Initially, the number of people affected grows exponentially and slows down toward the saturation point due to public health intervention. A regular time series with non-linearity over a time domain can be analyzed by making it linear using an appropriate transformation, such as logarithmic or square roots.

Analyzing irregular or non-linear time series requires a different set of tools, which is out of the scope of this book.

# Summary

In this chapter, you have learned about the components of a time series: trend, seasonality, cycles, and irregularity. The components could be combined in two ways, either additively or multiplicatively. We have discussed two important characteristics of time series, stationarity and autocorrelation. An autocorrelation plot is an important tool in time series modeling to inspect how strongly the current observation depends on the past values. You have learned how to interpret an ACF plot and have seen scenarios where one prominent component masked underlying patterns. Additionally, you learned that there are other important types of time series, intermittent, irregular, and non-linear, although these are not explored in detail in this book.

Real-world time series data comes in raw and messy formats that you need to clean, organize, and prepare for analysis. *Chapter 5* is dedicated to this topic, and you will learn with examples how to manipulate, summarize, and aggregate time series data.

# Join our community on Discord

Join our community's Discord space for discussions with the author and other readers:

`https://discord.gg/G4YmB6DQvf`

# 5

# Time Series Data Wrangling and Visualization

In this chapter, some common data wrangling and visualization techniques for time series are presented. Data wrangling is an essential yet often overlooked process of grappling with raw data, getting it into shape for further analysis. This pivotal stage includes cleaning, transforming, refining, and summarizing data in alignment with analysis objectives. In addition, visualization plays a significant role when working with time series data. It aids in recognizing existing patterns and understanding their combinations within the observed data. Skillful visualization is instrumental in selecting the most suitable forecasting model for the task at hand.

The complexity of time series data wrangling largely depends on two things: (a) the messiness of the raw data, and (b) the objectives of next-stage analysis. For example, if you have daily data and in the experiment phase, you do not know whether a model with weekly data or a model with monthly data will be more insightful. In such a situation, you need to clean the daily data as much as possible and aggregate it to the desired level just before modeling. In other scenarios, the data you have received could already be summarized at the desired level, but you may still need to clean and transform it appropriately.

This chapter is organized as follows:

- Section 5.1: Parsing date-time variables
- Section 5.2: Setting components of date-time
- Section 5.3: When to transform data into time series object
- Section 5.4: Wrangling functions
- Section 5.5: Visualization functions

- Section 5.6: Time plots
- Section 5.7: Seasonal plots
- Section 5.8: Visualizing relationships among time series
- Section 5.9: Interactive plots

Note

**Your purchase includes a free PDF copy + exclusive extras**

Your purchase includes a DRM-free PDF copy of this book, 7-day trial to the Packt+ library (no credit card required), and additional exclusive extras. See the *Free benefits with your book* section in the *Preface* to unlock them instantly and maximize your learning.

# Technical requirements

Clone the GitHub repo of the book, `https://github.com/PacktPublishing/Modern-Time-Series-Analysis-with-R.git` on your local machine. You will find the data files used in this chapter in the `data/data_chap5` folder.

Alternatively, you can manually download the files from the GitHub repo and save them under the `data/data_chap5` folder in the local R project on your machine.

In your R workspace, load the following packages to practice examples from this chapter:

```
#install.packages("pacman")
library(pacman)
p_load( "conflicted"
       ,"fable"
       ,"fabletools"
       ,"feasts"
       ,"forecast"
       ,"GGally"
       ,"lubridate"
       ,"Mcomp"
       ,"nycflights13"
       ,"patchwork"
       ,"plotly"
       ,"tidyverse"
       ,"tsibble"
       ,"tsibbledata")
```

```
# manage conflicts
conflicts_prefer(dplyr::filter)
```

The `conflicts_prefer()` function manages function conflicts between R packages. In this case, the `filter()` function from the `dplyr` package is preferred over the same-named function from the `stats` package.

First, we discuss how to properly parse a date-time variable. You will also learn about the issues that arise when dates are not properly parsed, especially when working with data coming from various sources.

# 5.1 Parsing date-time variables

During the data wrangling process, you need some kind of mechanism to check whether formatting date/date-time variables in R produces no error and gives desirable results. Unfortunately, there is no one-size-fits-all method. However, validating simple logic will save time and frustration later. For example, calculating unique years and sorting them enables us to check whether formatting performs as expected. Another option is to calculate the number of observations in each year. A too-low number or too-high number (more than 365 for daily data, or more than 12 for monthly data) would identify any potential issues.

Let's elaborate on this with a toy dataset. Load the `mismatch_date.csv` file using the `read_csv()` function and check the variables type:

```
mismatch_dt <- read_csv("data/data_chap5/mismatch_date.csv")
glimpse(mismatch_dt)
```

You will see the following output where the date column is loaded as character type:

```
Rows: 9
Columns: 2
$ date  <chr> "1/01/2015", "1/02/2015", "1/03…
$ value <dbl> 1630, 826, 567, 139, 256, 113, …
```

The data frame has nine rows and the date column could be in `DD-MM-YYYY` format. Now format the date and compute the number of observations in each year using the following code:

```
year_count <- mismatch_dt |>
  mutate(date = as.Date(date, format = "%d/%m/%Y")) |>
  mutate(year = year(date)) |>
```

```
  group_by(year) |>
  tally()
year_count
```

The following output showing two different year entries in date:

```
# A tibble: 2 × 2
   year     n
  <dbl> <int>
1    15     1
2  2015     8
```

The impact of even a small number of incorrect data entries in a large file can be significant, especially when summarizing data at higher levels, such as weekly, or monthly. This issue can escalate when data processing includes loading and merging files from multiple sources. It is unlikely that a unified date-time input is always followed across an organization. Similar challenges can arise when sourcing data from multiple schemas and tables within a database. These inconsistencies and discrepancies can lead to inaccurate analyses, misleading conclusions, and potentially costly decisions. It highlights the critical importance of rigorous data validation and cleaning procedures before proceeding with data aggregation and analysis.

The `lubridate` package offers a wide range of helper functions for easy and fast parsing of inconsistent date-time formats. The choice of function depends on the format in the source file, not how you expect them to appear in R. For illustration, apply the `dmy()` and `ymd()` to parse the date column in the `mismatch_dt` as shown in the following code:

```
mismatch_dt |>
  mutate(date_accurate = dmy(date),
         date_inaccurate = ymd(date))
```

The code generates the following output with a warning:

```
# A tibble: 9 × 4
  date      value date_accurate date_inaccurate
  <chr>     <dbl> <date>        <date>
1 1/01/2015  1630 2015-01-01    NA
2 1/02/2015   826 2015-02-01    NA
3 1/03/2015   567 2015-03-01    NA
4 1/04/2015   139 2015-04-01    NA
5 1/05/2015   256 2015-05-01    NA
6 1/06/2015   113 2015-06-01    NA
```

```
7 1/07/2015    166 2015-07-01     NA
8 1/8/15       315 2015-08-01     NA
9 1/09/2015    710 2015-09-01     NA
Warning message:
There was 1 warning in `mutate()`.
i In argument: `date_inaccurate = ymd(date)`.
Caused by warning:
! All formats failed to parse. No formats found.
```

In the `mismatch_dt.csv` file, the date column was entered as date/month/year style. The `dmy()` function parsed it accurately, whereas the `ymd()` function failed to parse and produced NA with a warning message.

The `ymd()` function expects formats such as "2015-08-01", "2015080", or "2015/08/01", not "1/8/15" or "1/08/2015". The `lubridate` package includes a variety of parsing functions: `ymd()`, `mdy()`, `dmy()`, `ymd_hms()`, and `dmy_hms()`. Use the appropriate function based on the date-time formatting style in the source (not how you want to see it in R). Any warning message or NA instance will indicate whether a parsing function is suitable and produces desirable results.

After correctly parsing a date-time variable while loading data in R, you may need to extract various parts from the date-time variable, such as weekdays from a date or week numbers in a calendar year. There are many helper functions that can be used for setting or extracting components of a date-time variable. In the next section, some of these helper functions are discussed.

# 5.2 Setting components of date-time

At times, you need to extract different components from a date-time variable, for example, extract the year or month from a date. The `lubridate` and `tsibble` packages provide helper functions to perform such operations.

Let's take a closer look at some functions from `lubridate` package:

- The `week()`, `month()`, and `year()` functions return the corresponding part of a date as an integer. The week/month name can be returned with the `label=TRUE` option.
- The `day()`, `wday()`, `mday()`, `qday()`, and `yday()` functions return the corresponding number of days.
- The `hour()`, `minute()`, and `second()` functions extract the corresponding part of a time index.

You can see some of the aforementioned functions in action in the following code snippet:

```
set.seed(45)
test_df <- tibble(
  date = as.Date(c("2019-01-01",
                   "2019-09-12",
                   "2020-01-01",
                   "2020-05-10")),
  value = runif(4) )

test_df |>
  mutate(year=year(date)
         ,day_year=yday(date)
         ,day_month=mday(date)
         ,month_name=month(date, label=TRUE))
```

You can examine the following output to understand the results from each function:

```
# A tibble: 4 × 6
  date       value  year day_year day_month month_name
  <date>     <dbl> <dbl>    <dbl>     <int> <ord>
1 2019-01-01 0.633  2019        1         1 Jan
2 2019-09-12 0.318  2019      255        12 Sep
3 2020-01-01 0.241  2020        1         1 Jan
4 2020-05-10 0.378  2020      131        10 May
```

Functions from the `tsibble` package include the following:

`yearmonth()`, `yearquarter()`, and `yearweek()` extract corresponding component and align it nicely with the year part. The following code snippet showing you how to apply these functions:

```
test_df |>
  mutate(year_month=yearmonth(date)
         ,year_quarter=yearquarter(date)
         ,year_week=yearweek(date))
```

The following output is generated:

```
# A tibble: 4 × 5
  date       value year_month year_quarter year_week
```

```
   <date>     <dbl>      <mth>         <qtr>    <week>
 1 2019-01-01 0.633   2019 Jan       2019 Q1   2019 W01
 2 2019-09-12 0.318   2019 Sep       2019 Q3   2019 W37
 3 2020-01-01 0.241   2020 Jan       2020 Q1   2020 W01
 4 2020-05-10 0.378   2020 May       2020 Q2   2020 W19
```

Observations may be collected against some units of time, but that does not necessarily mean that they are already a time series object by the definition of R. Will you transform any temporal data into a time series object as soon as you load it into R? The answer depends on the objectives of your analysis. Let's clarify this question in the next section.

# 5.3 When to transform data into time series object?

In an analysis, the need for a time series object arises when specific models require time series data as input. However, some models do not necessitate a time series object. In such cases, temporal data can be kept in a `tibble` object, allowing for wrangling tasks, such as cleaning and summarizing without the need for transformation into a time series object. Therefore, your choice of functions for data wrangling depends on the downstream analysis requirements.

Regardless of the future modeling choice, parsing temporal data into a `tsibble` object when data wrangling proves efficient, especially when handling numerous series simultaneously. This format retains the time series structure while offering flexibility for various analyses, making it a valuable option for managing temporal data.

In *Chapter 2*, we discussed various ways to transform data into a time series object. They are summarized in *Table 5.1* for your reference.

| Package | Function | Application |
|---|---|---|
| `forecast` | `ts()` | Transform a single series |
| `tsibble` | `tsibble()` or `as_tsibble()` | Transform many series at once, works with `ts`, `tibble` or `data.frame` format |

*Table 5.1: Functions to transform data into a time series object*

Both the `forecast` and `tsibble` packages provide helpful tools for managing time series data.

The `tsibble` package is constructed on top of tibble, which adheres to the tidy data principle by default. This means it provides an organized and efficient ways to store and wrangle temporal data. With tsibble, you can easily manage multiple time series in a structured manner, facilitating tasks, such as aggregation, transformation, and analysis.

On the other hand, the `ts()` function from the `forecast` package is designed to be applied on a single series. However, it can be extended to handle multiple series using techniques, such as *for loops* or functional programming with the `purrr` package. While these methods are effective, they might not always be as streamlined or intuitive as using a `tsibble` object for handling multiple time series simultaneously.

So, depending on the complexity of your time series data and the tasks at hand, `tsibble` offers a more integrated and efficient solution for managing and wrangling multiple time series, while the base `ts()` object can still be used effectively for single series or with additional programming effort for multiple series.

In the next section, some common data wrangling functions from the `dplyr` and `tsibble` package are presented. The functions from `dplyr` are available in `tsibble` too, whereas the latter also includes specialized functions needed to wrangle time series data.

# 5.4 Wrangling functions

The collection of helper functions that comes with the `dplyr` package makes any data wrangling process fast, logical and efficient. The functions are collectively known as **verbs**. All the functions from `dplyr` are applicable to a `tsibble` object. Verb functions from the `tsibble` package are optimized for temporal data. *Table 5.2* lists functions from both the packages that are commonly used for data wrangling in R.

| **Operation** | **package:dplyr** | **package:tsibble** |
|---|---|---|
| *Select variables/columns* | `select()` | `select()` |
| *Subset observations/rows by position* | `slice()` | `slice()` |
| *Subset observations/rows evaluating condition* | `filter()` | `filter()` |
| *Time index filtering* | | `filter_index()` |

| Operation | package:dplyr | package:tsibble |
|---|---|---|
| *Transform/create variables* | `mutate()` | `mutate()` |
| *Summarize each group to a row* | `summarise()` | `summarise()` |
| *Identify subgroups to operate on* | `group_by()` | `group_by_key()` |
| *Temporal subgrouping* | | `index_by()` |
| *Order/sorting* | `arrange()` | `arrange()` |
| *Rename columns* | `rename()` | `rename()` |
| *Unique rows* | `distinct()` | `distinct()` |
| *Identify implicit gaps in time* | | `has_gaps()` |
| *Scan for an implicit missing observations* | | `scan_gaps()` |
| *Count gaps* | | `count_gaps()` |
| *Expose implicit missing values* | | `fill_gaps()` |

*Table 5.2: Data wrangling functions from the dplyr and tsibble packages*

The `forecast` package comes with some handy functions as well, such as, `tsoutliers()` for identifying and replacing outliers; `tsclean()` for identifying and replacing outliers and missing values, and `na.interp()` to interpolate missing values in a time series. However, these functions are applied on a time series object that is created by the `ts()` function.

Appropriate use of these helper functions on a proper data object can transform time series data wrangling from a tedious process to a more efficient and rewarding one. The code and examples of these functions are omitted from this chapter as they are used throughout the book from which you will see their contextual application.

You may not have a crystal-clear idea about the objectives and the expected result at the beginning of a data wrangling phase. However, some clarity on the points listed next will save you time, effort, and frustration in the long run:

- Is it a one-off analysis or do you need to repeat it at regular frequency?
- Is it exploratory work versus potential to be part of a data and/or modeling pipeline in production?
- Is the code reproducible?

Maintaining reproducibility in the data preparation process is crucial, regardless of your role in an organization.

For a data analyst preparing data for monthly board meetings, it's essential to have a clear and reproducible workflow. This ensures that each month's data cleaning and transformation steps can be easily replicated with new data, allowing for consistent and reliable presentations. This might involve creating scripts or notebooks that document each step, from raw data to final presentation-ready tables and plots.

As a data scientist handling both data and modeling pipelines, reproducibility is of utmost importance. Your code should be organized in a way that anyone (including your future self) can follow the steps taken from initial data exploration to model development. This helps in understanding decisions made along the way and allows for easy adjustment or expansion of the analysis.

For a machine learning (ML) engineer working on productionizing a model, reproducibility ensures that the entire pipeline, including data cleaning, feature engineering, and model training, can be replicated in a production environment. This guarantees consistent results and easier troubleshooting if issues arise.

In all these scenarios, using version control systems, such as Git and documenting code with comments and clear explanations helps maintain reproducibility. Creating modular and reusable functions for common data cleaning and transformation tasks also streamlines the process and reduces the chances of errors.

By developing reproducible code, you not only ensure consistency and reliability in your analyses but also make it easier for others to understand and build upon your work.

In the next section, you will be introduced to some specialized time series visualization techniques.

# 5.5 Visualization functions

In base R, the `plot()` function with `type=l` is used to generate a time plot. The famous `ggplot2` package is widely used to create visualizations in R, and also has time series data covered. The `ggplot2` package is included in the `tidyverse` package suite. You can also load only the `ggplot2` package. In the rare scenario that you are using `ggplot2` for the first time, the brief overview that follows will be helpful.

When creating a plot using functions from the `ggplot2` package, you need to pass various types of information:

- **Data**: Usually long-format data works better (i.e., variables in columns and observations in row).
- **Mapping**: Assign variables to horizontal (x) or vertical (y) axis, apply groupings (if any), and indicate plotting characteristics, such as size, shape and color. This is done via the `aes()` (short for aesthetic) function.
- **Geom functions**: Define how to represent data graphically. There are many geoms available and new ones are introduced frequently. Some common geoms are `geom_point()` for scatter plot and dot plots, `geom_bar()` for bar plots, and `geom_line()` to create a line plot for a time series.
- **Theme**: Options to customize graphics.

First, a plotting object is created using the `ggplot()` function. Note the difference between this function's name and the name of the package, `ggplot2`. In the `ggplot()` function, pass the data and mapping along with at least one geom to create a plot.

Multiple geoms can be used in one plot. There are many options to customize a plot, such as using theme, scaling and faceting functions. The geom and the customization functions are considered as layers that are chained by a "+" in between each function. There is no limit on the number of layers used in a `ggplot` object. However, the ordering of layers may impact the final output.

The standard template for a plot created via `ggplot` is as follows:

```
ggplot(data = <DATA>, mapping = aes(x,y)) +
  <GEOM_FUNCITONS>() +
  <THEME_FUNCTIONS>()
```

The `autoplot()` function from the ggplot2 package is a flexible, short-hand, single command curated for plot for a particular class. Instead of passing raw data, you need to pass an *object with a defined class*, which determines the output plot.

The flexible nature of `autoplot()` makes it an attractive visualization tool to create default object-specific plotting in many packages. Both the forecast and `fabletools` packages have multiple versions of `autoplot()` and `autolayer()` that are applied on specific object to create a specific plot. For example, to plot a time series, `forecast::autoplot()` requires a `ts` class object whereas `fabletools::autoplot()` works with a `tsibble` object.

In short, the advantage of using `autoplot()` is that you need only one function to generate many different visualizations; all you need is to pass an appropriate object into it.

In the next section, you will use both the `autoplot()` and `ggplot()` to make a time plot.

## 5.6 Time plots

In time series visualization, the focus is on visually identifying the various components. You already saw some examples of time plots in *Chapter 4*, where we discussed the characteristics of a time series. Time plots are also known as line plots.

In a time plot, observations are plotted against time. The observations are joined by a line, reflecting the fact that time is a continuous variable. There are multiple ways you can create a time plot based on the class of your data:

- `tsibble class`: The `autoplot()` function automatically creates a time plot if the data is in tsibble format.
  In the following examples, we will be using the `aus_production` dataset from the `tsibbledata` package. The dataset is loaded along with the package. This dataset contains quarterly production figures for beer, tobacco, bricks, cement, electricity and gas in Australia from 1956 Q1 to 2010 Q2. First check the class of the data as follows:

```
class(aus_production)
# [1] "tbl_ts"     "tbl_df"     "tbl"        "data.frame"
```

  The `aus_production` is in a tsibble format. The `fabletools::autoplot()` function can be used in two different ways to visualize the gas production series. In the first example, the column to plot is explicitly specified and passed into the plotting functions as shown:

```
autoplot(aus_production, Gas)
```

In the second example, the time variable (Quarter) and the series (Gas) are selected, and the `autoplot()` function is chained without mentioning which variable to plot explicitly. The function is smart enough to identify the types of variables and plot them on appropriate axes:

```
aus_production |>
  select(Quarter, Gas)  |>
autoplot()
```

Both code examples will produce exactly the same plot presented in *Figure 5.1*. The rule of thumb for `autoplot()` is that the function requires users to explicitly call out the plotting time series when there are multiple variables of the same type available in the dataset.

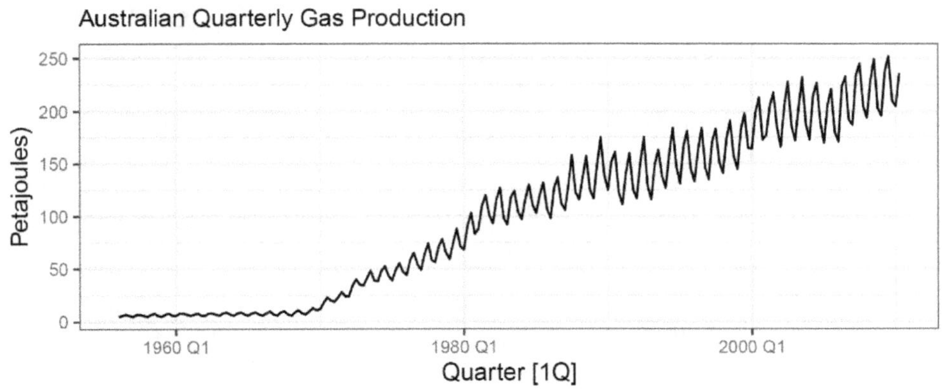


*Figure 5.1: Using autoplot() function on a tsibble object to create time plot*

- `ts class`: Use the `autoplot()` function from the `forecast` package. The other two variations should also work.
  When you pass a `ts` class object in the `forecast::autoplot()` function, the output will automatically choose a time plot. The figure can be customized further using the standard `theme()` and `scale_*()` functions if required:

```
class(sunspots)
# [1] "ts"
```

```
autoplot(sunspots) +
  labs(title = "Monthly Sunspots")
```

- `tibble` or `data frame`: Use the `ggplot()` function along with `geom_line()` to produce time plots. First, you need to transform the time variable into a proper date/date-time variant if you haven't already done so.

In the following example, the date variable, `sched_dep_date`, is obtained by extracting the year, month, day, hour, and minute components using the `make_datetime()` and `date()` functions. The data was recorded in an odd format, hence the arithmetic modulus is used. This is an example of handling odd formatting of date-time based variables in time series data wrangling. An indicator variable, `is_delayed`, is created to flag whether a flight was delayed or not:

```
flights <-
  nycflights13::flights |>
  filter(!is.na(dep_time), !is.na(arr_time)) |>
  mutate(sched_dep_datetime = make_datetime(
    year, month, day,
    sched_dep_time %/% 100,
    sched_dep_time %% 100)
  ) |>
  mutate(sched_dep_date = date(sched_dep_datetime)) |>
  select(carrier, origin, dest, sched_dep_date, dep_delay) |>
  mutate(is_delayed = ifelse(dep_delay > 0, 'y', 'n'))
```

The total number of delayed flights is calculated and plotted for two carriers, `American Airlines` (AA) and `Delta Airlines` (DL).

```
flights |>
  filter(carrier %in% c("AA", "DL"),
         is_delayed == 'y') |>
  count(sched_dep_date, carrier) |>
  ggplot(aes(x = sched_dep_date, y = n,
             group = carrier,
             linetype = carrier)) +
  geom_line() +
  labs(x = " " ,
       y = "delayed flights") +
  theme(legend.position = "top")
```

The result is shown in *Figure 5.2*:

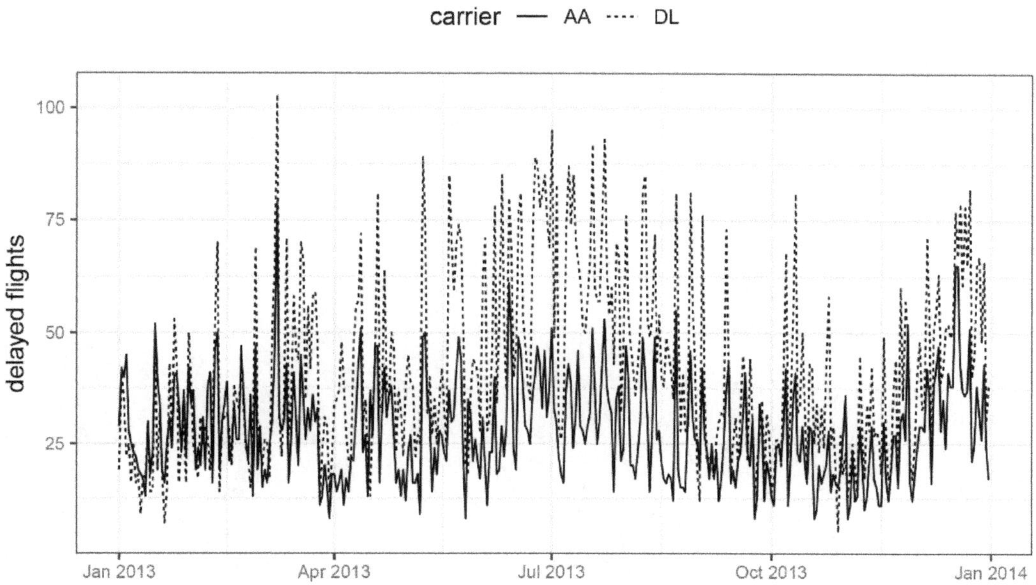


*Figure 5.2: Number of delayed flights from AA and DL carriers*

The plot reveals that, compared to AA, more DL flights (dotted line) were delayed on a daily basis. Both carriers had fewer delayed flights in September and October. However, the plot shows only one year's worth of data, and it is not possible to identify any monthly and weekly seasonality.

Let's see an example of how to do time-based aggregation and visualize the result. The following code computes the weekly number of delayed flights and plots them:

```
flights |>
  filter(carrier %in% c("AA", "DL"),
         is_delayed == 'y') |>
  mutate(week_end_date = ceiling_date(sched_dep_date, "week")) |>
  count(week_end_date, carrier) |>
  ggplot(aes(
    x = week_end_date,
    y = n,
    group = carrier,
    linetype = carrier
  )) +
  geom_line() +
```

```
    labs(x = " " , y = "delayed flights") +
    theme(legend.position = "top")
```

The result is shown in *Figure 5.3*:

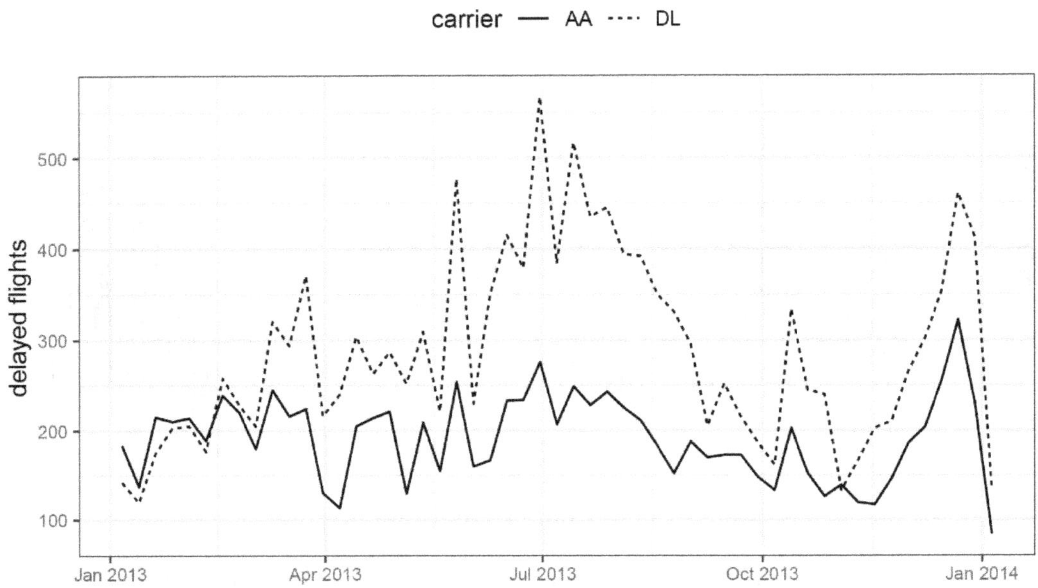


*Figure 5.3: Weekly number of delayed flights from AA and DL carriers*

Now, discrepancies in weekly delayed flights between the two airlines are better visible from the plot. Since we are showcasing ggplot functionality, we applied weekly aggregation in the tibble dataframe. The `lubridate::ceiling_date()` function is used to round a date up to the end date of the week (technically, the start of the next week). The `floor_date()` function rounds a date down to the start date of a week. Alternatively, you can transform the data to a `tsibble` format and then use the corresponding functions to calculate the week end date.

Another common aggregation level in time series data is statistical aggregation. For example, you may be interested in comparing the weekly average delayed time between these two carriers and examining any impact of the origin of the flights.

The following code first computes the weekly average delay time of all carriers that took off from John F Kennedy International Airport (JFK). Then, it computes the weekly average delay time for AA and DL:

```
# weekly average delayed time of all carriers from JFK
jfk_delay <- flights |>
  filter(origin == "JFK" ,
```

```
          is_delayed == 'y' ) |>
    mutate(sched_dep_week = ceiling_date(sched_dep_date, "week")) |>
    group_by(sched_dep_week) |>
    summarise(avg_jfk = mean(dep_delay, na.rm = T)) |>
    ungroup()


  flight2_delay <- flights |>
    filter(carrier %in% c("AA", "DL"),
          is_delayed == 'y' ,
          origin == "JFK"
    ) |>
    mutate(sched_dep_week = ceiling_date(sched_dep_date, "week")) |>
    group_by(sched_dep_week, carrier, origin, dest) |>
    summarise(avg_delay = mean(dep_delay, na.rm = T)) |>
    ungroup()
```

In the visualization part, a ggplot object is first created, and two datasets are passed separately to it with corresponding aesthetics. This offers flexibility to apply finer control on the individual datasets. The overall presentation layout, such as `facet_grid()` and `theme_bw()`, applies to the entire plot:

```
ggplot() +
  geom_line(
    data = flight2_delay |>
      filter( dest %in% c("LAX", "SFO")),
    aes(x = sched_dep_week, y = avg_delay),
    linetype = 3, size = 0.9
  ) +
  geom_line(
    data = jfk_delay,
    aes(x = sched_dep_week,  y = avg_jfk), size = 0.7
  ) +
  facet_grid(carrier ~ dest ) +
  scale_x_date(date_breaks = "3 months", date_labels = "%b\n%Y") +
  labs(x = "", y = "")
```

A visual comparison of AA and DL from JFK to two destination airports, (LAX and SFO) is presented in *Figure 5.4*.

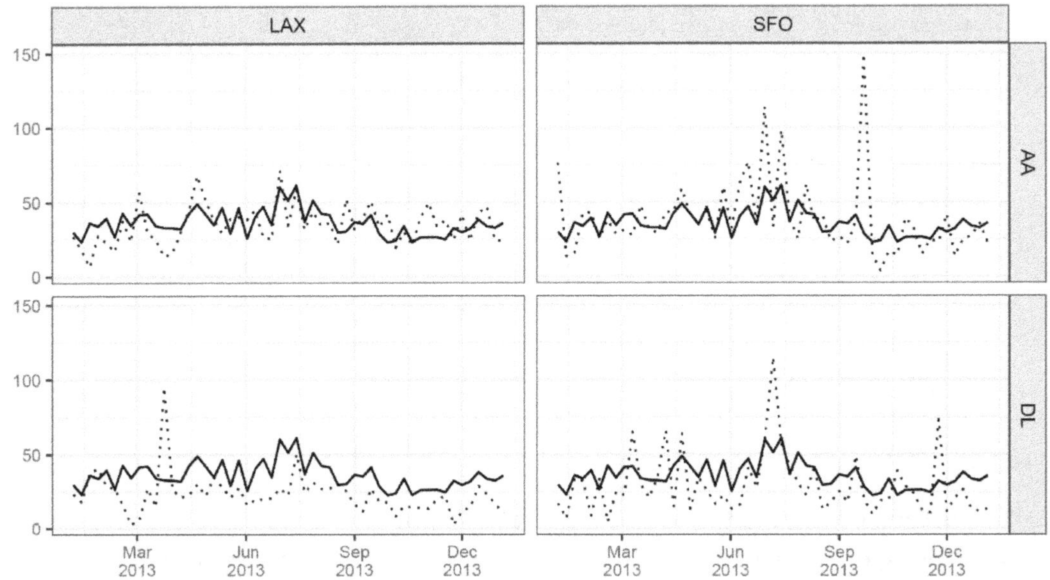


*Figure 5.4: Comparison of weekly average delay time of AA and DL flights against all carriers that took off from JFK to LAX and SFO*

This plot compares the individual weekly average against the JFK benchmark. Some insights can be drawn, such as the following:

- The DL weekly average is way below the JFK benchmark. Compared to DL, AA flights were delayed more often.
- There is a destination impact too: JFK-SFO routes have more delayed flights from both AA and DL carries compared to the JFK-LAX route.
- The spikes indicate weeks with extreme delays.

By utilizing appropriate aggregation levels and effective visualization layers, you can compare many aspects of time series and make insightful observations of this nature. In the next section, you will learn how to analyze seasonal data through visualization.

# 5.7 Seasonal plots

A seasonal plot is a time plot in which observations are plotted against the seasons. The seasons are the units and/or frequency at which observations are collected within a year.

The `gg_season()` function from the `feasts` package provides a seasonal plot for a `tsibble` data object as shown in the following code:

```
aus_production |>
  select(Quarter, Gas) |>
  gg_season(Gas)
```

The seasonal plot in *Figure 5.5* highlights that the highest amount of gas was produced in the third quarter of each year.

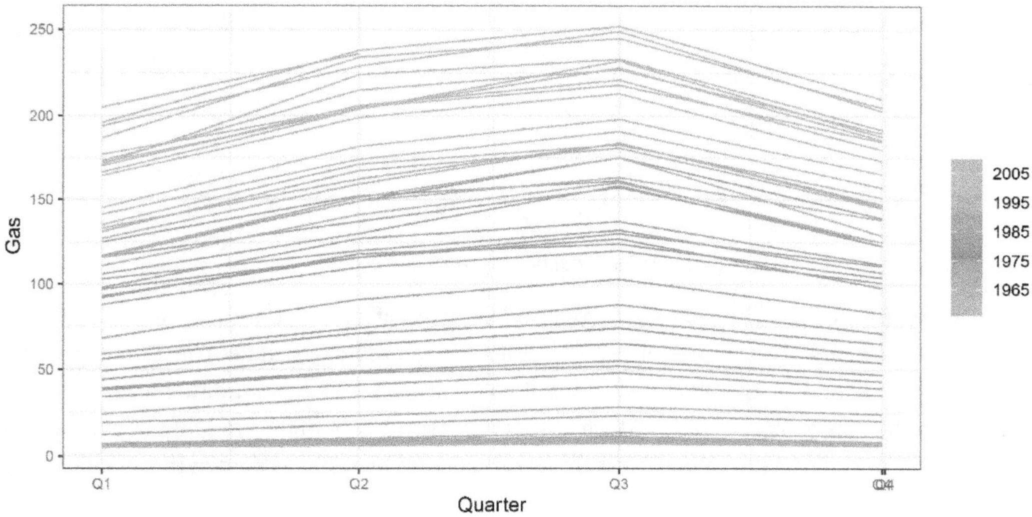


*Figure 5.5: Seasonal plot of quarterly gas production between 1956 Q1 and 2010 Q2*

This fact is much clearer when using a seasonal plot, compared to the time plot of the same data presented in *Figure 5.1*.

The period argument in the `gg_season()` function controls the display of seasonal periods. By specifying it, you can generate seasonal plots at various frequencies if the information is available. If you have monthly data for many years, the period should be "year" when producing a seasonal plot. Here's an example of this being applied on AirPassengers data:

```
AirPassengers |>
  as_tsibble() |>
  gg_season(value, period = "year") +
  labs(x = "",
       y = "Total passengers")
```

The output is presented in *Figure 5.6*:

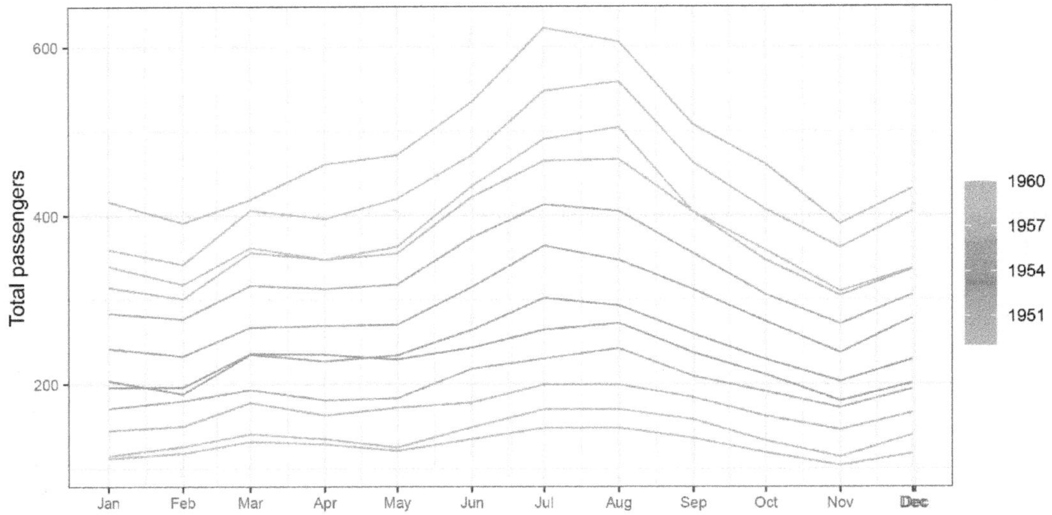


*Figure 5.6: Seasonal plot of AirPassengers data*

Most real business data do not have a long period of observation. A lot of the time you only have one or two years of data to explore and build a model. You can still use these visualization techniques to learn about the observed series. For example, the NYC flight data has daily observations for one year only. To produce a weekly seasonal plot of delayed flights irrespective of the carrier, origin, and destination, set the `period="week"` option as shown in the following code block.

```
flights |>
  filter(is_delayed == "y") |>
  count(sched_dep_date) |>
  as_tsibble(index = sched_dep_date) |>
  gg_season(n, period = "week" ) +
  theme(legend.position = "none") +
  labs(x = "",
       y = "number of delayed flights") +
  scale_x_date(
    date_breaks = "1 day",
    labels = date_format("%a"))
```

That provides the output shown in *Figure 5.7*:

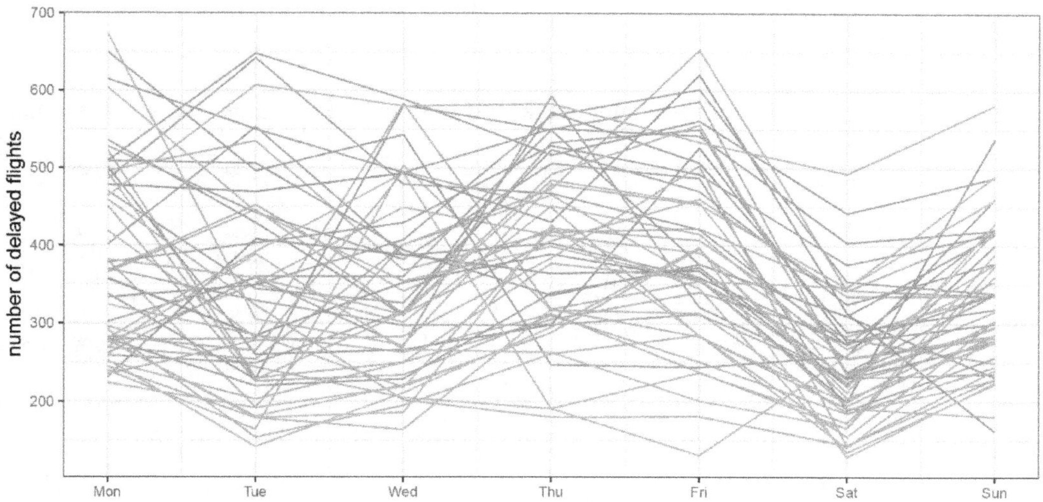


*Figure 5.7: Weekly seasonal plot of delayed flights*

Among the messiness of data, one key point is that the majority of flights on Saturdays departed on time compared to weekdays.

## 5.7.1 Seasonal subseries plots

An alternative way to visually examine seasonality is to plot observations of one seasonal period over multiple years together. The `gg_subseries()` function does exactly that. It also includes the average line for each season.

For the next example we will use Australian retail turnover data. We will use takeaway food service turnover for two states, Victoria and New South Wales (NSW) from January 2015 to December 2022. By visualizing data via a seasonal subseries plot, you can examine how patterns change over seasons and compare between series when there are multiple series:

```
takeaway_turnover <- readRDS("data/data_chap5/abs_retail.rds") |>
  filter(
    industry == "Takeaway food services",
    state %in% c("Victoria", "New South Wales"),
    year(month) >= 2015 ,
    year(month) <= 2022
  ) |>
  select(month, state, turnover)


takeaway_turnover |>
```

```
gg_subseries(turnover) +
labs(x = "", y = "Turnover $M")
```

*Figure 5.8* shows a monthly subseries plot of takeaway services in two large states in Australia.

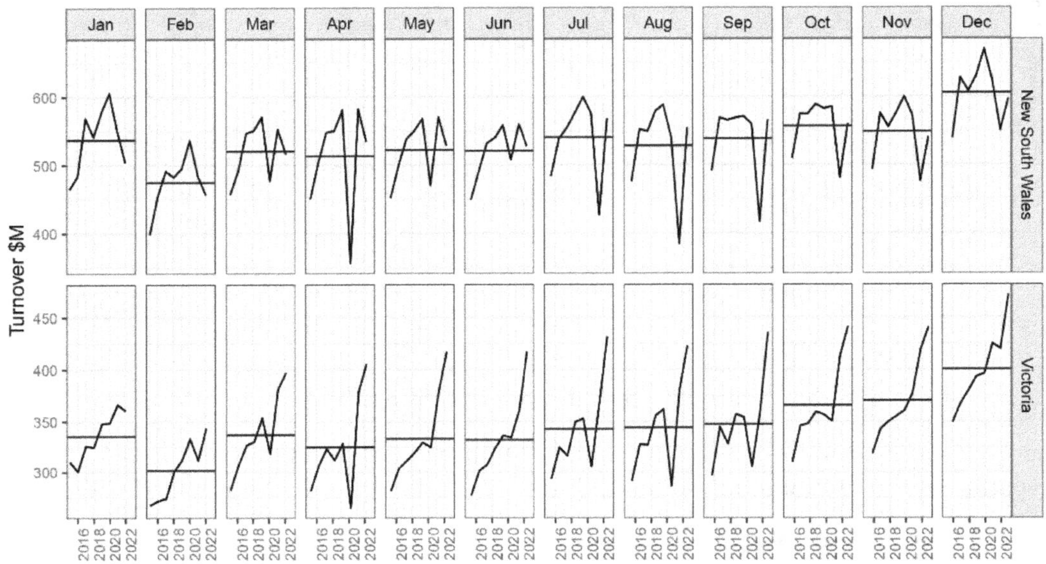


*Figure 5.8: Seasonal subseries plot of takeaway services for two states in Australia during 2015 -2022*

From the plot, you can visually identify the following:

- Victoria's monthly average was lower than NSW's monthly average.
- December is the month in which the highest number of people ordered takeaways.
- The drop in the February average could be due to the fact that it has a smaller number of days.
- Victoria recovered from the COVID-19 slowdown much faster than NSW.

Using these plotting functions can reveal interesting facts about the observed time series. The `ggseasonplot()` and `ggsubseries()` functions are two functions from the `forecast` package that provide the same functionality in case you have data in the `ts` class.

So far, we have explored visualization techniques for a single time series. Next, we look at how to visually inspect relationships among multiple time series.

# 5.8 Visualizing relationships among time series

Correlation is a common technique to identify the strength of a liner relationship among variables. A standard practice in statistical and ML modeling is to compute the correlation between variables to check whether to include them in a model. In time series modeling, you could check the correlation between a series and other explanatory variables, such as electricity demand and temperature. Similarly, you could analyze the correlation among multiple time series to assess how closely they manifest patterns. A scatter plot is the simplest way to visually inspect the correlation among variables. Now you will learn to make scatter plots for time series.

## 5.8.1 Scatter plots

Usually, a scatter plot is used to examine the correlation between two variables, one dependent and one explanatory. The `GGally` package offers the `ggpairs()` function to plot a pairwise scatter plot for multiple series. The same plot also prints the pairwise correlation value along with its significance in the top diagonal panels, and distribution in the diagonal line.

In the following example, you will use the turnover from six industries between 2015 and 2022 for Victoria from Australian retail data. The long industry names are replaced by extracting the corresponding first word using `sapply` after applying the `str_split()` function and a regex (split in white space or comma) pattern:

```
vic_retail <-
  readRDS("data/data_chap5/abs_retail.rds") |>
  filter(
    industry %in% c(
      "Clothing retailing",
      "Food retailing",
      "Electrical and electronic goods retailing",
      "Hardware, building and garden supplies retailing",
      "Newspaper and book retailing",
      "Supermarket and grocery stores"
    ),
    state == "Victoria" ,
    year(month) >= 2015 ,
    year(month) <= 2022
  ) |>
  select(month, industry, turnover) |>
  mutate(industry = sapply(
    str_split(string = industry,pattern = "[, ]+") ,'[', 1))
```

Let's see the turnover from individual industries over the years as presented in *Figure 5.9*:

```
vic_retail |>
  ggplot(aes(x = month, y = turnover)) +
  geom_line(size = 0.7) +
  facet_grid(vars(industry), scales = "free_y") +
  labs(x = " ")
```

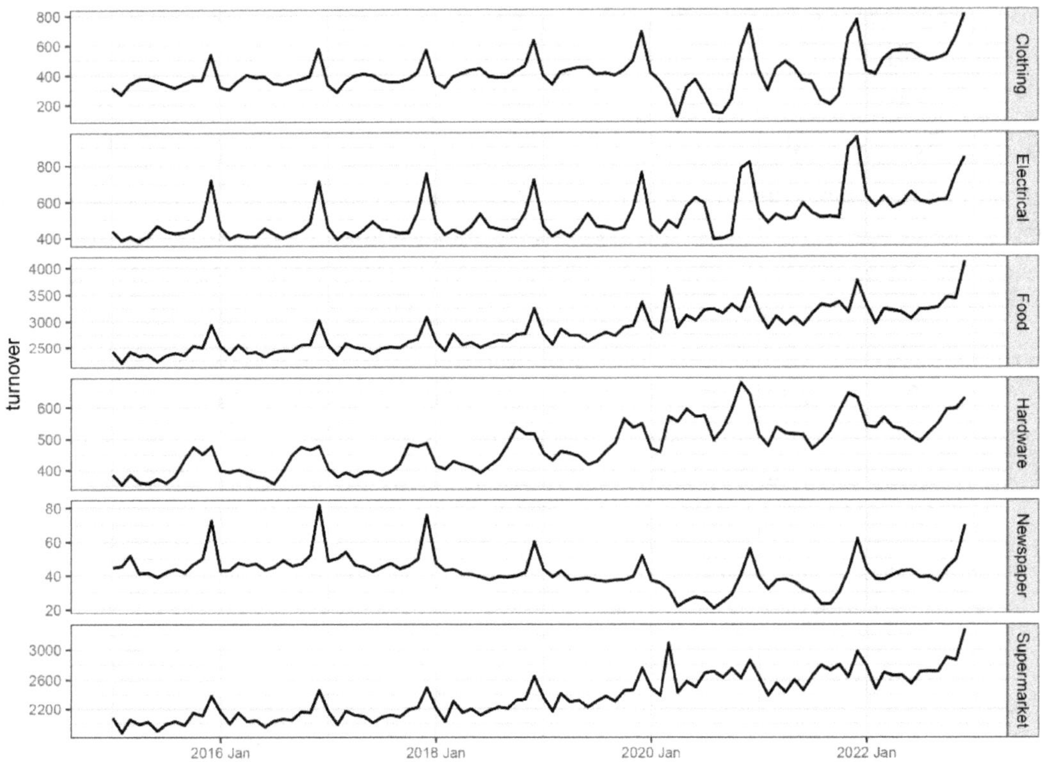


*Figure 5.9: Turnover from six industries between 2015 and 2022*

The `ggpairs()` function expects the data to be in a wider format, which is done using the `pivot_wider()` function. You then pass the column numbers as an argument. The `progress=F` instance controls the printing of the computation process:

```
vic_retail |>
  pivot_wider(names_from = industry,
              values_from = turnover) |>
  GGally::ggpairs(columns = 2:6, progress = F)
```

This will give us the following image:

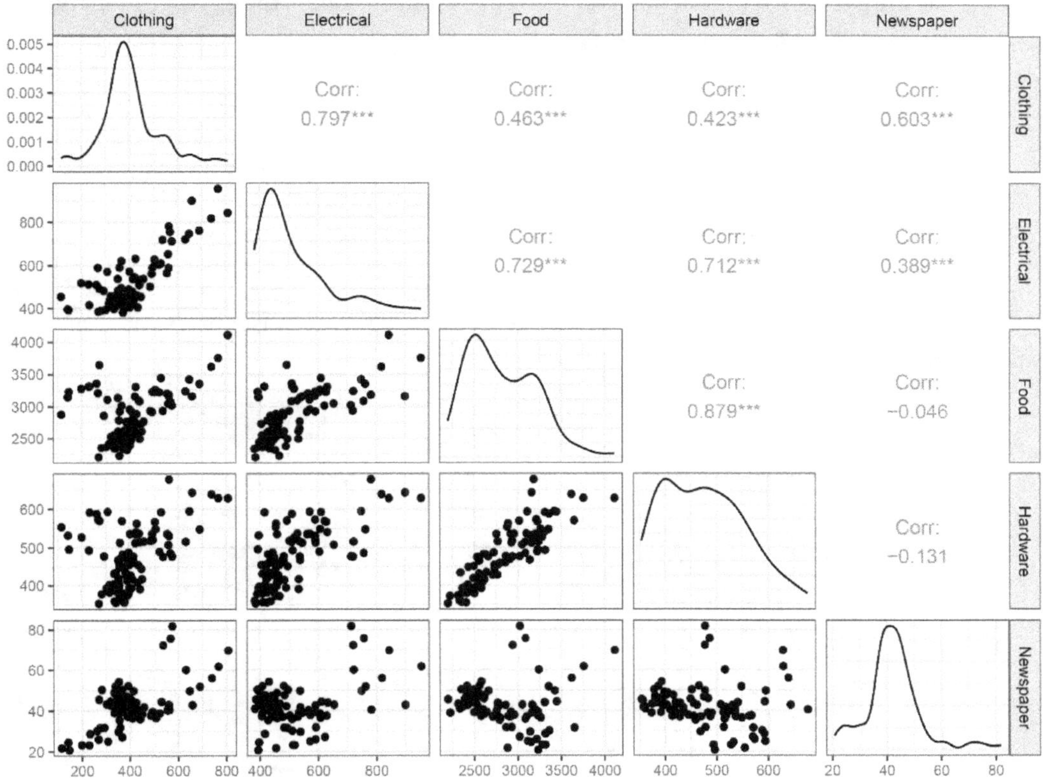


*Figure 5.10: Scatter plot of retail turnover from six industries between 2015 and 2022*

The food-hardware pair has the strongest positive linear correlation, and the food-newspaper pair shows a negative correlation. This may suggest that when spending on food increases, spending on newspapers and books tends to decrease, possibly reflecting shifts in consumer priorities.

## 5.8.2 Lag plots

You learned about autocorrelation in *Chapter 4* when examining serial dependency on the past values of a time series. Scatter plots with a lagged value, also known as lag plots, are graphical representations used in time series analysis to visualize the relationship between a variable and its lagged observations. In a lag plot, the current value of a variable is plotted against its previous value(s) at a certain time lag. This helps reveal patterns, trends, or correlations that might exist in the data over time. It is another way to explore the autocorrelation or dependence structure of a time series.

There are functions in the two packages that you can use to create lag plots. The `forecast` package has the `gglagplot()` function and the `feasts` package offers the `gg_lag()` function. The latter is used in the following example:

```
vic_retail |>
  filter(industry == "Clothing") |>
  gg_lag(turnover, lags = 1:12, geom = "point") +
  labs(x = "lag(Turnover,1)",
       y = "Turnover")
```

*Figure 5.11* shows a strong positive dependency at lag 1 and lag 12, as most observations are gathered around the upward line.

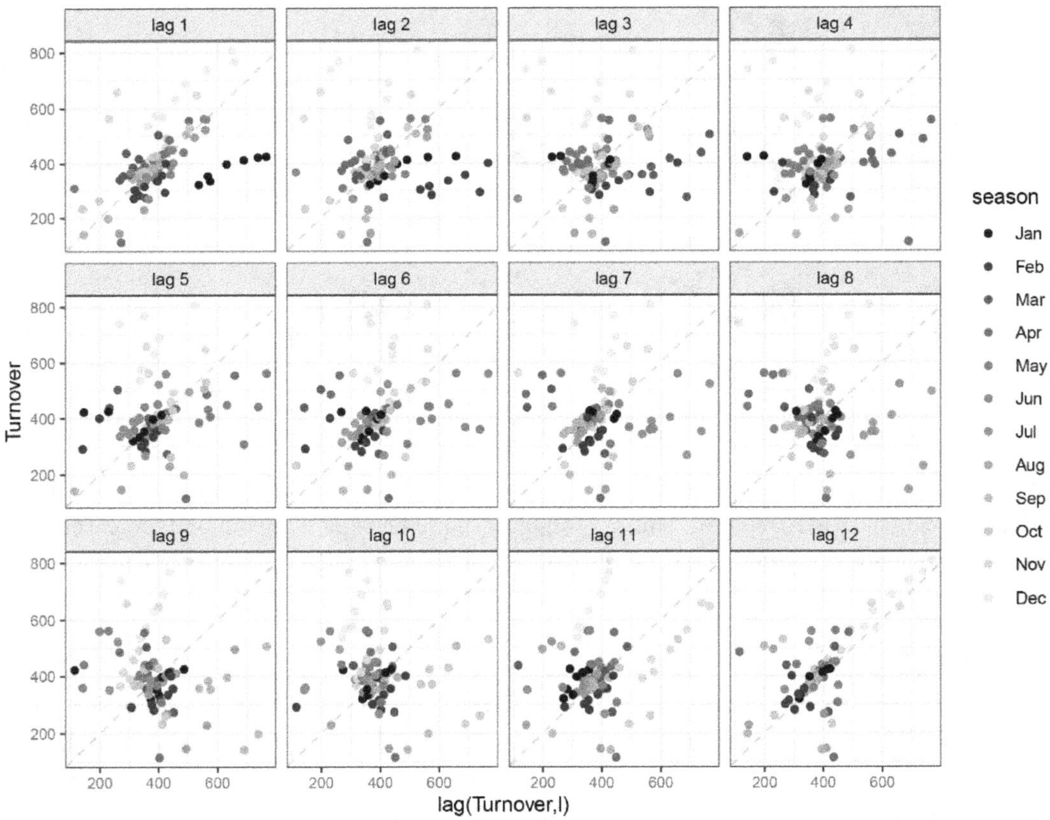


*Figure 5.11: Lag plot of Clothing turnover in Victoria during 2015-2022*

The observations are scattered in other lags and do not indicate any strong dependency.

In case you are curious to examine the autocorrelation function of the same data to cross-check the dependency pattern, you could plot that by either using the `ggAcf()` function from the `forecast` package or applying `autoplot()` on a `feasts::ACF` object:

```
vic_retail |>
  filter(industry == "Clothing") |>
  ACF(turnover) |>
  autoplot() +
  labs(title = "" )
```

The autocorrelation function in *Figure 5.12* shows significant spikes at lags 1 and 12.

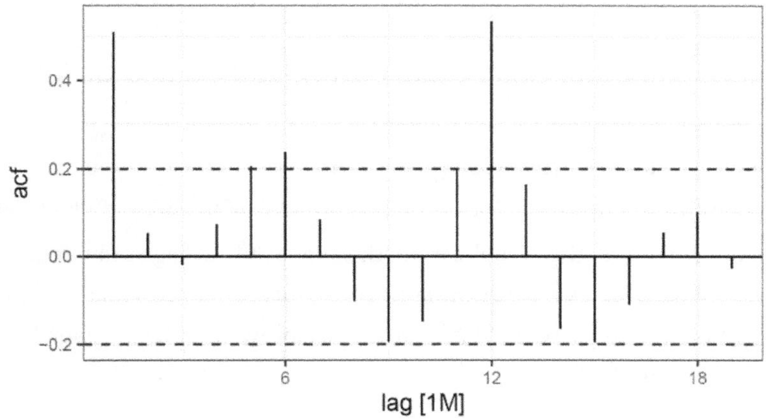


*Figure 5.12: Autocorrelation function of clothing turnover in Victoria between 2015 and 2022*

A strong spike at lag 1 suggests that the current month's value is strong and positively correlated with the previous month's value. The significant spike at 12 indicates a strong seasonal pattern; that is, the current month's value is positively correlated with the same month in the previous year's value.

By using scatter plots and lag plots on time series data, you can examine the relationship between multiple time series as well as dependency within a series. There are a handful of R packages that offer interactive plotting functionality. In the next section, you will see how the `Plotly` package is used to turn a static graphic into an interactive one.

## 5.9 Interactive plots

The `Plotly` package is well renowned for its capability in producing interactive plots. The `ggplotly()` function from the `Plotly` package turns any static `ggplot` object into an interactive and dynamic visualization. To do this, you need to save `ggplot` into an object and then pass it inside the `ggplotly()` function, as shown in the following example:

```
lag_turnover <- vic_retail |>
  filter(industry=="Clothing") |>
  gg_lag(turnover, lags = 1:12, geom = "point") +
  labs(x="lag(Turnover,1)",
       y="Turnover")

ggplotly(lag_turnover)
```

The code will generate an interactive plot in your R environment where you can pan, zoom, and hover over data points. This enables users to get engaged and access more information.

For those interested in interactive visualization, the `dygraphs` package provides a highly effective method for creating dynamic time series charts using the `dygraphs()` function. This allows users to pan and zoom through data directly in a browser or IDE viewer. You can explore such functionality on your own.

## Summary

In this chapter, you learned about two important steps: wrangling and visualizing time series data in R.

First, you learned how to properly parse a date-time variable along with helper functions to get and set several components from a time index. You learned about how to transform data into a time series object and a wide class of functions that can be used when wrangling data. Regardless of your specific role or the next phase in your analysis, building reproducible code when cleaning and transforming data is essential.

You also learned about standard and specialized visualization for time series data, with examples. An array of visualization functionality was discussed ranging from simple time plots to time series scatter plots, seasonal plots and subseasonal plots.

You now have basic knowledge of time series. Time series components, and their characteristics were covered in *Chapter 4*, while how to wrangle time series data and examine it visually was the topic of this chapter.

As you get prepared to begin your journey into time series analysis, it's crucial to understand the scope of problems that time series analysis can address. Also, there are analytical problems that fall beyond the boundaries of the time series realm. These topics are discussed in detail in the following chapter.

# Get this book's PDF version and more

Scan the QR code (or go to `packtpub.com/unlock`). Search for this book by name, confirm the edition, and then follow the steps on the page.

UNLOCK NOW

*Note: Keep your invoice handy. Purchases made directly from Packt don't require an invoice.*

# 6

# Business Applications of Time Series Analysis

Time series analysis can be applied to solve a wide range of real-life problems. By time series applications, we refer to the domain or field of study that involves analyzing and predicting patterns, trends, and behavior in data indexed by time. The goal of time series analysis is to extract meaningful insights, understand the underlying dynamics of the data, and use them accordingly. This chapter highlights time series applications in various industries and disciplines, including finance, economics, business, medicine, retail, public health, tourism, healthcare, manufacturing, and more.

Identification of a time series analysis problem is crucial because the proper choice of analytics methods depends on the nature and objectives of a problem. A misalignment between a problem and the method adopted for a solution will cause confusion rather than being a solution. By consulting the problem spaces covered in this chapter and the high-level concept of the method, you will be able to assess your problem and which methods to use for a solution.

From the definition of time series, you must have realized that time series data are numerical data in nature. Standard textbooks on the analysis of numerical data cover four key analytics objectives:

- Visualization
- Summary/descriptive analysis
- Inferential analysis
- Predictive analysis

Time series analytics also encompass challenges commonly faced in numerical data analysis. However, due to the incorporation of a time index alongside numerical observations, this problem space necessitates the application of extended analytics methods that often go beyond standard numerical data analysis. These specialized methods will be explored in upcoming chapters. In the current chapter, you will gain insights into when time series analysis techniques are employed to tackle real-life problems.

It's crucial to highlight that this chapter focuses exclusively on problem spaces that directly leverage time series characteristics. However, it's worth noting that numerous inferential and predictive analytics methods within numerical data analysis offer the capability to incorporate time series data attributes, particularly serial dependency. These methods will be discussed in later chapters of the book.

This chapter is organized as follows:

- Section 6.1: Time series characteristics level problem domains
- Section 6.2: Inference and attribution problem domains
- Section 6.3: Impact measurement problem domains
- Section 6.4: Forecasting problem domains

Note

**Your purchase includes a free PDF copy + exclusive extras**

Your purchase includes a DRM-free PDF copy of this book, 7-day trial to the Packt+ library (no credit card required), and additional exclusive extras. See the *Free benefits with your book* section in the *Preface* to unlock them instantly and maximize your learning.

# Technical requirements

Clone the GitHub repo of the book, `https://github.com/PacktPublishing/Modern-Time-Series-Analysis-with-R.git` in your local machine. You will find the data files used in this chapter in the `data/data_chap6` folder.

Let's load the following packages into your R workspace:

```
#install.packages("pacman")
library(pacman)
p_load("conflicted",
       "dataseries",
       "fpp3",
       "tidyverse",
       "tsibble")

conflicts_prefer(
  dplyr::filter,
  fabletools::components
)
```

Let's start by understanding some common business problems that can be solved by analyzing time series characteristics.

# 6.1 Time series characteristics level problem domains

Understanding the inherent characteristics of time series is crucial for effective analysis, modeling, and interpretation. The time series characteristics-based problem space focuses on understanding and describing fundamental properties and attributes of time series data. Recognizing these characteristics is essential for drawing meaningful insights. Three major problem spaces exist around time series characteristics:

- Trend-cycle analysis
- Seasonality analysis
- Anomaly detection

We will now discuss the nature of these problems in detail. First, you will learn how trend analysis can be very helpful to businesses.

## 6.1.1 Trend (trend-cycle) analysis

Trend analysis is one of the most encountered problems in time series analysis. This analysis can be performed on time series data with or without seasonality. Trend analysis can be a part of visual, descriptive, inferential, or predictive analytics depending on specific requirements. Trend analysis involves examining historical data over a period to identify and understand long-term patterns and changes. It helps in uncovering underlying trends that can provide insights into the direction of a phenomenon, predicting future behavior to aid decision-making. Some examples include the following:

- **Economics**: Analyzing trends in economic indicators, such as gross domestic product (GDP), unemployment rates, and inflation, to understand the health of an economy.
- **Finance**: Studying trends in stock prices, commodity prices, and exchange rates to make investment decisions in finance.
- **Inventory management**: Using trend analysis to anticipate demand for products or services and make informed decisions about production, inventory, services, and marketing strategies.
- **Public health**: Identifying trends in patient data to predict disease outbreaks and allocate medical resources effectively in public health.

Trend analysis is also known as trend-cycle analysis. Both trend and cycle are long-term behaviors, and that's why they often need to be analyzed together. Short-term trends are often part of and could be influenced by economic cycles, business cycles, or other external factors. A common example of trend-cycle analysis is a change in economic indicators such as GDP, which is subject to movement of the official interest rate, which often follows a cycle due to monetary policy to tackle inflation.

Trend analysis in time series involves uncovering meaningful patterns and changes within historical data to gain insights into long-term behaviors. This process is crucial for making accurate predictions, informed decisions, and strategic planning in various domains. Intuitively, a trend analysis is as simple as fitting a trend line on a scatter plot, which is achievable using common software such as Excel, however, in the time series context, it requires additional care, and this is covered in more detail in the time series smoothing and decomposition chapter.

Use of trend analysis is widespread in business decision-making via the monitoring of key business metrics. Trend analysis is the most presented visual in the media when covering news about economics and finance.

To illustrate the idea, we will use the monthly retail sales index for Switzerland between 2000 and 2020. You can download the original index series, and the underlying trend using the `ds()` function from the `dataseries` package (`Sax, 2017`). The two series are downloaded from www.dataseries.org using the R code shown next. Then column bind them into one `tsibble` object:

```
# Swiss retail sale index with 2015 base
swiss_retail_sale <- ds("ch_fso_rtt.brut.n.noga0801", "ts")
swiss_retail_sale_trend <- ds("ch_fso_rtt.tr.n.noga0801", "ts")

swiss_retail <- cbind(
  sale_index = swiss_retail_sale,
  index_trend = swiss_retail_sale_trend ) |>
  as_tsibble()
swiss_retail
```

Let's take a look at the data. The data remains in a long format despite earlier use of `cbind` (column bind). This is because we downloaded them as `ts` objects and used `cbind` and `as_tsibble` together to convert them into a `tsibble` object containing monthly observations for two keys (variables): `sale_index` and `sale_trend_index`. The date-time index variable is in year-month format:

```
# A tibble: 480 × 3
   index    key         value
   <chr>    <chr>       <dbl>
 1 2000 Jan sale_index   81.5
 2 2000 Feb sale_index   81.4
 3 2000 Mar sale_index   84.0
 4 2000 Apr sale_index   85.0
 5 2000 May sale_index   92.6
 6 2000 Jun sale_index   79.4
 7 2000 Jul sale_index   83.5
 8 2000 Aug sale_index   81.3
 9 2000 Sep sale_index   87.5
10 2000 Oct sale_index   88.6
# i 470 more rows
```

Let's visualize the sales index and examine the time plot presented in *Figure 6.1*:

```
swiss_retail |>
  filter(key == "sale_index") |>
  ggplot(aes(x = index, y = value)) + geom_line() +
  labs(x = "", y = "Index")
```

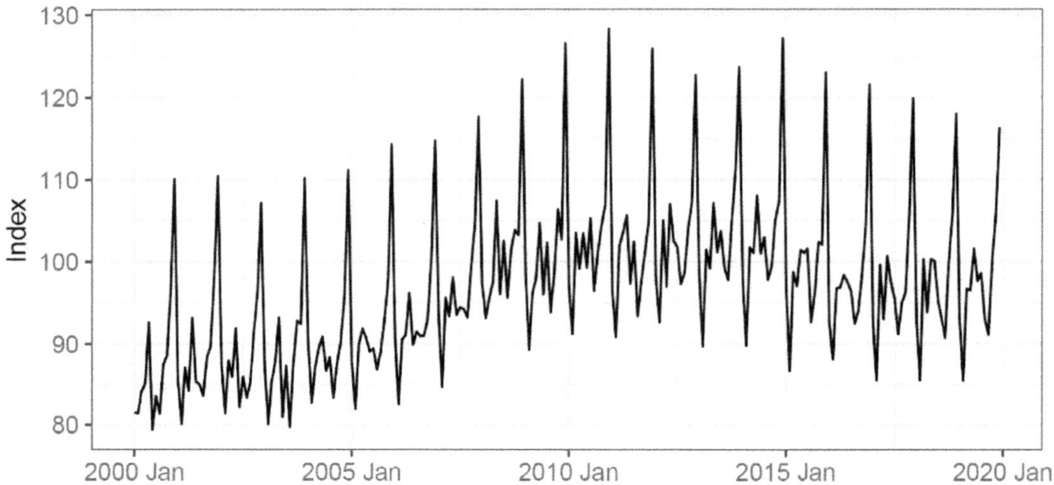


*Figure 6.1: Monthly Swiss retail sales index with base (2015 = 100)*

Business metrics are frequently benchmarked for making comparisons. The Swiss retail sales series in *Figure 6.1* were also benchmarked to show how sales compared with the year 2015. This benchmarked series is often called an indexed series.

Despite the presence of seasonality and random fluctuations, just by visual inspection, one will be able to identify the gradual upward movement of retail sales from 2005 to 2010. Retail sales remain stable till 2015, after which they go down before stabilizing. Proper identification of such underlying smooth movement can help decision makers understand the past and current state of the retail sector compared to 2015.

The trend analysis draws exactly the same conclusion using appropriate mathematical tools instead of visual inspection. We will not go into the details of the methods for the extraction of the underlying trend right now. Throughout the book, you will learn several techniques and methods to perform trend analysis and how to use them in time series modeling.

At this point, you might be wondering why you need a mathematical tool that you can identify just by visual inspection. Possible reasons could be (but are not limited to) the following: (i) many real metrics in business are not as well behaved as this example dataset, (ii) without quantifying the underlying trend, further analysis and modeling are not feasible, and (iii) visual inspection is not suitable when you need to analyze many series regularly.

Now, plot the sales index and index trend. Before plotting, the key column is transformed into a factor to keep the sales index variable on top of the chart:

```
swiss_retail |>
  mutate(key = factor(key, levels = c("sale_index", "index_trend"))) |>
  autoplot(linewidth = 0.9, color = "black") +
  facet_wrap(key ~., ncol = 1, scales = "fixed") +
  labs(x = "", y = "Index") +
  theme(legend.position = "none")
```

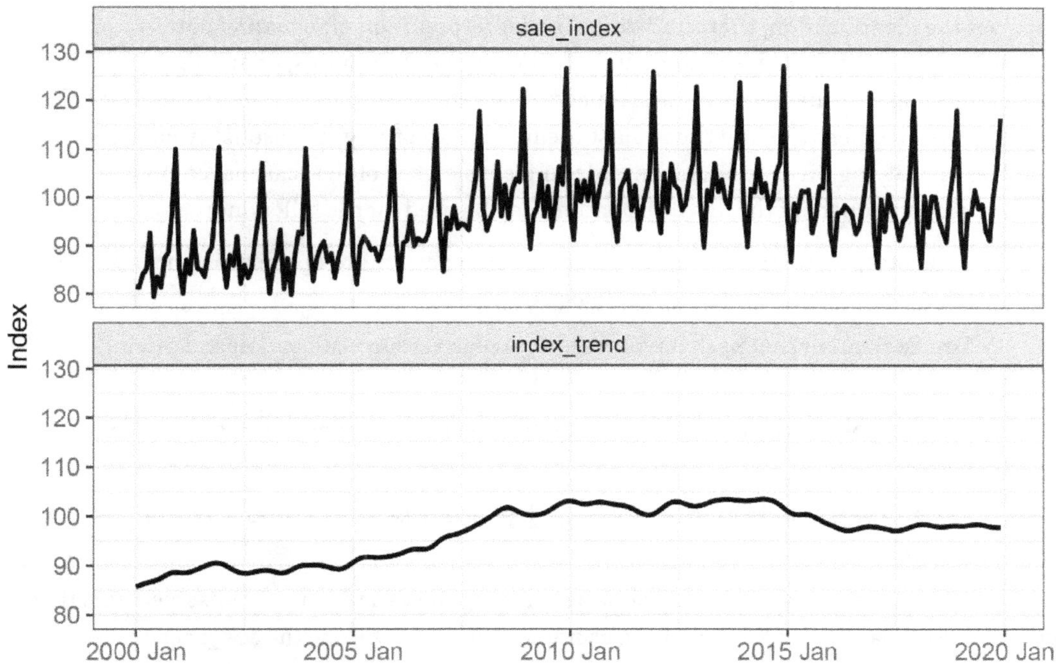


*Figure 6.2: Sales index and underlying trend for the Swiss retail sales index with base (2015 = 100)*

*Figure 6.2* shows the original series and the underlying trend after removing seasonality and random variation. Now the trend series reveals long-term movement in the data and is also useful for identifying short-term movement peaks and troughs to better understand the potential impact of some short-term policy decisions.

In the preceding example, seasonality is removed to reveal the underlying trend. You will have realized that the extraction of underlying trends from an observed series could be quite insightful for making short- and long-term business decisions. Trend analysis is an important part of forecasting future directions. You will see how trend analysis is incorporated into forecasting from the discussion presented in *Chapters 10* to *14*.

After detecting trends, identifying seasonality in time series data helps businesses or sectors analyze data that depends on seasonal patterns. Let's see how seasonality analysis helps in such a scenario in the following section.

## 6.1.2 Seasonality analysis

Similar to trend analysis, seasonality analysis is a common type of time series analysis used in visual, descriptive, inferential, or predictive analytics. Seasonality analysis in time series involves the identification, characterization, and interpretation of repeating patterns or fluctuations that occur at fixed intervals such as days, weeks, months, or quarters. Understanding seasonality is essential for accurate trend analysis and forecasting, as it helps capture repeated behavior and thus adjust predictions accordingly. Seasonality analysis involves identifying the pattern, magnitude, and consistency of the seasonal component. As a systematic component of the data, this pattern can be used for the following:

- **Marketing**: Understanding seasonal shopping patterns to optimize inventory management and marketing campaigns.
- **Tourism**: Analyzing peak travel seasons to plan resource allocation and pricing strategies.
- **Demand planning**: Predicting electricity or heating demand based on seasonal usage patterns.
- **Agriculture**: Anticipating planting and harvesting seasons to optimize crop management.

One key element of seasonality analysis is quantifying the variation in the observed data that occurs due to seasonality. To provide an example of this, we will use the `aus_accommodation` data from the `fpp3` package. This dataset contains the quarterly percentage of room occupancy rates from short-term accommodation from 1998 Q1 to 2016 Q2 in Australia. Let's first plot New South Wales data, and then we'll estimate the magnitude of variations in each quarter:

```
aus_accommodation |>
  filter(State == "New South Wales") |>
  select(Date, Occupancy) |>
  autoplot() + labs(x = "", y = "% of room occupancy")
```

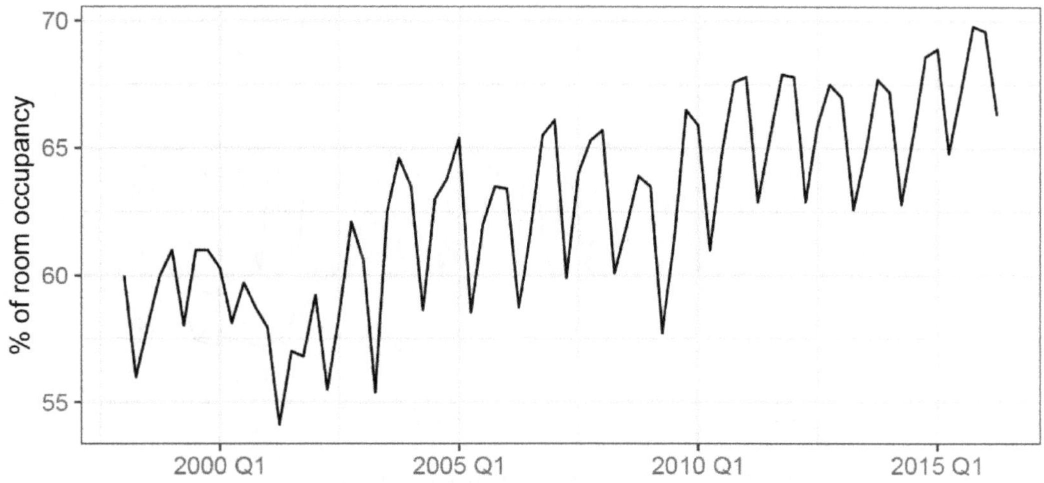


*Figure 6.3: New South Wales (Australia) tourist accommodation room occupancy rate, 1998 Q1-2016 Q2*

From *Figure 6.3*, the upward trend and seasonal variation in every quarter are visually identifiable. However, due to random variation, quarter-to-quarter movement is not fixed and is hard to measure visually. Next, we calculate the magnitude of changes in each quarter from the data. This is done via decomposing the observed series into its components and then computing the variations in every season. Decomposing methods are elaborately discussed in *Chapter 7*.

In this section, we only present the outputs to illustrate the concept with an example.

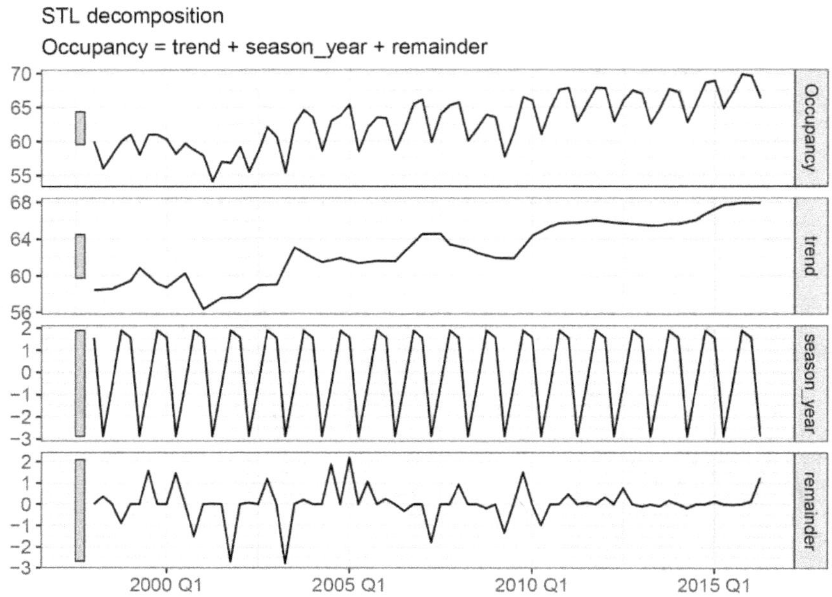


*Figure 6.4: Decomposition of tourist room occupancy rate*

*Figure 6.4* presents the decomposed room occupancy rate and its components. The second panel shows the underlying trend obtained after seasonal variations were removed. The third panel shows the underlying seasonal variations after the trend was removed. The quarterly variations are now more consistent. Within a year, a peak occurred in Q4, followed by Q1, and room occupancy was lower in Q2 or Q3.

*Table 6.1* presents the computed magnitude of seasonal variations. You can interpret the values as follows: the highest room occupancy rate was in Q4, at 1.89% higher than the annual average, and the lowest was in Q2, which was 2.78% below the annual average.

| Quarter | Seasonal index |
|---|---|
| 1 | 1.56 |
| 2 | -2.87 |
| 3 | -0.57 |
| 4 | 1.89 |

*Table 6.1: Seasonal variation in the room occupancy rate*

This is an example of using seasonality analysis from time series data, which is commonly required in business decision-making. Throughout the book, you will see other applications of seasonality analysis. After seasonality, another key time series application is anomaly detection, which is explored in the next section.

### 6.1.3 Outlier/anomaly detection

Time series anomaly detection involves identifying observations or patterns in time series data that deviate significantly from expected or normal behavior. Anomalies can signify unusual events, errors, faults, or other irregularities that require attention or investigation. Solving time series anomaly detection problems is crucial for ensuring the reliability and quality of data-driven processes. Some examples of time series anomaly detection problems are the following:

- **Cybersecurity**: Detecting unusual patterns in network traffic that might indicate potential cyber-attacks or unauthorized access.
- **Industrial IoT**: Identifying anomalies in sensor data to detect equipment malfunctions or deviations from expected performance.
- **Finance**: Detecting fraudulent transactions or activities that deviate from typical financial behavior.
- **Healthcare**: Identifying anomalies in patient vital signs or medical sensor data to alert medical professionals to critical conditions.
- **Energy**: Detecting abnormal energy consumption patterns that might indicate system malfunctions or inefficiencies.

Solving time series anomaly detection problems involves methods and models that identify deviations from normal patterns. Accurate anomaly detection is crucial for ensuring the integrity, security, and efficiency of various systems and processes that rely on time-ordered data. Various statistical and machine learning (ML) algorithms can be applied too. This problem space can work with data with or without trend and/or seasonality. Time series-based anomaly detection and imputation techniques are presented in *Chapter 17*.

Time series characteristics-based problem spaces are a foundational step in time series analysis. Recognizing trends, seasonality, and other attributes helps in the proper selection of techniques, models, and strategies for extracting valuable insights from time series data.

Next, we look at inference and attribution problems that can be solved by time series analysis.

## 6.2 Inference and attribution problem domains

A time series attribution model refers to the analytical framework that is used to distribute the impact of different factors on changes observed in time-evolving series of outcomes. This problem space often arises in scenarios where you want to understand how multiple independent variables contribute to changes in a dependent variable that is structured as a time series. Examples of such problems are the following:

- **Marketing attribution**: Understanding how various marketing channels (e.g., social media, email, paid ads) impact a time series metric such as sales or website visits.
- **Economic indicators**: Assessing the influence of economic factors (e.g., interest rates, GDP, inflation) on time series outcomes such as stock prices or consumer spending.
- **Healthcare**: Analyzing how different treatments, patient demographics, or external factors influence patient recovery rates over time.
- **Supply chain**: Examining the impact of different factors (e.g., raw material costs, shipping delays) on inventory levels or production output.

Time series regression techniques can be used to meet the analytical needs of this problem space. This involves building regression models where time series data is used as both independent and dependent variables. This approach captures the effects of various factors on the time series outcome. An alternative could be using ML algorithms that can handle complex interactions and patterns in the data, such as random forests or gradient boosting.

When a statistical regression model is used, examining the significance of coefficients helps identify which variables have a statistically significant impact on the time series outcome. However, an ML model will achieve only attribution with a quantified direction without inference.

Accurate attribution can lead to better decision-making, improved resource allocation, and a deeper understanding of dynamics influencing time series outcomes. The methods discussed in *Chapter 15* are useful for solving inference and attribution problems involving time series.

Next, you will see how time series can be used to measure the causal impact of a policy or treatment.

## 6.3 Impact measurement problem domains

An extension of attribution modeling is causal impact estimation using time series. Causal impact analysis in time series data involves assessing the impact of a specific intervention or event on a time series variable. This analysis aims to determine whether the intervention caused a significant change in the observed time series and to quantify the magnitude of that impact. The problem space of causal impact analysis using time series data can include the following:

- **Policy evaluation**: Assessing how a policy change (e.g., tax reform, regulation) affects economic indicators such as GDP, unemployment rates, or consumer spending.
- **Treatment/drug testing**: Analyzing whether the introduction of a new medical treatment or drug influences patient outcomes over time.
- **Product launches**: Measuring the effect of introducing a new product on overall sales or market share.
- **Social media trends**: Investigating whether a viral social media post or event affects user engagement metrics on a platform.

A critical challenge is estimating what the time series would have looked like in the absence of intervention, which serves as a baseline for comparison. This can be achieved by following a variety of avenues, such as before-and-after comparison. In such problems, we can compare time series behavior before and after the intervention to identify any meaningful changes. Causal effect can also be estimated through post-intervention change by assessing whether there is a statistically significant change in time series behavior after the intervention. Such analysis can both detect and measure the magnitude of causal impact.

Causal impact analysis is vital for making informed decisions, understanding the effectiveness of interventions, and distinguishing between causation and correlation in time series data. *Chapter 15* discusses time series based methods for causal impact analysis.

The most important use of time series data, forecasting, is kept as the last item to discuss. Next, we explore how forecasting is applied across a range of modern use cases.

## 6.4 Forecasting problem domains

Forecasting is perhaps the most intuitive problem space for time series data. Forecasting is often confused with prediction, and we distinguish forecasting from prediction by adopting the definition that forecasting is a process of estimating the future path of a variable. In contrast, prediction covers the problem space where the user wants to know the conditional mean of a variable based on given values of a set of correlated variables. Generally, trend (trend-cycle) line fitting with or without adjusting for seasonality is a predictive modeling problem space instead of forecasting.

The time series forecasting problem space entails predicting future values of a time-dependent variable based on historical observations. This problem is central to numerous domains, as accurate forecasts facilitate informed decision-making, resource allocation, and strategic planning. The time series forecasting problem space spans a wide range of domains, including the following:

- **Business and finance**: Forecasting sales, revenue, marketing expenditure, and financial indicators to support budgeting and investment decisions.
- **Supply chain management**: Forecasting demand for products or resources to optimize inventory levels and production scheduling.
- **Economics**: Forecasting economic indicators such as GDP, inflation, and unemployment rates to guide policy decisions, such as setting official interest rates.
- **Energy**: Forecasting energy consumption and demand patterns to ensure efficient resource allocation for uninterrupted supply.
- **Demography**: Forecasting the future population of a country to support infrastructure planning.
- **Meteorology**: Forecasting weather conditions and natural phenomena to enable preparedness and response efforts.

Accurate forecasts empower organizations to make proactive decisions, allocate resources effectively, and adapt to changing trends and conditions.

For illustration, a forecasting model was fitted on the Swiss retail sales index where forecasts are generated for the next 5 years with 95% intervals. *Figure 6.5* shows the observed data, the forecasts, and the 95% intervals around the forecast values. This visualization helps to validate the results produced by a forecasting model.

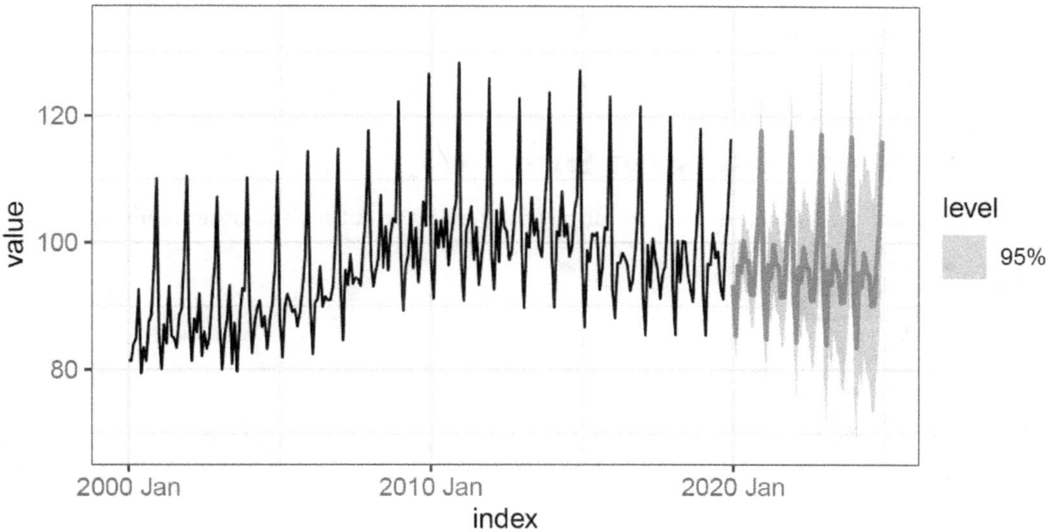


*Figure 6.5: Forecasts of the Swiss retail sales Index*

Forecasting is a powerful analytical tool that helps make better business decisions when applied with a clear understanding of a problem's objective and context. *Chapter 10* to *14* are dedicated to forecasting models, tools, and techniques in depth. You will learn about a wide range of forecasting models from these chapters, and importantly, when and why a particular forecasting technique should be used.

# Summary

This chapter accounted for a wide range of problem spaces where you can apply time series-based methods for descriptive, inferential, predictive, and forecasting analysis. We cover examples from various disciplines and avenues of analysis that will help you identify an appropriate approach to solve your problem. Of course, there are many other usages of time series methods, tools, and techniques, and those can even be used in conjunction with other analytics techniques. As you develop a better understanding of models and methods, the examples provided in this chapter will help to draw those potential links.

From this chapter, you came to know the broader scopes of applying time series-based methods and models to solve real-life problems. Now, it is time to learn about those techniques in detail. *Chapter 7* covers various methods of time series adjustment and transformations, an important step to prepare your data for modeling.

## Further reading


- Sax, C. (2017). *Dataseries: Switzerland's data series in one place.*


## Join our community on Discord

Join our community's Discord space for discussions with the author and other readers:

# Part 3

# The Makeover: Adjusting the Appearance

This part of the book focuses on techniques to process time series for analysis. To extract information properly, a time series may need to be processed through multiple adjustments and transformations. Decomposition is a crucial step when analyzing time series by breaking it down into its components. Feature extraction is another perspective for extracting information from an observed series. Smoothing and filtering techniques remove noise from a time series to unmask the underlying characteristics.

*Part 3* comprises the following chapters:

# 7

# Time Series Adjustments, Transformations, and Decomposition

Time series data analysis plays a pivotal role in various fields, from finance to healthcare, meteorology, and economics. Time series data from businesses comes in various lengths (short to long), shapes (e.g., exponential growth), frequencies, and a combination of components. In order to extract meaningful insights and make accurate forecasts from time series data, it is essential to employ robust techniques for adjustments, transformations, and decomposition. This chapter presents a comprehensive exploration of advanced methodologies and best practices in time series adjustments, transformations, and decomposition.

The chapter is organized as follows:

- Section 7.1: Summary of differences: Adjustments, transformations, and decompositions
- Section 7.2: Rationales for time series adjustments and transformations
- Section 7.3: Time series adjustments
- Section 7.4: Time series transformations
- Section 7.5: Decomposition plots
- Section 7.6: Decomposition using MAs
- Section 7.7: Trend and seasonal decomposition with the classical method
- Section 7.8: Model-based decomposition methods
- Section 7.9: STL decomposition

Note

**Your purchase includes a free PDF copy + exclusive extras**

Your purchase includes a DRM-free PDF copy of this book, 7-day trial to the Packt+ library (no credit card required), and additional exclusive extras. See the *Free benefits with your book* section in the *Preface* to unlock them instantly and maximize your learning.

# Technical requirements

You need the following packages to follow the examples in this chapter. Let's load them into your R workspace:

```
#install.packages("pacman")
library(pacman)
p_load("conflicted"
       ,"dataseries"
       ,"fable"
       ,"fabletools"
       ,"feasts"
       ,"here"
       ,"readr"
       ,"seasonal"
       ,"slider"
       ,"tidyverse"
       ,"tsibble")

#manage conflicts
conflicts_prefer(
  dplyr::filter,
  fabletools::components )
```

The data used in this chapter is available in the `data/data_chap7` folder in the GitHub repo, `https://github.com/PacktPublishing/Modern-Time-Series-Analysis-with-R.git`. The code file is also available in the same repo for you to download and practice.

# 7.1 Summary of differences: Adjustments, transformations and decompositions

Adjustments, transformations and decompositions serve fundamentally different analytical purposes in time series analysis.

Adjustments are about ensuring the data accurately reflects the underlying characteristics by correcting for measurable, external influences (e.g., holidays, trading-day effect).

Transformations are mathematical functions applied to the data to improve statistical properties (i.e., reducing volatility) required by further modeling.

Decomposition methods are procedures applied to the data to separate inherent time series components (trend, seasonality, and remainder) for better structural understanding, modeling and forecasting.

A brief comparison of time series adjustment, transformation, and decomposition techniques is presented in *Table 7.1*. You will learn the usage of these techniques and when to apply them in an analytical framework:

| Criteria | Adjustments | Transformations | Decomposition |
|---|---|---|---|
| Primary goal | Data cleaning: Remove external biases or normalize scale | Model prep: Stabilize variance and normalize distribution for modeling assumptions | Analysis: Break series into inherent components (trend, season, remainder) |
| Nature of change | Value modified by external factors (e.g., day counts, population size) | Mathematical function applied to values (e.g., log, power function) | Statistical model that separates deterministic parts of the series |
| Methods included | Calendar, population, indexing | Logarithmic, Box-Cox | MA-based methods, classical, X-13-ARIMA, STL |

*Table 7.1: Comparison of time series adjustment, transformation, and decomposition methods*

In summary, while each of these three techniques manipulates data differently, they collectively form an essential analytical toolkit in the time series analysis framework. Depending on the problem at hand, you may need to apply only one technique or all three combined.

Let's look at scenarios where adjustments and transformations are required for time series data.

# 7.2 Rationales for time series adjustments and transformations

Adjustments and transformations make time series data suitable for analysis and modeling. The purposes of time series adjustment and transformations are the following:

- **Removing known sources of variation**: Some data variation can occur solely due to calendar effects. Calendar adjustments are needed before identifying any other existing pattern in the data.
- **Handling outliers and anomalies**: Outliers are large measurement errors, and anomalies are data points that do not belong to the assumed data distribution. They can distort the analysis and modeling of time series data, and may be caused by errors in data collection or unusual events. Adjustments may involve identifying and handling outliers by winsorizing or applying time series smoothing techniques.
- **Transforming data to a stationary state**: Stationarity is a property of time series data that means the mean, variance, and autocorrelation structure do not change over time. The validity of many analyses and models depends on the stationarity of the input data. Differencing or detrending is used to make non-stationary data stationary.
- **Seasonality adjustment**: Seasonality is a pattern in data that repeats at fixed intervals. Often this masks the magnitude of other components. Seasonal differencing and seasonal decomposition are two common techniques for removing seasonal patterns from data, making it easier to analyze underlying trends and irregular components.
- **Missing observations**: Missing observations are a common scenario you will face when dealing with business time series data. Imputation and interpolation methods are used to fill in missing data points, ensuring a complete series for modeling.
- **Data non-normality**: Many statistical analyses and inferences assume that the input data follows a normal distribution. If the observed data does not follow normality, then a logarithmic or Box-Cox transformation is applied to make the data more normally distributed.

- **Heteroscedasticity**: Heteroscedasticity refers to the changing variability of residuals from a time series model over time. It affects the reliability of statistical analyses. This will also happen when non-linear data is passed into a modeling framework that expects a linear series. A logarithmic function is the most common way to transform data in such a scenario. Square root or cube root functions are also used, depending on the nature of non-linearity. It is important to apply a reverse transformation on the analyzed or model output to bring data back to its original scale.
- **Noise-reduction**: Too much noise or random variability in the data obscures underlying patterns. Time series smoothing or filtering methods are used to reduce noise and highlight underlying components.
- **Comparability**: When multiple time series are compared or data is combined from different sources, adjustments are often needed to ensure that the data is on a common scale and unit of measurement.
- **Interpretability**: Adjustments and transformations often simplify complex patterns or relationships, which makes the data more interpretable. This helps in making informed business decisions.
- **Improving forecast accuracy**: Sometimes, adjustments and transformations are performed solely to improve the forecast accuracy of a model. These adjustments are based on business rules and largely depend on the problem at hand.

Now that you have learned about the reasons for applying adjustments and transformations to time series data, let's look at some common adjustment techniques.

## 7.3 Time series adjustments

There are many different types of adjustments that can be used for time series data. The most appropriate adjustment will depend on the specific characteristics of the data. Each of the time series adjustment techniques has its own strengths, limitations, and best use cases. *Table 7.2* summarizes the three common adjustment techniques covered in this book. We discuss them in detail in subsequent sections.

| Method | Assumptions | Strengths | Limitations | Use case |
|---|---|---|---|---|
| Calendar adjustment | Stable relationship between day counts and the measured variable (e.g., sales) | Removes distorting calendar effects (holidays, trading days) | Requires accurate historical day counts; complexity with moving holidays (Easter) | Retail sales, production data |
| Population adjustment | Linear relationship between population growth and the time series variable | Normalizes data by a constant unit (e.g., per capita) for comparison | Ignores demographic structure changes (e.g., age distribution shifts) | Macroeconomic metrics (GDP, unemployment) |
| Base adjustment (indexing) | The base period is representative and stable; comparisons are relative | Simplifies comparisons of growth rates across diverse series; highly interpretable | Base period changes disrupt long-term consistency; sensitive to base choice | Benchmarking economic indicators or stock market indices |

*Table 7.2: Summary of common time Series adjustments*

Now, you will learn how to apply these adjustment techniques to time series data.

### 7.3.1 Calendar adjustments

These adjustments are used to remove the effects of known calendar events, such as holidays and trading days. For example, you are analyzing the monthly total number of visitors to a website over 2 years, and February is expected to be lower than the other months solely because of the smaller number of days. You can adjust this variation by using the average monthly visitors rather than the totals. This is known as calendar adjustments. There may be other reasons why visitors in February could be lower. Calendar adjustments also help you separate known facts from the unknown when analyzing data.

For illustration, we will use a synthetic time series of total daily visitors to a website. The daily website visitor count is generated using a Poisson distribution with an average of 950 visitors per day. The `total_visitor` and `avg_visitor` are computed per month using the following code:

```
# Set the seed for reproducibility
set.seed(3302)
visitors <- tibble(visit_date = seq(as.Date("2021-01-01"),
                                    as.Date("2022-12-31"),
                                    by = "1 day")) |>
    mutate(visitor = rpois(n = n(), lambda = 950)) |>
    as_tsibble() |>
    group_by_key() |>
    index_by(year_month = ~yearmonth(.)) |>
    summarize(total_visitor = sum(visitor),
              avg_visitor = mean(visitor))
```

Let's visually inspect the time plots of these two series that are presented in *Figure 7.1* presents:

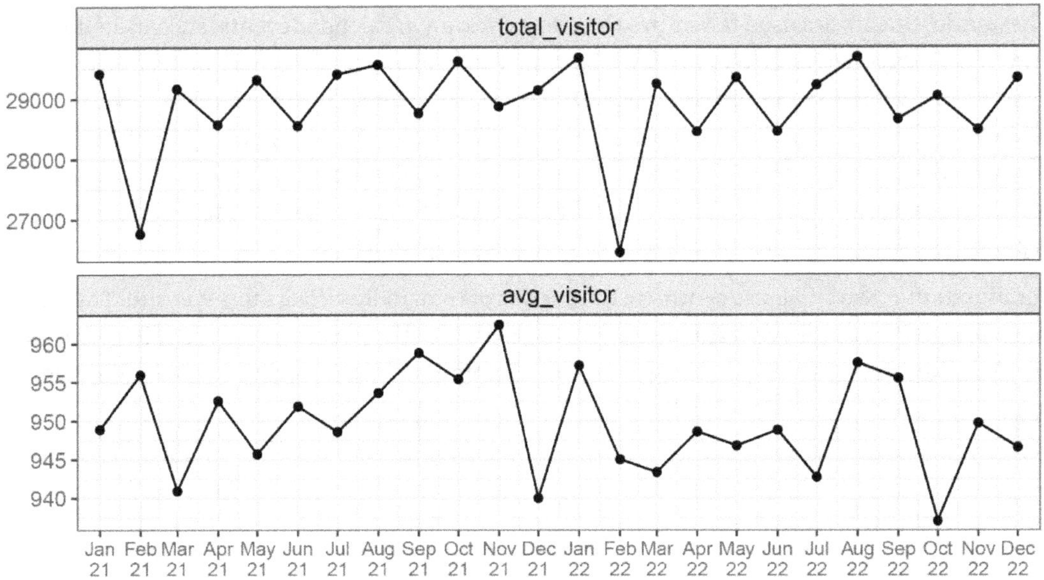


*Figure 7.1: Calendar adjustment on a simulated website visitor series*

In *Figure 7.1*, the top panel shows the total number of visitors, and the bottom panel shows the average number of visitors per month. In the original `total_visitor` series, February was the lowest month in every year. However, after the calendar adjustment (taking the monthly average), it is identified that December 2021 and October 2022 had the lowest average visitors.

Calendar adjustments, therefore, reveal patterns that could be hidden due to the data collection unit of measurement. It is a good idea to perform calendar adjustments on time series that are highly likely to be impacted by holidays, such as hotel occupancy rate, retail sales, air travelers, traffic volume, and so on.

## 7.3.2 Population adjustments

Population adjustments may require any data that is impacted by changes in the population. A large number of metrics are always reported after population adjustments, such as birth and death rates, crime rates, school enrollment rates, unemployment rates, inflation rates, and so on. Population adjustments are usually applied to raw measurements, such as total births, deaths, and number of unemployed, which would increase with a larger population.

Let's illustrate the concept with a simulated scenario where you are analyzing the clicks of a certain product on a website to assess its popularity. The simulation runs as follows:

1. Simulate a base time series (`stable_clicks`) for a stable population (`base_visits`).
2. Introduce an increasing population bias: 20% annual growth in raw visits.
3. Compute daily raw clicks and population-adjusted clicks normalized to the average population level.
4. Compute monthly unadjusted raw clicks (`unadjusted_clicks`) and population adjusted clicks (`adjusted_clicks`) by taking the monthly average to accommodate calendar adjustment and population adjustment at the same time.
   Let's see a full breakdown of the code:

```
set.seed(3302)
days_in_sim <- 730
clicks_data <- tibble(
  visit_date = seq(as.Date("2021-01-01"),
                   as.Date("2022-12-31"),
                   by = "1 day"),
  stable_clicks = rpois(n = days_in_sim, lambda = 100),
  base_visitors = rpois(n = days_in_sim, lambda = 5000)) |>
  # introduce annual population growth
  mutate(
    date_index = as.numeric(visit_date - min(visit_date)),
    raw_visits = base_visitors * (1 + 0.2 * (date_index/365))
  ) |>
  mutate(
    raw_clicks = stable_clicks * (1 + 0.2 * (date_index/365)),
    pop_adjusted_clicks = raw_clicks/raw_visits * mean(raw_visits)
  ) |>
  as_tsibble() |>
  index_by(year_month = ~ yearmonth(.)) |>
  summarize(
    unadjusted_clicks = mean(raw_clicks),
    adjusted_clicks = mean(pop_adjusted_clicks)
  )
```

5. Plot the results to visually highlight the differences. *Figure 7.2* presents time plots of the two series in one graphical window:

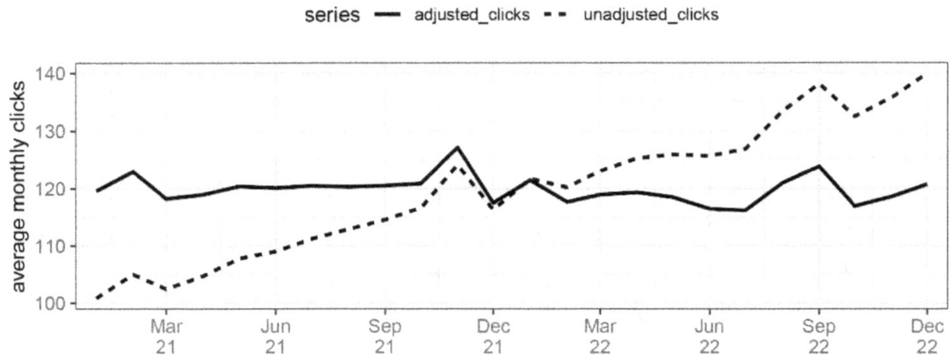


*Figure 7.2: Effect of population and calendar adjustments to a simulated product click data*

The visual comparison of `unadjusted_clicks` and `adjusted_clicks` clearly illustrates the importance of population adjustment in time series analysis. The unadjusted series (dashed line) exhibits a strong upward trend, reflecting the 20% annual visitor growth on the website. On the other hand, the adjusted series (solid line) removes this growth bias, remaining stable and flat over a two-year period. Based on the unadjusted clicks, some might interpret that the product has become popular, whereas the adjusted clicks tell the opposite story.

Population adjustments are important because they isolate the true underlying behavior of observed data, free from the confounding influence of external population changes, and ensure that the data is comparable over time.

Next, you will see how to perform base adjustments on time series data.

## 7.3.3 Base adjustments

Base adjustment is the process of designating a specific observation within a given time as the base, and then subsequently expressing all other observations relative to that base. Base adjustments are also known as reference point adjustments.

For example, you compare sales before, during and after a promotion trial. To identify the impact of the promotion, it is important to adjust all observations to a reference point. You might use a day's sales as the base before your analysis period begins. If there is strong weekly seasonality in your data, then you can use seven days' sales (one for each day of the week) as a base instead of a single day's figure and adjust the data using the corresponding weekdays' base value. This would allow you to compare the actual impact of promotion, even if the population size or the sales rate had changed.

Base adjustments can be done using a variety of methods, including but not limited to the following:

- **Indexing**: This is the most common method for base adjustments. It involves dividing each data point by the value of the data point in the base period and then multiplying by 100. Indexing transforms all series to a common scale (base=100), which makes comparisons among series simpler.
- **Deflating**: By deflating, you can remove inflation from the data. This can be done by dividing each observation by a price index. The **Consumer Price Index** (**CPI**) and the **Produce Price Index** (**PPI**) are commonly used for adjusting inflation in economic data.
- **Chaining**: Chaining involves linking together successive periods of data by calculating the percentage change in each period relative to the previous period. This technique is commonly used in business to compare changes over time, as percentage changes can be more informative than measuring changes in actual values.

We will explain how to perform indexing on time series data using the Switzerland quarterly total employment data for females and males. The data are downloaded from `www.dataseries.org` using the `dataseries` package and filtered to have full-year data from 2000 to 2019. Here, the yearly total employees are calculated for both groups and adjusted using their respective values in 2000 as a base:

```
swiss_employment <- cbind(
  male = ds("ch_comb_emp.ttl.m.tot", "ts"),
  female = ds("ch_comb_emp.ttl.f.tot", "ts")
) |>
    as_tsibble() |>
    filter_index("2000 Q1" ~ "2019 Q4") |>
    index_by(year = ~year(.)) |>
    group_by_key() |>
    summarize(employee = sum(value)) |>
    group_by_key() |>
    mutate(employee_base_2000 = round(100 * employee/employee[1]))
```

The unadjusted and base adjusted series (using 2000 as base) for both the genders are plotted in *Figure 7.3*:

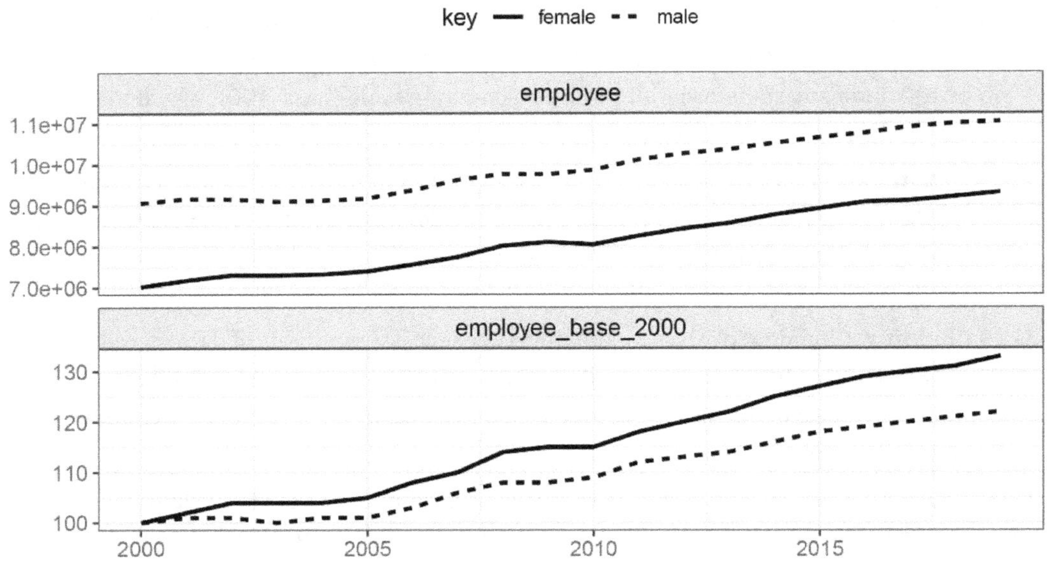


*Figure 7.3: Switzerland female and male employee during 2000-2019 with and without base adjusted*

The absolute number of male employees remained consistently higher than that of female employees (top panel in *Figure 7.3*). After adjusting, the indexed series shows that female employment increased more rapidly than male employment from 2000 onward, resulting in a widening gap between the two over time.

When analyzing multiple time series of varying magnitudes observed during the same time period, indexing proves to be a valuable tool. By converting scales to a common one, indexing enables meaningful comparisons and inferences among multiple series with various magnitudes. One advantage of base adjustments is ensuring that the underlying trend is not obscured by other factors. Whichever method you use for base adjustments, be aware of its impact on modeling and interpretation of forecast values.

Next, we discuss two mathematical transformations that are commonly applied to time series: logarithmic and Box-Cox transformations.

# 7.4 Time series transformations

Mathematical transformations are performed when time series data exhibits heteroscedasticity; this refers to when the variance of the data changes along with the level of the data over time. These types of transformations also bring non-normal data to normality—an important assumption to hold for statistical analysis and modeling.

We will look at logarithmic transformations first.

## 7.4.1 Logarithmic transformation

A logarithmic (in short, log) transformation can be used in multiple scenarios. The two most useful cases are to transform data when it is not normally distributed, and to stabilize the variance.

Logarithmic transformations are widely used in practice because they make it easy to understand the relative changes in a set of data. In base-10 logs, an increase of 1 unit corresponds to a 10× increase in the original value. This makes it easy to evaluate how the data is changing over time, even if the original values are very large or very small. However, log transformation cannot be used when there are zero or negative numbers in the data.

If the original observations are denoted as, $y_1, y_2, \ldots, y_T$ then you can obtain the log transformed series as $z_t = \log(y_t)$.

Let's use monthly Indian tourist arrival data to explain why logarithmic transformation is required to stabilize variance.

First, download the series and perform logarithmic transformation on it as shown in the following code:

```
indian_tourism <- ds("ch_fso_hesta.62.1", "ts") |>
  as_tsibble() |>
  filter_index("2005 Jan" ~ "2019 Dec") |>
  rename(arrivals = value) |>
  mutate(log_arrivals = log(arrivals))
```

Visualize the original and transformed series to inspect the effect of the logarithmic transformation. The plot is presented in *Figure 7.4:*

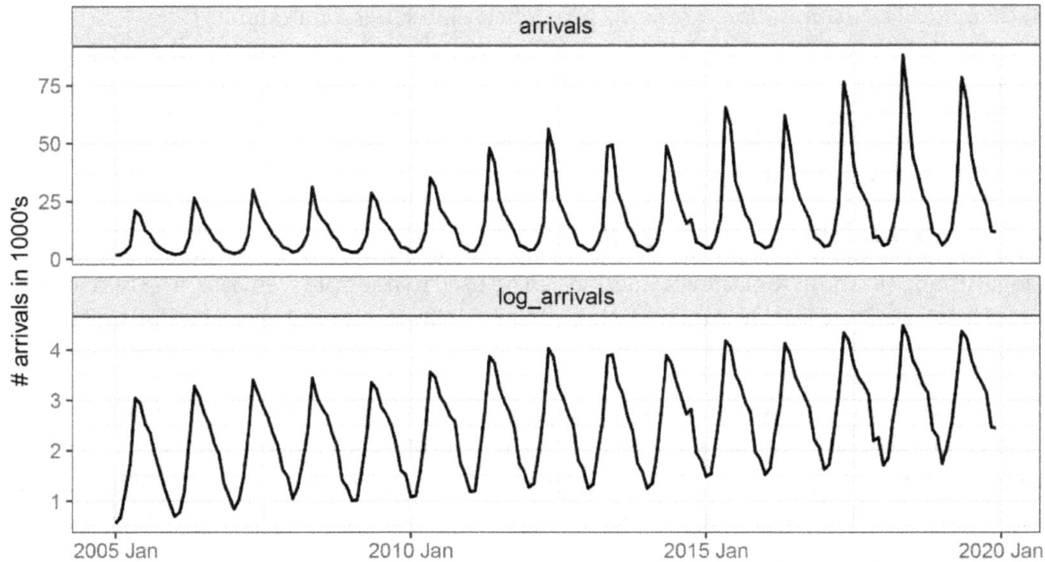


*Figure 7.4: Monthly tourist arrivals in India 2005-2019*

The original series in the top panel of *Figure 7.4* reflects increasing exponential growth, which may jeopardize the proper estimation of seasonality and trend. The log-transformed series in the bottom panel shows the consistent seasonality and slight upward trend over time. The logarithmic transformation makes the non-linear series into a linear one, which clearly reveals underlying seasonal peaks and troughs.

Some other commonly used transformations are the following:

- Square root transformation: $z_t = \sqrt{(}y_t)$ or $z_t = y_t^{1/2}$
- Cube root transformation: $z_t = y_t^{1/3}$

They are also known as power transformations and have the general form $z_t = y_t^p$. One disadvantage of power transformations is that they are not quite interpretable.

This leads to a more general family of transformations widely used in time series analysis.

## 7.4.2 Box-Cox transformation

Box-Cox transformations is a family of mathematical transformations that were introduced by Box et al. (1964). These include both logarithmics and power transformations that depend on a parameter, $\lambda$. A modified Box-Cox transformation is presented in Bickel et al. (1981), and the mathematical form is as follows:

$$z_t = \begin{cases} \log(y_t) & \text{if } \lambda = 0; \\ \dfrac{y_t^\lambda - 1}{\lambda} & \text{if } \lambda \neq 0 \end{cases} \qquad (7.1)$$

Here, the log is a natural logarithm (i.e., to base $e$).

The transformation in (7.1) uses a natural logarithm (base $e$) when $\lambda = 0$. When $\lambda \neq 0$, a power transformation is used, and some simple scaling is done.

This transformation also changes the shape of the transformed data except when $\lambda = 1$. If $\lambda = 1$, then $z_t = y_t - 1$, and the transformed data shifts downwards without any impact on shape. This particular form of Box-Cox transformation allows negative values in the original series, when the parameters $\lambda > 0$.

There are multiple implementations of Box-Cox transformations in several R packages:

- The `box_cox()` function in the `fabletools` package which implements the form presented in (7.1). The function also provides an inverse Box-Cox transformation, `inv_box_cox()`.
- The `BoxCox()` function in the `forecast` package implements a slightly different formula in which the sign of the observation is considered.

You can select the value of $\lambda$ automatically following a model-independent method proposed in Guerrero (1993). The optimal value is chosen because it minimizes the coefficient of variation across several subseries. You can select the optimal transformation for a series using either the `forecast::BoxCox.lambda()` function, or the `guerrero()` function from the `feasts` (Feature Extraction and Statistics for Time Series) package. Functions from two packages may produce slightly different results depending on whether the data contains complete observations for the full seasonal period or not.

We can find the optimal transformation required ($\lambda$ values) for the `arrivals` series applying the function as follows:

```
lambda_auto_selection <- feasts::guerrero(indian_tourism$arrivals)
```

The function returns $\lambda = 0.4$ as optimal for the arrivals series:

```
round(lambda_auto_selection,1)

# lambda_guerrero
#             0.4
```

The value indicates that the `arrivals` series requires a transformation between a logarithm ($\lambda = 0$) and a square root ($\lambda = 0.5$) to stabilize its variance.

Along with the automated selection, let's apply two other manual values, $\lambda = 0$ and $\lambda = 0.7$, via the Box-Cox transformations to the Indian tourism data. These transformations on the arrivals series are shown in the following code:

```
indian_tourism_bc <- indian_tourism |>
  mutate(
    `lambda 0(manual)` = box_cox(arrivals, lambda = 0),
    `lambda 0.7(manual)` = box_cox(arrivals, lambda = 0.7),
    `lambda 0.4(auto)` = box_cox(arrivals,
                                 lambda = feasts::guerrero(arrivals)))
```

We have three transformed series obtained from three different levels of transformation. *Figure 7.5* presents the original and three transformed series that is generated using the following code:

```
indian_tourism_bc |>
  pivot_longer(
    cols = c(
      arrivals,
      `lambda 0(manual)`,
      `lambda 0.7(manual)`,
      `lambda 0.4(auto)`
    ),
    names_to = "lambda_value",
    values_to = "value"
  ) |>
  ggplot(aes(x = index,y = value)) +
  geom_line() +
  facet_wrap(
    ~lambda_value,
    ncol = 2,
    scales = "free_y"
  ) +
  labs(x = "", y = "")
```

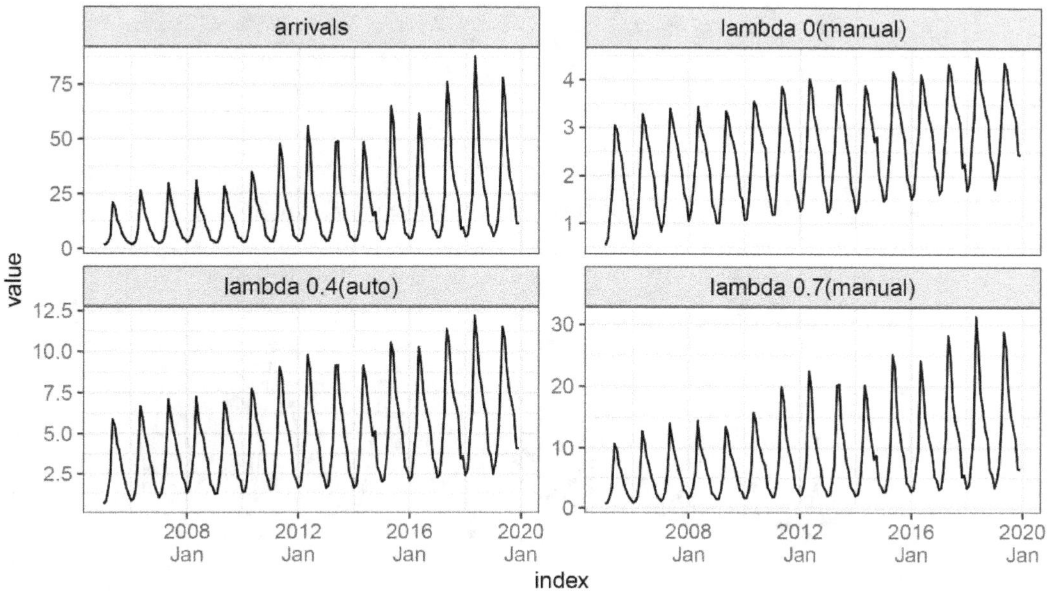


*Figure 7.5: Effect of three different Box-Cox transformations of Indian tourist arrivals 2005-2019*

In *Figure 7.5*, you can see that different values of $\lambda$ have different impacts on the original data. The logarithmic transformation ($\lambda = 0$) over-corrects the data, flattening the structure too aggressively. When $\lambda = 0.7$, the data is only transformed a little. The auto selection, $\lambda = 0.4$, reduces the exponential nature of the data without compromising the actual nature of the series, that is, variability increased along with trend. This is crucial for performing proper analysis as the transformation did not hide the actual pattern in the data.

If Box-Cox transformations are used, it is important to apply inverse Box-Cox transformations on forecast values to bring them to the original scale to ensure that the forecasts are interpretable and comparable.

Now, we will deep dive into time series decomposition techniques, starting with decomposition plots in the next section.

# 7.5 Decomposition plots

We will use the monthly Australian takeaway food services turnover between 2015 and 2022 to explain decomposition plots. First load the ABS retail data (`abs_retail.rds` is available in the `data/data_chap7` folder in the GitHub repo). After loading, filter the data as follows:

```
abs_data <- read_rds(here("data", "data_chap7","abs_retail.rds"))
```

```
takeaway_turnover  <- abs_data  |>
  filter(state == "Total (State)",
         industry == "Takeaway food services") |>
  filter(year(month) >= 2015, year(month) <= 2022 ) |>
  select(month, turnover)
```

Let's examine the series visually from the time plot presented in *Figure 7.6*:

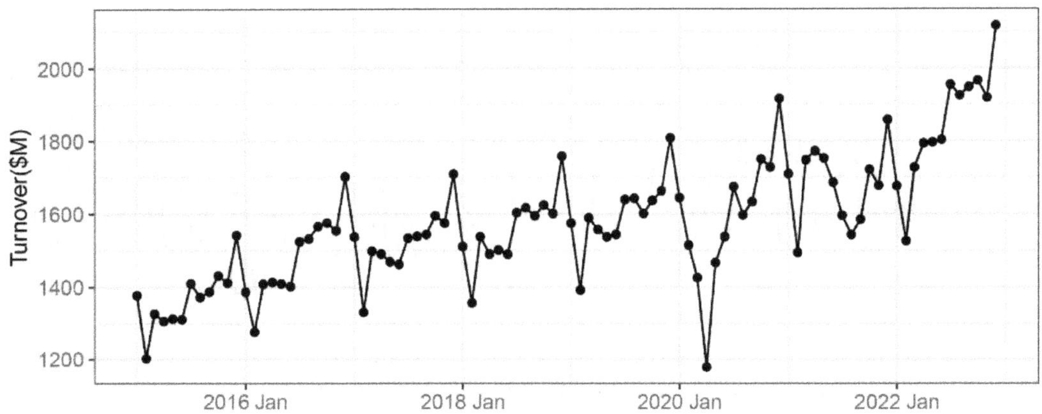


*Figure 7.6: Monthly Australian takeaway food services turnover: 2015-2022*

The series exhibits trend or trend-cycle ($T$), seasonality ($S$) and remainder ($E$) components. Now, apply STL decomposition on this data to illustrate decomposition plots. A detailed discussion on the STL method is included in *Section 7.9* later in the chapter.

We will use the `STL()` function from the `feasts` package:

```
turnover_dcomp <- takeaway_turnover |>
  model(stl = STL(turnover))
turnover_dcomp
```

The code produces the following output:

```
# A mable: 1 x 1
      stl
  <model>
1   <STL>
```

The output is a `mable` (model table) object which is nothing but a list-column `tibble` object. It is called mable because it contains a model, which is STL in this case. If you apply multiple models on one data object in a tidyverse fashion, they will all appear in one `mable` object in different columns.

Now, you need to extract the contents of the `mable` object using the `components()` function from the `fabletools` package as follows:

```
components(turnover_dcomp)
```

The function will produce the following output:

```
# A dable: 96 x 7 [1M]
# Key:     .model [1]
# :        turnover = trend + season_year + remainder
   .model    month turnover trend season_year remainder season_adjust
   <chr>     <mth>    <dbl> <dbl>       <dbl>     <dbl>         <dbl>
 1 stl    2015 Jan     1378 1354.        4.13    20.4           1374.
 2 stl    2015 Feb     1203. 1357.    -163.        8.97          1366.
 3 stl    2015 Mar     1328. 1360.      -27.2     -4.74          1355.
 4 stl    2015 Apr     1307. 1363.      -67.4    11.5            1374.
 5 stl    2015 May     1314. 1366.      -52.6     -0.341         1366.
 6 stl    2015 Jun     1312. 1370.      -57.5     -0.724         1370.
 7 stl    2015 Jul     1410. 1374.       34.6      1.60          1376.
 8 stl    2015 Aug     1373. 1379.       19.5    -25.2           1353.
 9 stl    2015 Sep     1388. 1383.       21.6    -16.9           1366.
10 stl    2015 Oct     1432  1388.       62.2    -17.8           1370.
# i 86 more rows
```

The extracted components are presented in a `dable` (decomposition table) object. It is basically a `tibble` object with extra information about the STL model fit.

The turnover = trend + season_year + remainder tells us that the STL model used an additive model for decomposition. Along with the data column the `dable` object includes the following columns:

- Estimated trend (`trend`)
- Estimated seasonal effect (`season_year`)
- Remainder (`remainder`)
- Seasonally adjusted observations (`season_adjust`)

From the additive model formulation, it is clear that any component can be obtained by subtracting the other two from the original observation at any time point. You can confirm the computation to significant digits by using `View()` or `print()` function that reveals the decimal places on any `tibble` object.

*Figure 7.7* shows the decomposition plot, obtained by applying the `autoplot()` function directly on the `dable` object:

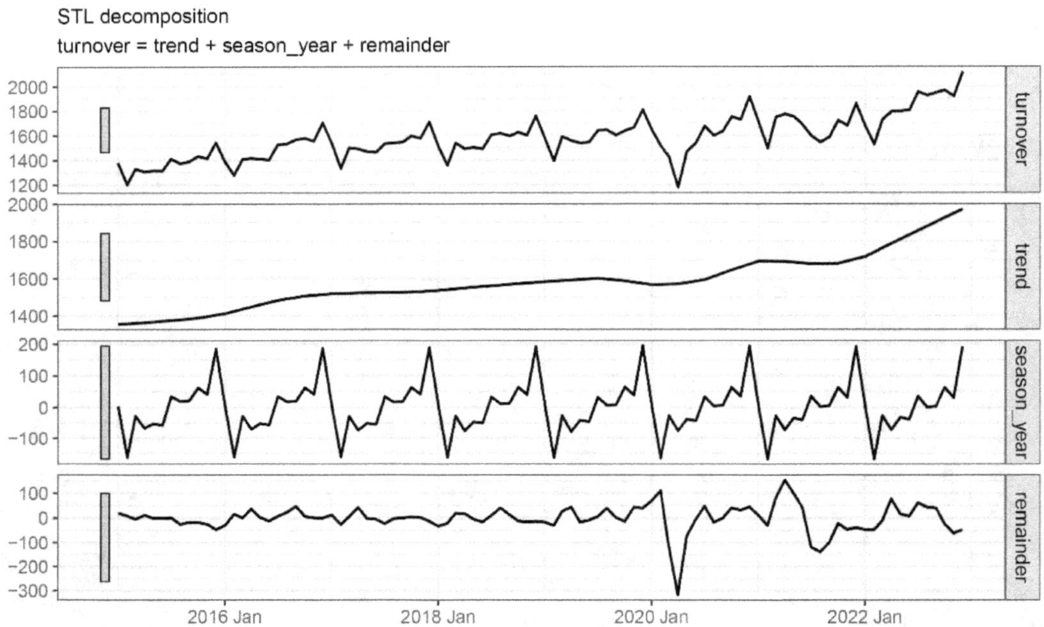


*Figure 7.7: STL decomposition of Australian takeaway food services turnover 2015-2022*

The decomposition plot shows the original series in the top panel, followed by the estimated trend, seasonality and remainder. There is an overall upward trend in the data with slight bumps that correspond to the large dip in the remainder components. The seasonality pattern is consistent from year to year. The gray bars on the left side serve as scale indicators. They standardize the vertical scale across panels, making it easier to visually compare the relative magnitude of each component to the original series. The seasonal component has the lowest variation relative to the actual series, followed by the remainder and then the trend. You may consider how much each panel has been zoomed on y-axis scale.

Sometimes we overlay the estimated components with the actual observation for comparison. You can extract the corresponding column from the `dable` object into a `tsibble` object for plotting as shown in the following code:

```
components(turnover_dcomp) |>
  as_tsibble() |>
```

```
autoplot(turnover, colour = "gray", linewidth = 1.2) +
geom_line(aes(y = trend), colour = "black", linewidth = 1.2) +
labs(x = "" ,y = "Turnover")
```

The preceding code produces *Figure 7.8*, in which one can evaluate the effect of decomposition by comparing decomposed components with the original data:

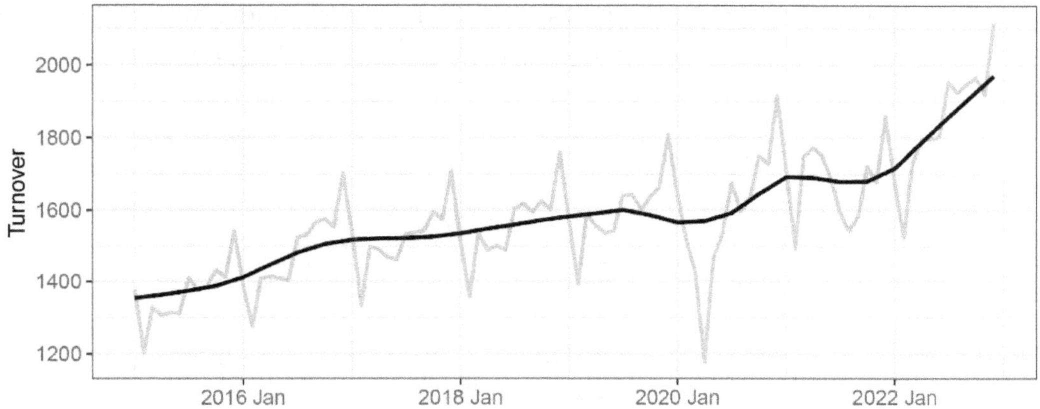


*Figure 7.8: Takeaway food services turnover and decomposed trend from STL*

A decomposition plot presents information of a complex time series by breaking it into three distinct, simpler components. It helps us to see patterns that are otherwise hidden in the raw data.

Next, we discuss several decomposition techniques and will use a decomposition plot where applicable.

# 7.6 Decomposition using MAs

The MA takes the average of a fixed number of observations by moving the time horizon. The number of observations is fixed, and observations leave and enter as the time horizon changes. The MA removes noise in the data and reveals underlying trends, trend-cycles, or seasonal components. That is why the technique is known as a time series smoothing technique than a decomposition method.

The MA method has a parameter, $m$, which denotes the number of observations of the averaging window. For example, a 3-MA uses three data points to calculate the average. The MA value is aligned with the center of the window (for odd $m$ value). Naturally, there are no MA values at the beginning or at the end of a series. The higher the $m$ value, the larger the information loss is. Some allow a shorter window of observations at the end of the series. Even if there is information loss, the technique works really well for short-to-medium series to depict any underlying components.

Next, we'll see how to decompose a series with trends using MA smoothing and then we'll decompose a series with seasonality using the same technique.

## 7.6.1 MA smoothing of data with trends

You will find an example of smoothing annual takeaway food retail turnover with a 3-MA and a 5-MA. The helper functions from the `slider()` package (`Vaughan, 2022`) provide an easier way to implement MAs in your code:

```
smooth_turnover <- takeaway_turnover |>
  index_by(year = ~year(.)) |>
  summarise(turnover = sum(turnover)) |>
  mutate(`3-MA` = slide_dbl(.x = turnover,
                            .f = mean,
                            .before = 1,
                            .after = 1,
                            .complete = FALSE),
         `5-MA` = slide_dbl(.x = turnover,
                            .f = mean,
                            .before = 2,
                            .after = 2,
                            .complete = FALSE ))
```

Now, visualize the two MA series with the original series to inspect the effect of smoothing. Both the original and the MA series are plotted in *Figure 7.9*:

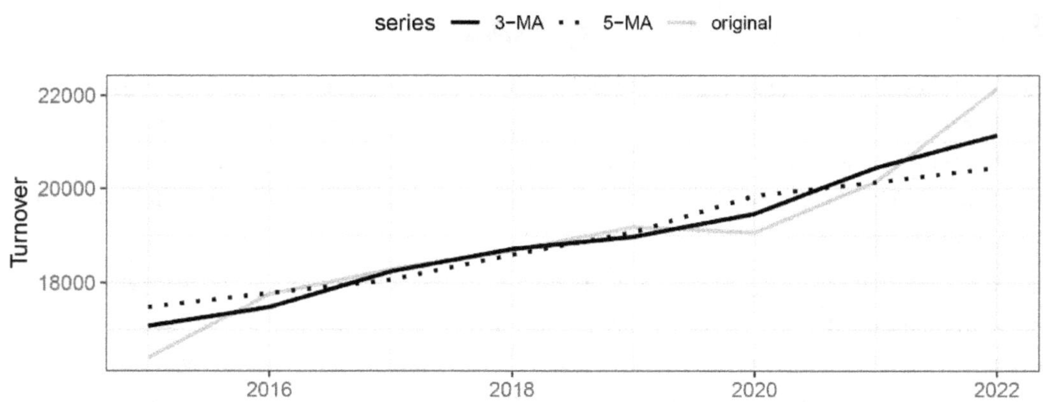


*Figure 7.9: MAs of annual Australian takeaway food services turnover 2015-2022*

The parameter of MA controls the amount of noise reduction. There is a possibility of introducing new noises if the data is over-smoothed. As we see in *Figure 7.9*, the 3-MA series has smoothed the annual series to reveal the trend, whereas the 5-MA series changes the magnitude of the trend at the end by over-smoothing observations around 2020.

## 7.6.2 MA smoothing of data with seasonality

For seasonal series, it is best practice to use an MA that matches the frequency of the seasonal data; for example, using a 12-MA for a monthly series. As you will notice, it is not possible to align MA values exactly to a middle observation when the sliding window is even. Taking a further 2-MA on the 12-MA helps mitigate the issue as shown in the following code:

```
turnover_monthly_ma <- takeaway_turnover |>
  mutate(`12-MA` = slide_dbl(.x = turnover,
                              .f = mean,
                              .before = 5,
                              .after = 6,
                              .complete = TRUE),
         `2x12-MA` = slide_dbl(.x = `12-MA`,
                                .f = mean,
                                .before = 1,
                                .after = 0,
                                .complete = TRUE))
```

The smoothed data should reveal the underlying trend or trend-cycle component, if exists. Some data points are lost as we used the `.complete=TRUE` argument while calculating MAs.

Let's visualize the original series with the 2x12-MA smoother series using the following code:

```
turnover_monthly_ma |>
  ggplot(aes(x = month)) +
  geom_line(aes(y = turnover,
                color = "original"),
            linewidth = 0.8 ) +
  geom_line(aes(y = `2x12-MA`,
            color = "2x12-MA"),
            linewidth = 0.8 )+
  scale_color_manual(name = "series",
                     values = c("original" = "gray",
                                "2x12-MA" = "black")) +
  labs( x = "", y = "" ) +
  theme(legend.position = "top")
```

The preceding code generates *Figure 7.10*:

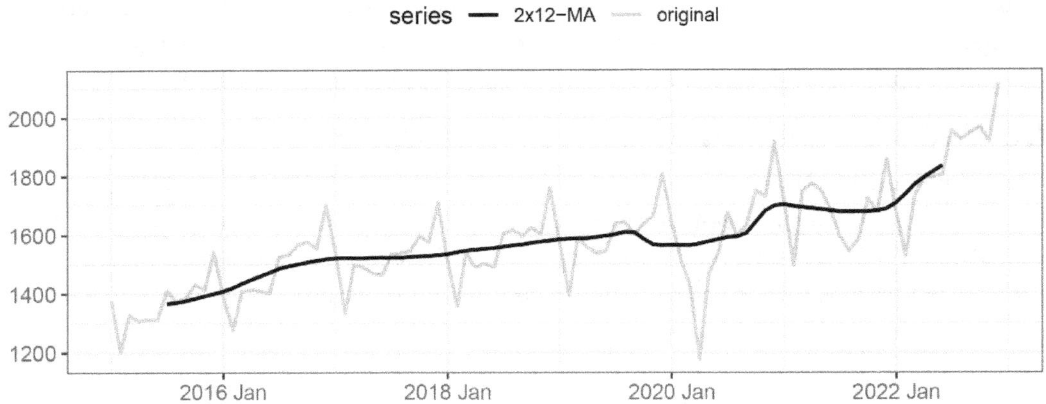


*Figure 7.10: Seasonal MA of the monthly Australian takeaway food services turnover*

You may notice that 12-MA smoothing removed noise and seasonality to reveal the underlying trend or trend-cycle component, even though there is information loss at both ends in the observation period. Another limitation of the MA smoothing method is that it does not estimate any seasonal component.

The classical decomposition method provides an estimate for both the trend and the seasonal component which is discussed in the next section.

# 7.7 Trend and seasonal decomposition with the classical method

The classical decomposition method uses MA smoothing to estimate trends, and seasonal components using de-trended values.

In *Chapter 4*, we saw that time series components can be combined either additively or multiplicatively. Classical decomposition is of two types based on how components are presented in the data. Additive decomposition is used when the seasonal component is constant from year to year, while multiplicative decomposition is used when the seasonal component varies over time. The technique follows the steps described next.

1. The trend-cycle ($\hat{T}_t$) is estimated using MA smoothing. The smoothed values are used to de-trend the observed value ($Y_t$). The de-trended series contains the seasonal ($S_t$)

and remainder ($R_t$) terms. Depending on how the components are combined, one of the following two equations is used to de-trend:

- Additive: $\widehat{Y}_t - \widehat{T}_t$
- Multiplicative: $\widehat{Y}_t / \widehat{T}_t$

2. The seasonal indices $\widehat{S}_i$,for $i = 1, \ldots, m$ are obtained by taking the average of the de-trended value at each seasonal period $m$:

   - Additive: The seasonal indices are adjusted so that $\sum \widehat{S}_i = 0$
   - Multiplicative: The seasonal indices are adjusted so that $\sum \widehat{S}_i = m$

   The seasonal components $\widehat{S}_t$ are obtained by chaining these seasonal indices and repeating for each year.

3. The remainder component is then obtained as follows:

   - Additive: $\widehat{R_t} = \widehat{Y}_t - \widehat{T}_t - \widehat{S}_t$
   - Multiplicative: $\widehat{R_t} = \widehat{Y}_t / (\widehat{T}_t \times \widehat{S}_t)$

You can apply the preceding method using the `classical_decomposition()` function from the `feasts` package. The type argument allows you to pass which method of decomposition should be used, additive or multiplicative.

An additive model is used on the takeaway food services turnover as the seasonal component stays constant over the years. The following code fits the classical decomposition on turnover series and generates a decomposition plot:

```
takeaway_turnover |>
  model(classical_decomposition(turnover, type = "additive"))  |>
  components() |>
   autoplot(linewidth = 1) +
  labs(x = "")
```

Now, you can visually examine the classical decomposition method to the takeaway turnover series from *Figure 7.11*:

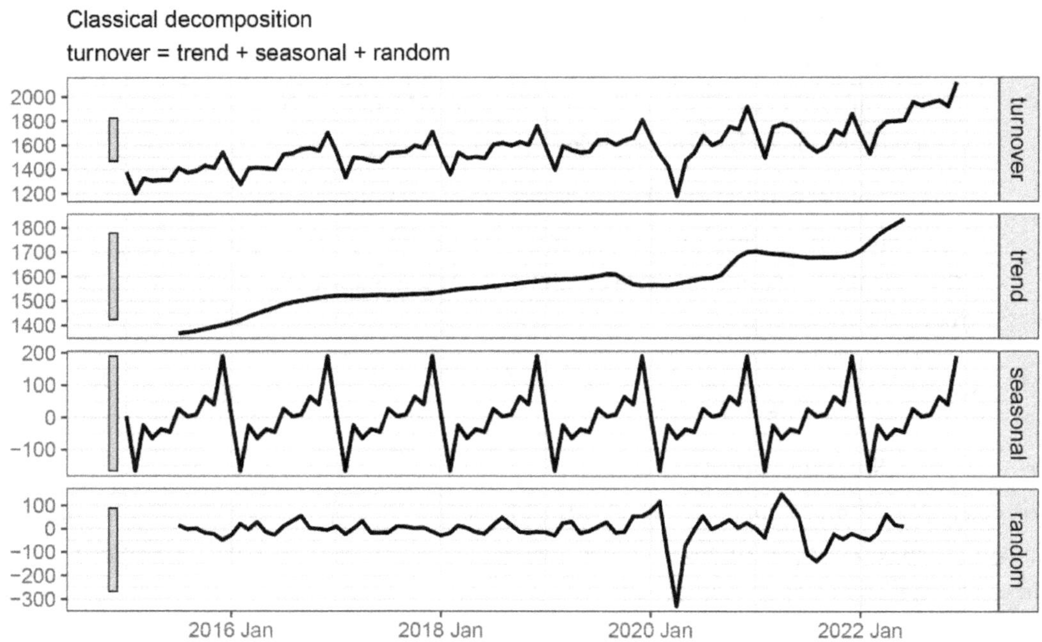


*Figure 7.11: Classical decomposition of monthly Australian takeaway food services turnover*

The decomposition plot shows that the estimated trend component has increased over time with a couple of dips during 2020 and 2022. The seasonal pattern was consistent throughout the observed period.

Classical decomposition is applicable in situations where the seasonal pattern does not change over time. Information loss at the beginning and end due to the MA, the tendency of over-smoothing of a trend-cycle, and the inability to handle sudden shocks in the data are some disadvantages of the classical method.

In the following section, you will learn about model-based decomposition methods, which are mainly developed and evolved to adjust economic and financial time series data.

# 7.8 Model based decomposition methods

Many financial and economic time series require proper seasonal adjustment. For policy makers it is crucial to compare seasonally adjusted data to evaluate existing policies and design new ones if needed. From that need, model-based decomposition methods, such as X-11 and Seasonal Extraction in ARIMA Time Series (SEATS) were developed and implemented in software. The X-11-ARIMA method was first developed by the US Census Bureau in 1965 and has evolved in multiple versions, the most recent one being X-13-ARIMA. The SEATS method was developed by the Bank of Spain and implemented in two kinds of software. Many government agencies, statistical bureaus, and financial institutions around the world widely use these model-based decomposition methods. However, detailed discussion on these topics is out of the scope of this book. For further reference see Dagum et al. (2016).

Advantages of these model-based decomposition techniques include, but not limited to, the following:

- No information loss due to MA smoothing
- Better seasonality estimates in quarterly and monthly data (cannot handle weekly or daily seasonality, though)
- Handling of trading day variations
- Handling holiday effects and outliers robustly
- Use of external predictors

Both the X-11 and SEATS methods for seasonal decomposition are implemented in the feasts package via the `X_13ARIMA_SEATS()` function. This function calls the seasonal package, which offers further controls and helper functions.

You can apply the methods to the takeaway food turnover data using the following code:

```
turnover_model_decomp <- takeaway_turnover |>
  model(x11 = X_13ARIMA_SEATS(turnover ~ x11()),
        seats = X_13ARIMA_SEATS(turnover ~ seats()))
```

Extract the components estimated using X-11 and make a decomposition plot as follows:

```
turnover_model_decomp |>
  select(x11) |>
  components() |>
   autoplot(linewidth = 1) +
  labs(x = "")
```

Let's examine the X-11 decomposition on the takeaway turnover presented in *Figure 7.12*:

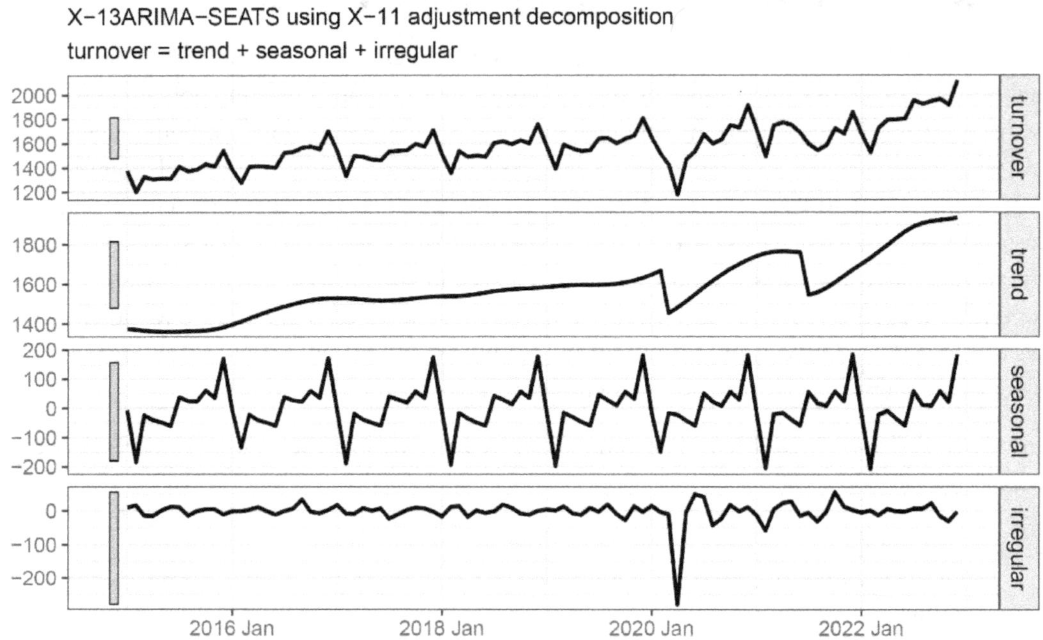


*Figure 7.12: X-13-ARIMA decomposition on monthly Australian takeaway food services turnover 2015-2022*

In contrast to the classical decomposition shown in *Figure 7.11*, the X-11 model incorporates the outlier in the trend-cycle component to provide an estimate of the seasonal component. This seasonal estimate can be used to de-seasonalize the observed series to estimate long-term trends.

If you are looking for a versatile and robust decomposition method that considers all seasonal frequencies, unlike the model-based methods, or you are wishing for a fully automated method with flexible user controls, then you should use STL decomposition. You have already seen STL mentioned in several sections. Now, you will learn more about it in the next section.

# 7.9 STL decomposition

**Seasonal-Trend decomposition using Loess** (**STL**) was introduced by Cleveland et al. (1990). The method uses **Locally Estimated Scatterplot Smoothing** (**LOESS**) techniques in the estimation of the time series components. It overcomes all the major disadvantages of the classical and model-based decomposition methods, as follows:

- Handles all types of seasonal frequency ranges, from daily to monthly
- Captures variations in seasonal components over the years

- More user control over trend-cycle smoothness, which reduces the possibilities of over-smoothing
- Outliers are estimated via a robust method

However, STL assumes an additive model by design and cannot automatically handle trading-day variation or holiday effects. The STL method can be extended to allow the multiplicative nature of time series after proper transformation.

The method is implemented in the `forecast::stl()` and `feasts::STL()` functions. We will use the latter version as it fits with the tidyverse ecosystem. STL decomposition uses multiple parameters under three broad categories. The defaults are set for each parameter, which works well for the majority of the series. As a user, you may need to control the following parameters depending on the observed series:

- `trend(window=?)`: Determines how many nearest observations are to be considered in the LOESS method to estimate the trend-cycle component. A larger window value allows for greater smoothness. For monthly data the default is 21. Defaults are calculated for each seasonal frequency. Should always be an odd value.
- `season(window=?)`: How many consecutive years to be considered in seasonal indices estimation. The default value is 13 and should be an odd value. To consider a consistent seasonal pattern across the years use 'periodic' instead.
- `lowpass(window=?)`: The number of observations used in the low-pass filter for each subseries. Uses an odd number matching the seasonal period of the data by default.

The robust option also handles data with outliers. We will use the same takeaway turnover data as we used before to compare across several decomposition techniques. Let's extract time series components from monthly turnover using STL:

```
turnover_stl <- takeaway_turnover |>
  model(
    stl_dcomp = STL(
      turnover ~ trend(window = 12) +
                 season(window="periodic"),
      robust = TRUE
    )
  )
```

Earlier in *Section 7.5*, we showed the use of `components()` function to extract the estimated components from an `STL` object and generate a decomposition plots from an STL fit. You can experiment by passing different parameter values in the `STL` function and then compare the outputs by generating decomposition plots.

By default, STL assumes components are combined additively in the observed series. You can use STL to decompose data even when components are combined multiplicatively after transforming data either using a logarithmic or Box-Cox transformation. Taking the logarithmic route makes the data to an additive model:

$$\begin{aligned} Y_t &= T_t \times S_T \times R_t \\ log(Y_t) &= log(T_t) + log(S_t) + log(R_t) \end{aligned}$$

Box-Cox transformation provides greater flexibility as it enables robust decomposition when the data has issues, such as non-constant variance. When the optimal transformation parameter, $\lambda$, is between 0 and 1 ($0 < \lambda < 1$), the series must first be transformed before applying the `STL()` function. It is crucial to back-transform the resulting decomposed series to the original unit of measurement for valid interpretation.

We have covered four different time series decomposition techniques. Each of them has its own strengths and limitations, applied widely in current business practices. In the next section, a summary of these techniques is presented.

## 7.9.1 Summary of decomposition techniques

Decomposition is an integral part of analyzing time series data. Time series decomposition methods, ranging from simple to complex, are discussed with examples in *Table 7.3*:

| Method | Assumptions | Strengths | Limitations | Use case |
|---|---|---|---|---|
| MA | Seasonality is strictly stable/ constant (additive or multiplicative) | Simple calculation; intuitive understanding; good for quick visualization | Highly sensitive to outliers; information loss at start/ end; cannot handle moving seasonality | Simple, stable data for quick, exploratory analysis |

| Method | Assumptions | Strengths | Limitations | Use case |
|---|---|---|---|---|
| Classical decomposition | Seasonal component and trend component are constant/stable over time | Simple and fast; provides clear separation of components | Cannot handle moving seasonality; poor outlier handling | Historical economic data with very stable patterns; initial exploration |
| X-13ARIMA / SEATS | The underlying time series can be adequately modeled by an ARIMA model structure | Industry standard in official statistics (e.g., Census Bureau); handles outliers and trading days automatically | Complex and resource-intensive; black box for many users; difficult to integrate automated pipelines | Official government statistics (e.g., GDP, unemployment rates) where high accuracy is required |
| STL | Seasonality can move slowly over time (LOESS smoothing handles evolution) | Very robust to outliers; handles all types of seasonal frequencies | Default is additive decomposition (requires pre-transformation for multiplicative data); does not handle trading days/holidays automatically | Applicable to most time series analysis |

*Table 7.3: Summary of time series decomposition methods*

# Summary

This chapter presented you with a collection of methods for performing adjustments, transformations, and decomposition on time series. All three techniques are used as data pre-processing steps and are simultaneously important for identifying patterns.

You learned about calendar adjustments, population adjustments and base adjustments. Logarithmic and the Box-Cox transformations are widely used when data does not conform to model assumptions. The main idea behind adjustments and transformation is to make data comparable, increase interpretability, and improve forecast accuracy. The time series decomposition methods are an important toolkit to help you understand the nature of time series data.

The techniques you learned in this chapter provided you with results, graphs and stats about the presence and strength of time series components. Apart from analyzing the four main components, there are many other features that can be derived from a time series to study its characteristics. In the next chapter, *Chapter 8*, you will learn about time series features and how to extract them using a faster R implementation.

# Further reading


- Bickel, P. J., & Doksum, K. A. (1981). An analysis of transformations revisited. *Journal of the American Statistical Association, 76(374).*
- Box, G. E. P., & Cox, D. R. (1964). An analysis of transformations. *Journal of the Royal Statistical Society. Series B, Statistical Methodology, 26(2).*
- Cleveland, R. B., Cleveland, W. S., McRae, J. E., & Terpenning, I. J. (1990). STL: A seasonal-trend decomposition procedure based on loess. *Journal of Official Statistics, 6(1).*
- Dagum, E. B., & Bianconcini, D. (2016). *Seasonal adjustment methods and real time trend-cycle estimation.* Springer.
- Guerrero, V. M. (1993). Time-series analysis supported by power transformations. *Journal of Forecasting, 12(1).*
- Vaughan, D. (2022). *Slider: Sliding window functions* (Version 0.3.0).

# Get this book's PDF version and more

Scan the QR code (or go to `packtpub.com/unlock`). Search for this book by name, confirm the edition, and then follow the steps on the page.

UNLOCK NOW

*Note: Keep your invoice handy. Purchases made directly from Packt don't require an invoice.*

# 8

# Time Series Features

Feature extraction is a common machine learning (ML) technique in which different attributes are calculated from a series that are used to analyze data characteristics. Feature extraction can be performed on time series data too. Many different attributes can be extracted from a univariate time series. These features can then be used to analyze the series itself or can be collectively used to compare a large collection of time series.

The chapter is organized as follows,

- Section 8.1: Time series features
- Section 8.2: Statistical summary-based features
- Section 8.3: Autocorrelation based features
- Section 8.4: STL-based features
- Section 8.5: Tiling window features
- Section 8.6: Sliding window features
- Section 8.7: Statistical test-based features
- Section 8.8: Other useful features
- Section 8.9: Clustering using time series features

We will discuss a wide range of time series features, each of which provides valuable insights for observed data. When we move to modeling, you will see that information provided by these features is automatically embedded in a large number of time series forecasting models, and you don't need to manually incorporate it. However, understanding these features helps you choose a plausible class of forecasting models for a given dataset. Later in the chapter, we present an example use case on how a time series feature set can be used in clustering when many series need to analyze and/or forecast.

Note

**Your purchase includes a free PDF copy + exclusive extras**

Your purchase includes a DRM-free PDF copy of this book, 7-day trial to the Packt+ library (no credit card required), and additional exclusive extras. See the *Free benefits with your book* section in the *Preface* to unlock them instantly and maximize your learning.

# Technical requirements

Load the following packages into your R console.

```
library(pacman)
p_load("bundle"
       ,"conflicted"
       ,"factoextra"
       ,"fabletools"
       ,"feasts"
       ,"fpp3"
       ,"fracdiff"
       ,"fs"
       ,"here"
       ,"readr"
       ,"tidyverse"
       ,"tidymodels"
       ,"urca"
       ,"uwot" )

conflicts_prefer(
  dplyr::filter,
  dplyr::lag,
  fabletools::components,
  fabletools::accuracy )

conflict_prefer_all("feasts", "tsfeatures")
```

The data used in this chapter is available in the `data/data_chap8` folder in the GitHub repo at `https://github.com/PacktPublishing/Modern-Time-Series-Analysis-with-R.git`. You will also find the code file in the same GitHub repo.

Let's start by learning what time series features are and why they are important in time series analysis.

# 8.1 Time series features

The trend, seasonality, cyclical, and remainder components define an observed time series and provide insights into the underlying data generation process. Other attributes, such as mean, variance, and autocorrelation, are calculated from a series and provide information about the observed system. Just as we can compare the mean and variance of two series to comment on their population distributions, time series features, such as the strength of trend or tiling window variance can be used to compare different time series in a similar way.

In other words, time series features can be used to quantify the characteristics of a time series in a way that is independent of specific values in the series. This allows us to compare different time series even if they have different scales or units of measurement.

For example, you could compare two sales series using their strength-of-trend feature. A higher strength of trend would indicate that the sales are increasing or decreasing more rapidly. You could also use the tiled-window-variance feature to compare the variability of two different time series. A higher tiling window variance would indicate that the values in the time series are more spread out.

By comparing time series features, you can gain a better understanding of the similarities and differences between various time series. This can be helpful for a variety of tasks, such as forecasting, anomaly detection, and time series classification.

The `feasts` (which stands for **Feature Extraction and Statistics for Time Series**) package (O'Hara-Wild et al., 2023) provides a collection of feature computations for time series data. You have already used functions from this package, such as `ACF()`, `gg_season()`, `gg_subseries()`, and so on, in previous chapters.

The computed features and statistics from the `feasts` package can be arranged into several groups for explanation and usage purposes:

- Autocorrelation-based features
- STL-based features
- Tiling window features
- Sliding window features
- Statistical test-based features
- Statistical summary-based features
- Other useful features

In the examples of this chapter, most features are computed using default parameters and are optimized where required. However, you can test these arguments and find optimal settings for your data and use case. The important part is to understand what these features offer and how to use them appropriately.

Alongside learning how to compute feature sets from time series data, we would also like to know how to interpret them and use them appropriately. For illustration, we will use two example series from the `abs_retail` data, available in the `data/data_chap8` folder in the GitHub repo. Let's load the data into your R workspace, filter two series, and organize them:

```
# Load data
abs_retail <- read_rds("data/data_chap8/abs_retail.rds")

retail_df <- abs_retail |>
  filter(
    state == "Victoria",
    industry %in% c("Department stores", "Supermarket and grocery stores")
  ) |>
  mutate(
    industry = ifelse(
      industry == "Supermarket and grocery stores",
      "Supermarket stores", industry)
  ) |>
  filter_index("2000 Jan" ~ "2022 Dec") |>
  select(industry, month, turnover)
```

Let's visually examine them from their time plots presented in *Figure 8.1*:

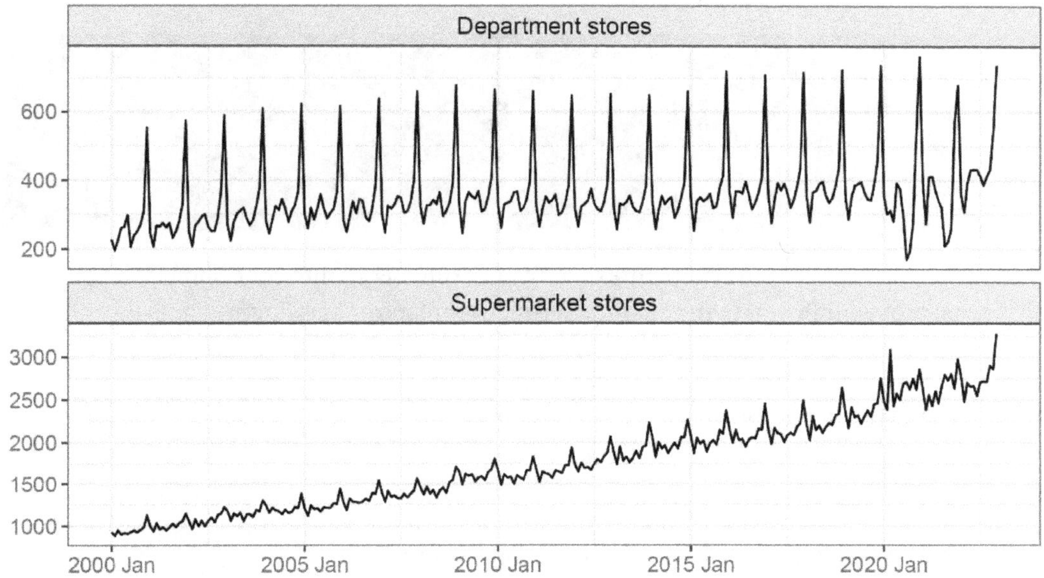


*Figure 8.1: Monthly turnover from two industries in Victoria, Australia*

The monthly turnovers from two industries in Victoria (Australia) from 2000 to 2022 are plotted in *Figure 8.1*. From a visual inspection, the two series look quite different in nature, magnitude, trend, and seasonality. We will extract various features from these two series and learn what insights can be drawn about them.

We will start with some simple statistical summary-based features in the next section.

## 8.2 Statistical summary-based features

When we talk about features in data analytics, the first thing that comes to mind is descriptive statistics, such as, mean, variance, quantiles and so on. You can compute these features by applying the `features()` function from the `feasts` package.

Let's compute the mean and variance from the observed turnover from two industries, department stores and supermarket/grocery stores:

```
stat_feat <- retail_df |>
  features(turnover, list(mean=mean, var=var))

stat_feat
```

The result is shown in the following output:

```
# A tibble: 2 × 3
  industry               mean     var
  <chr>                 <dbl>   <dbl>
1 Department stores      356.  11900.
2 Supermarket stores    1751. 305344.
```

The quantile features provide cut points that divide a time series into continuous intervals with equal probabilities. You can compute the quantile features as follows:

```
quant_feat <- retail_df |>
  features(turnover, quantile)

quant_feat
```

The quantile-based features produce the following output:

```
# A tibble: 2 × 6
  industry             `0%` `25%` `50%` `75%` `100%`
  <chr>               <dbl> <dbl> <dbl> <dbl>  <dbl>
1 Department stores    170.  298.  334.  370.   763.
2 Supermarket stores   886  1266. 1676. 2149.  3278.
```

The statistical summary-based features provide information on the scale and distribution of an observed series. Supermarket stores operate on a significantly larger scale compared to department stores. The supermarket series' mean turnover was substantially higher (1751 vs 356), and variance (305K) is over 25 times greater than for department stores (11.9K). The quantile features highlights that the difference is consistent across all percentile levels, from a minimum of 0% to a maximum of 100%. In summary, supermarket and grocery stores generate higher sales volumes and exhibit greater absolute volatility than department stores during the observed period.

Next, we will discuss the autocorrelation-based features that will provide information on a time series dependency structure.

# 8.3 Autocorrelation based features

The **autocorrelation functions** (**ACF**) measure the long-term strength and direction of the linear relationship between the current values and the lagged values in a time series. The autocorrelation coefficient at various lags from an original series indicates the strength and direction of trend. The ACFs computed from a first-differenced series indicate the effectiveness of first differencing, whether that has stabilized the mean and variance over time. The ACFs from second differencing indicate whether double differencing introduces an unwanted dependency structure due to over-differencing. The **Partial Autocorrelation Function** (**PACF**) shows the overall strength of short-term relationships.

*Table 8.1* provides a short description of ACF and PACF features. In general, they provide insights into how strong the trend in the original series is and what level of non-seasonal and seasonal differencing is required to make a series stationary:

| Feature | Description |
|---|---|
| `acf1` | The autocorrelation at lag 1 of the original series |
| `acf10` | The sum of squared autocorrelations for lags 1 through 10 of the original series |
| `diff1_acf1` | The autocorrelation at lag 1 of the first-differenced series |
| `diff1_acf10` | The sum of squared autocorrelations for lags 1 through 10 of the first-differenced series |
| `diff2_acf1` | The autocorrelation at lag 1 of the second-differenced series |
| `season_acf1` | The autocorrelation at the first seasonal lag 1 of the original series (lag 12 for monthly data) |
| `pacf5` | The sum of the squared partial autocorrelations for lags 1 through 5 of the original series |

| Feature | Description |
|---|---|
| `diff1_pacf5` | The sum of the squared partial autocorrelations for lags 1 through 5 after one round of non-seasonal differencing |
| `diff2_pacf5` | The sum of the squared partial autocorrelations for lags 1 through 5 after two rounds of non-seasonal differencing |
| `season_pacf` | The partial autocorrelation at the first seasonal lag (lag 12 for monthly data) |

*Table 8.1: Description of ACF and PACF based features*

The `feat_acf()` and `feat_pacf()` functions provide ACF and PACF-based features. Let's compute these features from the turnover series. The values are listed in *Table 8.2*:

| Features | Department stores | Supermarket stores |
|---|---|---|
| `acf1` | 0.266 | 0.952 |
| `acf10` | 0.207 | 8.026 |
| `diff1_acf1` | -0.229 | -0.437 |
| `diff1_acf10` | 0.232 | 0.661 |
| `diff2_acf1` | -0.482 | -0.630 |
| `diff2_acf10` | 0.297 | 1.170 |
| `season_acf1` | 0.889 | 0.858 |
| `pacf5` | 0.129 | 1.043 |
| `diff1_pacf5` | 1.581 | 1.106 |
| `diff2_pacf5` | 1.880 | 1.512 |

| Features | Department stores | Supermarket stores |
|---|---|---|
| `season_pacf` | 0.864 | 0.191 |

*Table 8.2: ACF and PACF features computed from Department and Supermarket stores turnover*

Based on the computed features, the following insights can be drawn on the two series:

- **Department stores** have a much lower `acf1` value (0.266). While still positive, the dependency structure is weak and decays faster. This series appeared closer to stationary as the overall dependency structure is weak (`acf10` = 0.207). The slightly higher values of `diff1_acf10` (0.231) and `diff2_acf10` (0.297) indicate that non-seasonal differencing is likely to introduce an unwanted dependency structure, evident by the first 10 ACFs are increasing in the first- and second-differenced series. The seasonality is quite strong, indicated by a high `season_acf1` value (0.889), pointing to the fact that a seasonal differencing will be required to reduce seasonal dependency in order to make the series stationary before further analysis or forecast modeling.
- **Supermarket and grocery stores** have a high positive acf1 value (0.952) and a very large `acf10` value (8.03). This indicates a strong, persistent positive upward trend. A much lower `diff1_acf10` value (0.661) suggested first differencing is required to stabilize the mean and variance (i.e., stationary). However, a relatively higher `diff2_acf10` value (1.17) indicates that a second difference will introduce unwanted dependency due to over-differencing. This series has a strong seasonal pattern with `season_acf1` (0.858). A high `pacf5` value (1.04) indicates a dominant, non-stationary trend, and double differencing is not required (`diff2_pacf5` is higher than `diff1_pacf5`). Seasonal PACF is relatively weaker (`season_pacf` = 0.19) suggesting that seasonal differencing is not required; rather, inclusions of seasonal moving averages (MAs) could model this series better.

You have learned about the STL decomposition which is used to extract time series components. In the following section, you will learn how to compute informative features from the output of STL decomposition.

# 8.4 STL-based features

**Seasonal-Trend decomposition using Loess** (**STL**) is a time series decomposition technique that was discussed in *Section 7.9* of *Chapter 7*. Features based on the STL method quantify the strength of trend and seasonal components by comparing the variability of the estimated components.

Before explaining how the features are computed, let's briefly recall how STL works. Considering $T_t$ is the trend component, $S_t$ is the seasonal component, and $R_t$ is the remainder component, the underlying model in STL decomposition at time $t$ is the following:

$$y_t = T_t + S_t + R_t.$$

After separating the observed series into its components by STL, you can measure the variance of each estimated component, such as the following:

- Variance of the seasonally adjusted series: $Var(T_t + R_t)$
- Variance of the trend adjusted series: $Var(S_t + R_t)$
- Variance of the remainder terms: $Var(R_t)$

The seasonally adjusted series removes the seasonal component from the series which leaves the trend and remainder terms. Similarly, the trend adjusted series removes the trend components leaving seasonal and remainder terms in the decomposed series. In the remainder series, both the trend and seasonality are removed. What information might these variances provide in terms of the relative strength of trend and seasonality?

In the seasonally adjusted series, a weak-to-no trend should have a similar value for $Var(R_t)$ and $Var(T_t + R_t)$. That means the variation in the trend and remainder components is basically dictated by the variation in the remainder component alone when there is no trend component. Hence, for a weak trend, the ratio $\frac{Var(R_t)}{Var(T_t+R_t)}$ approaches to $1$. On the other hand, for data with a strong trend, the same ratio will approach to $0$ as the $Var(R_t)$ becomes smaller compared to $Var(T_t + R_t)$.

The `trend_strength` feature, $F_T$ is defined as follows:

$$F_T = max\left(0{,}1 - \frac{Var(R_t)}{Var(T_t + R_t)}\right)$$

Here, $F_T = 0$ indicates there is no trend, and a higher value indicates the strong presence of a trend component, either positive or negative.

In a similar fashion, the `seasonal_strength` feature is defined as follows:

$$F_S = max\left(0{,}1 - \frac{Var(R_t)}{Var(S_t + R_t)}\right)$$

A value close to $1$ suggests strong seasonal dependency whereas a value close to $0$ indicates little to no seasonality.

When the STL decomposition cannot identify an autocorrelation at a particular lag, that information ends up in the remainder component, $R_t$. The spikiness feature is the leave-one-out-variances of $R_t$ which identifies any prevalence of spikes that were not captured by STL decomposition.

The `feat_stl()` functions provide other features along with the three discussed previously. A full list is presented in *Table 8.3*:

| **Feature** | **Description** |
|---|---|
| `trend_strength` | Measures the presence of the trend component using a variance ratio |
| `seasonal_strength_year` | Measures the presence of the seasonal component using a variance ratio |
| `seasonal_peak_year` | Identifies the specific seasonal observation (e.g., month or quarter) with the highest average value |
| `seasonal_through_year` | Identifies the specific seasonal observation (e.g., month or quarter) with the lowest average value |
| `spikiness` | Quantifies the variance of the remainder component of the time series decomposition |
| `linearity` | Assesses the linearity of the trend component in the decomposed series |
| `curvature` | Assesses the curvature of the trend component in the decomposed series |
| `stl_e_acf1` | Computes the first autocorrelation coefficient (ACF1) of the remainder series |
| `stl_e_acf10` | Computes the sum of squares of the first 10 autocorrelation coefficients of the remainder series |

*Table 8.3: Description of STL based features*

All these features are computed via one function, `feat_stl()`. The following code computes the STL based features for the turnover data:

```
stl_feat <- retail_df |>
  features(turnover, feat_stl)
```

Let's look at some important STL based features presented in *Table 8.4*:

| STL Features | Department stores | Supermarket stores |
|---|---|---|
| `trend_strength` | 0.77 | 0.99 |
| `seasonal_strength_year` | 0.96 | 0.85 |
| `linearity` | 526.70 | 8909.52 |
| `curvature` | -103.14 | 878.25 |
| `spikiness` | 21.70 | 1442.79 |
| `stl_e_acf1` | 0.20 | 0.10 |
| `stl_e_acf10` | 0.23 | 0.14 |

*Table 8.4: STL features of two retail series*

A value closer to 1 in `trend_strength` indicates that the variation in the series is dominated by the trend component. The trend in the supermarket series (0.99) is strong and consistent. Department stores also have a strong trend (0.77), but seasonality (0.96 vs 0.85) plays a relatively large role in their overall variation compared to the supermarket series.

The linearity and curvature feature measure the shape of the trend component. The high linearity suggests the supermarket trend has a very strong linear movement. The high curvature for supermarkets also suggests that while it is strongly linear, there are also significant changes in the growth rate over time. The lower linearity and negative curvature in the department stores series captured the fact that the trend was flat and moved slightly downward near the end of the observed period.

The spikiness measures the variance of the remainder component. The supermarket series has a much higher spikiness (1442.79 vs 21.70), indicating a much higher degree of irregular, short-term volatility or unpredictable spikes. The `stl_e_acf1` and `stl_e_acf10` values measure the autocorrelation structure remaining in the remainder (error) series. Ideally, these should be close to zero, suggesting all systematic structures have been captured by the trend and seasonal components.

Remember that the STL features are computed from the components extracted after an STL decomposition. Let's perform the STL decomposition on the supermarket turnover series and produce the decomposition plot as shown in the following code:

```
retail_df |>
  filter(industry == "Supermarket stores") |>
  model(STL(turnover)) |>
  components() |>
  autoplot() +
  labs(x="")
```

*Figure 8.2* presents the plot of STL decomposed supermarket turnover:

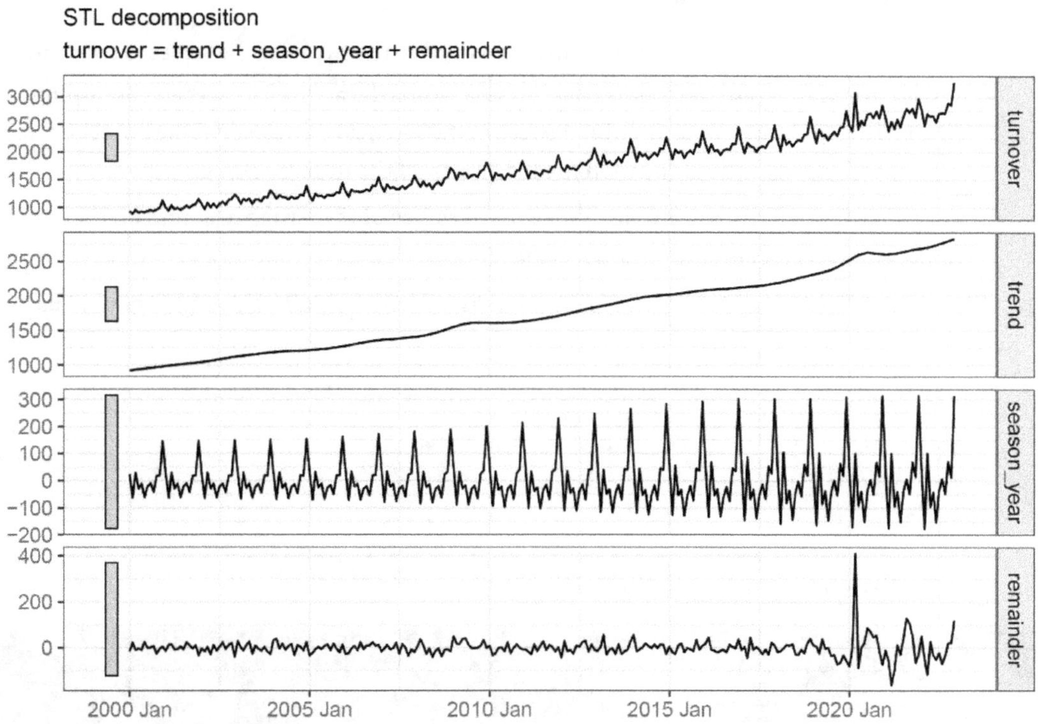


*Figure 8.2: STL decomposition of Supermarket monthly turnover series*

As you will notice, the high spikiness in the supermarket turnover was mainly due to the COVID-19 lockdown impact during the first half of 2020. Later, the noise settled down, but the volatility did not come back to the original level, even after two years.

Most of the facts, such as the shape of the trend component, changes in the growth rates, and changes in the volatility, are visible in the original series in *Figure 8.1*. However, comparison of different time series based solely on STL decomposition is more of a manual task. On the other hand, the quantified features from STL make comparing different series an objective, simplified, and potentially an automated task. Some features are computed on equally sized blocks of a time series. Two ways the blocks can be separated are through non-overlapping (tiled) and overlapping (sliding) windows. We will look at features that are computed over non-overlapping windows in the following section.

## 8.5 Tiling window features

The `var_tiled_mean` ("stability") and `var_tiled_var` ("lumpiness") features analyze the time series by dividing it into a series of non-overlapping "tiled" windows to assess how the central tendency and volatility change over time:

- `var_tiled_mean`: This feature measures how stable the mean of the series is over the full length of the observed period. For non-seasonal data, the size of the overlapping tile is 10. For a seasonal series, the tile size is the same as the seasonal period. Then, it calculates the variance of the tiled means across all the different tiles.
- `var_tiled_var`: This feature calculates the variance of the variance across all the different non-overlapping windows (tiles).

To calculate the tiling window features, pass them as a list into the `features()` function, as shown next:

```
tiled_feat <- retail_df |>
  features(turnover, list(var_tiled_mean, var_tiled_var ))

tiled_feat
```

The generated output is presented next:

```
# A tibble: 2 × 3
  industry           var_tiled_mean var_tiled_var
  <chr>                       <dbl>         <dbl>
1 Department stores           0.113         0.105
2 Supermarket stores          0.997         0.00125
```

A high var_tiled_mean value indicates the time series is unstable because the **average level** of the data changes significantly from one time window to the next. A low value suggests a stable average level over time. The department store's turnover is a stable series with a low var_tiled_mean value (0.113), indicating the average turnover did not change much, whereas the high value (0.997) in supermarket turnover indicates the average level in the series changed consistently across time.

The department store's turnover (0.105) is less lumpy, meaning the volatility in the data remains substantially the same over time. That is, there were no periods of high noise followed by periods of low noise. The supermarket turnover is even less lumpy (0.00125), except for the large volatility in the last two years.

These features are important for identifying non-stationary behavior and structural changes in a time series beyond simple trend and seasonality. Next, you will learn about features that are computed on sliding windows.

# 8.6 Sliding window features

Three features are computed on sliding windows (over-lapping windows) that measure when the maximum shift occurs at the average level, variance and in data distributions. They also report the time index when those shifts occurred:

- shift_level_max: Measures the largest mean shift between consecutive sliding windows. This is useful for identifying sudden jumps or drops in the average level (i.e., structural breaks) of the time series.
- shift_var_max: Calculates the largest variance shift between consecutive sliding windows. This feature is useful for detecting when a stable process becomes noisy, or vice versa, by capturing changes in the volatility.
- shift_kl_max: The Kullback-Leibler (KL) divergence measures the difference between two probability distributions. By computing the KL divergence on sliding windows of a time series, the shift_kl_max feature identifies the time and magnitude of the largest distributional shift. It is the most comprehensive measure of change as it detects shifts in the overall shape of the data distribution (average level, variance, and shape) between different periods.

Let's compute the sliding windows features for the example series:

```
sw_feat <- retail_df |>
  features(turnover,
           list(shift_level_max,
```

```
                    shift_var_max,
                    shift_kl_max))
```

The sliding window features computed on the turnover data are organized for comparison in *Table 8.5*:

| Features | Department stores | Supermarket stores |
|---|---|---|
| shift_level_max | 67.68 | 275.48 |
| shift_level_index | 265 | 242 |
| shift_var_max | 15559.6 | 40674.85 |
| shift_var_index | 264 | 244 |
| shift_kl_max | 2.77 | 0.16 |
| shift_kl_index | 262 | 237 |

*Table 8.5: Sliding window features of two retail series*

Before explaining the features, let's extract the time periods when the maximum shifts were detected using the following code:

```
series_idx <- retail_df |>
  filter(industry %in% c("Department stores", "Supermarket stores")) |>
  group_by(industry) |>
  mutate(row_index = row_number()) |>
  ungroup()

sw_feat |>
  select(c("industry", ends_with("index"))) |>
  pivot_longer(cols = ends_with("index")) |>
  left_join(
    series_idx,
    by = c( "industry" = "industry", "value" = "row_index")) |>
  arrange(industry, month)
```

The following output shows the month when the maximum shifts happened:

```
# A tibble: 6 × 5
  industry    name              value    month turnover
  <chr>       <chr>             <dbl>    <mth>    <dbl>
1 Department  shift_kl_index      262 2021 Oct      260
2 Department  shift_var_index     264 2021 Dec      680.
3 Department  shift_level_index   265 2022 Jan      359.
4 Supermarket shift_kl_index      237 2019 Sep     2304.
5 Supermarket shift_level_index   242 2020 Feb     2385.
6 Supermarket shift_var_index     244 2020 Apr     2428.
```

Department stores had the maximum changes in average level (`shift_level_max` = 67.68) in January 2022, and the volatility (`shift_var_max` = 15559.6) in December 2021. The distribution shift (`shift_kl_max` = 2.77) was detected in October 2021.

Supermarket turnover had the maximum shift in average level (`shift_level_max` = 275.48) in February 2020, and in volatility (`shift_var_max` = 40674.85) in April 2020. However, the distributional shift (`shift_kl_max` = 0.16) was much smaller compared to the Department stores series and was detected in September 2019.

Overall, supermarket turnover primarily reacted with large magnitude shifts in level and variance, likely to reflect the large shock due to COVID-19. The department store series, while also affected, experienced a more qualitative change in the distribution itself.

Next, we discuss features that are based on statistical properties of a time series.

# 8.7 Statistical test-based features

In time series analysis, we use different statistical tests to decide whether a series has unit roots, or how many differences are needed to make a series stationary. An important model fitting diagnostic is to check whether the residual from a model is white noise.

A unit root is a property in a time series where the average level shifts randomly over time. This makes the series level wise non-stationary and causes problems in parameters estimation and statistical inference. If a time series has unit root, then it needs to be differenced (i.e., made stationary) before applying any model.

There are various tests available to identify unit roots. The following features compute the test statistic and corresponding p-value to determine the existence of unit roots in a time series. In other words, these features identify whether a series is stationary or not:

- `unitroot_kpss`: This function computes the statistic from the Kwiatkowski et al. (1992) unit root test with linear trend and lag 1
- `unitroot_pp`: This function computes the statistic for the $Z_\tau$ version of Phillips and Perron unit root test with constant trend and lag 1

There are various arguments in the functions that you can study further and apply to your scenario.

Once you have identified that a time series is not stationary, the next step is to figure out how to make it stationary. A non-stationary series becomes stationary via a simple operation-taking the difference from its consecutive values for a non-seasonal series or taking the difference at seasonal periods for a seasonal series. Sometimes a double difference is required to make a series stationary. The following two features provide you with the optimal number of differencing required to obtain a stationary series:

- `unitroot_ndiffs()`: Identify the optimal differencing order that lies between 0 to 2 by applying the Kwiatkowski et al. (1992) unit root test.
- `unitroot_nsdiffs()`: Identify the optimal seasonal differencing order by evaluating the seasonal strength of an STL decomposition. Even though this method does not depend on a statistical test, it works well in identifying the order of seasonal differencing by evaluating the strength of the seasonal component in the data.

Let's check the stationarity status of the two example series, and what level of differencing is needed to make them stationary. You can perform all the unitroot tests using the following code:

```
uroot_feat <- retail_df |>
  features(turnover,
           list(unitroot_kpss,
                unitroot_pp,
                unitroot_ndiffs,
                unitroot_nsdiffs))
```

The output is summarized in *Table 8.6*:

| Feature | Department stores | Supermarket stores |
|---|---|---|
| `kpss_stat` | 2.1460 | 4.6127 |
| `kpss_pvalue` | 0.0100 | 0.0100 |
| `pp_stat` | -11.7199 | -0.2483 |
| `pp_pvalue` | 0.0100 | 0.1000 |
| `ndiffs` | 1 | 1 |
| `nsdiffs` | 1 | 1 |

*Table 8.6: Features to test the existence of a unitroot in department and supermarket turnover*

Both the turnover series constantly changed (strong trends and seasonal patterns) over time. A forecasting model assumes the average level in a time series should be stable. The low p-values (all `0.01` or `0.1`) from KPSS and Phillips and Perron indicates that the tests are very confident that neither set of turnover data is stable, that is, they are non-stationary. The series will become stationary after taking non-seasonal differencing once (subtract the previous consecutive value from each observation) and seasonal differencing once (subtract values of the same previous season from an observation). The trend adjustment (`ndiffs=1`) and the seasonal adjustment (`nsdiffs=1`) helps a forecast model to accurately capture data characteristics.

Once a forecast model is fitted, it is expected that the residuals from the forecast model should be white noise. That means, the residuals should be serially uncorrelated, have zero mean, constant variance (usually `1`), and be independent and identically distributed. A forecasting model diagnostic usually involves testing the autocorrelation structure in residuals. If the residuals are not white noise, then it suggests that the model could not capture the data dynamics well enough and needs improvement.

Two statistical tests, Ljung-Box and Box-Pierce, are commonly used to test white noise. Both tests examine the null hypothesis of independence in a given time series by computing autocorrelation at different lags and testing whether the sum of squared autocorrelation coefficients equals `0`. The main difference between these two tests is that they use slightly different test statistics.

Let's check whether the remainder series from the STL decomposition of the turnover series is white noise, using the following code:

```
retail_df |>
  model(stl = STL(turnover)) |>
  components() |>
  features(remainder,
           list(ljung_box, box_pierce)) |>
  mutate(across(3:6, \(x)round(x ,2))) |>
  select( - .model)
```

The result is given in the following output (lb stands for Ljung-Box and bp stands for Box-Pierce):

```
# A tibble: 2 × 5
  industry            lb_stat lb_pvalue bp_stat bp_pvalue
  <chr>                 <dbl>     <dbl>   <dbl>     <dbl>
1 Department stores     10.9       0       10.8       0
2 Supermarket stores    2.87      0.09    2.84      0.09
```

Usually the p-values from Statistical test are compared against a standard threshold of $0.05$ (significance level) for inference. For the department stores remainder, the p-values from both tests are way below $0.05$ ($0 < 0.05$), suggesting the remainder from this series is white noise. On the other hand, all the p-values for the supermarket series are higher than the $0.05$ significance level ($0.09 > 0.05$), indicating the remainder series is not white noise. That means, the supermarket remainder series exhibits some degree of autocorrelation.

The features discussed so far are commonly used in time series analysis. There are also some other features that are useful for understanding a time series. We present them in the following section as they cannot be grouped under a certain topic.

## 8.8 Other useful features

Apart from what has already been discussed, the features listed in this section provide valuable insights about data characteristics:

- `guerro()`: Identify the optimal $\lambda$ value for a Box-Cox transformation using the Guerrero method. You used this function in *Chapter 7, Section 7.4* when transforming a non-linear time series to a linear one.
- `coef_hurst()`: The Hurst coefficient measures the degree to which a time series exhibits long-term memory. A value of $0.5$ or below indicates that the time series is a

random walk, which means there is no long-term memory and that is difficult to forecast. A value greater than 0.5 indicates that the time series is likely to exhibit patterns that persist over a long period of time and could be easier to forecast.

- `n_crossing_points`: Computes the number of times a time series crosses the median.
- `longest_flat_spot`: Flat spots are computed by dividing the sample space of a time series into 10 equal-sized intervals and computing the maximum width within any single interval. This feature returns the number of intervals for which the data is relatively unchanged.
- `stat_arch_lm`: Checks the Autoregressive Conditional Heteroscedasticity (ARCH) condition based on a Lagrange Multiplier (LM) test.
- `feat_spectral`: This feature provides information about the forecastability of a time series. Based on the Shannon spectral entropy, this measures how much or how little noise there is in the data. A value close to 0 indicates that a series is easy to forecast as it has a strong trend and seasonal pattern. A value close to 1 indicates that the series is noisy and will be difficult to forecast.
- `feat_intermittent`: This feature provides useful information about intermittent time series. It returns four values: the average interval between non-zero observations (zero run mean), the squared coefficient of variation of non-zero observations (nonzero squared), the proportion of data that starts with zero (zero start prop), and the proportion of data that ends with zero (zero end prop).

All these features provide important information about different characteristics of time series. They provide insights into time series that are helpful for either understanding time series data or applying appropriate forecasting models. However, not all features are needed in every scenario. The choice of features will depend on what type of data you have and how you plan to analyze them.

The primary goal of time series features is to examine data characteristics from various perspectives and to suggest which kind of model would be suitable for analysis and forecasting. They can also be utilized to group similar time series together. In the next section, you will learn how some of these features are used to group a large collection of time series data.

If this seems important, you will be able to compute these important features from your time series data following the examples presented in this chapter.

## 8.9 Clustering using time series features

Time series features become a useful tool when you have a large collection of data and would like to summarize its characteristics. You may want to group series that have similar characteristics together. In literature multiple methods for time series clustering are available that incorporate features calculated from time series; see Wang et al. (2006) for details.

In the following example, you will find an application of feature-based time series clustering. The steps involved are the following:

- Compute important features for each series.
- Apply a dimension reduction technique, such as Principal Component Analysis (PCA) or Unifold Manifold Approximation and Projection for Dimension Reduction (UMAP), to reduce the dimension without compromising data variability.
- Apply a clustering technique, such as K-means or hierarchical clustering, to identify the optimal number of clusters for the entire dataset.

Each of these clusters consists of a series that shares common properties but is different from other clusters.

For illustration, let's use the 1,428 monthly time series from M3-competition (Makridakis et al., 2000). The dataset is available in the `M3comp` package in `ts` format.

We will download the same data in `tsibble` format from Monash Time Series Forecasting Repository (`https://forecastingdata.org/`) using the `monash_forecasting_repository()` function and reference code 4656298. After downloading the dataset, save it in `data/data_chap8` folder as shown in the following code:

```
m3_monthly <- monash_forecasting_repository(4656298)
saveRDS(m3_monthly,  here("data", "data_chap8", "m3_monthly.rds"))
```

The series are named T1, T2, up to T1428, and values are recorded against the first date of a month. Change `thestart_timestamp` column to a year-month type and format into a `tibble` object using the following code:

```
m3_monthly <- m3_monthly |>
  mutate(month = yearmonth(start_timestamp)) |>
  as_tsibble(index = month) |>
  select(-start_timestamp)
```

Now, we will calculate features for each of these series. You can generate selective features using respective functions as shown in the previous sections in this chapter. Alternatively, full set (48 of them) of features can be generated at once using a single line of code:

```
monthly_features <- m3_monthly |>
  features(value, feature_set(pkgs = "feasts"))
# save locally
saveRDS(monthly_features,  here(
  "data", "data_chap8", "monthly_features.rds"))
```

In business scenarios, the number and applicability of features should depend on your specific use case. For example, seasonality features are not required for annual series. Similarly, including the output of multiple similar statistical tests provides redundant information; therefore, features based on only one such tests may be sufficient.

Next, we will apply PCA on the features computed from the M3 monthly dataset to analyze all the time series together.

## 8.9.1 PCA on time series features

PCA is a dimensionality reduction technique that is commonly used on a dataset with a large number of interrelated variables. PCA transforms a large set of potentially correlated variables into a smaller set of uncorrelated components, known as the principal components (PCs). The PCs are statistically independent from one another. They provide a linear transformation of the original variables while capturing maximum variance (i.e., information) in the data.

So, why do we apply the PCA on the M3 dataset? In analytics, performing an explanatory analysis before modeling is a crucial step. PCA on features from M3 data will simplify the explanatory analysis. Instead of analyzing and visualizing each series individually, PCA will identify the most important patterns of variation and help us visualize this high-dimensional data (48 features reduced to 4 PCAs). Among other things, it will identify any outliers (if any) and may highlight inherent clusters by combining series that exhibit similar patterns into a group.

A number of PCA implementations are available in various R packages. We will use the `tidymodels` implementation of PCA. The `tidymodels` framework (`https://www.tidymodels.org/`) is an ecosystem of R packages designed to streamline and enhance the process of modeling and statistical analysis based on `tidyverse` principles. The `tidymodels` include packages such as `parsnip` for model specification, `recipes` for pre-processing, `tune` for hyper parameter tuning, and `yardstick` for model evaluation. A detailed discussion on `tidymodels` framework is beyond the scope of this book.

Let's apply PCA on the data by performing the following steps:

- Data processing steps are defined using the `recipe()` function.
- The ~. formula specification denotes the use of PCA as an unsupervised technique, leveraging all available features in `monthly_features` dataset.
- The `series_name` column is designated as an identifier using the `update_role()` function
- The `step_nzv()` function will identify features with near zero variance. Since all the predictors in the data are numeric, the `all_predictors()` function selects them at once.
- The `step_normalize()` function is applied to normalize all predictors.
- The `step_pca()` function is used to transform the features into four PCs.

The code described in the preceding steps is shown here:

```
pca_recp <- recipe(~., data = monthly_features) |>
  update_role(series_name, new_role = "id") |>
  step_nzv(all_predictors()) |>
  step_normalize(all_predictors()) |>
  step_pca(all_predictors(), num_comp = 4)
```

Apply the `prep()` function on the recipe object to performs the preprocessing steps, and fit the PCA model using the training data, `monthly_features`:

```
pca_prep <- prep(pca_recp, training = monthly_features)
```

Variations across the 48 features are now being summarized by just four PCA components. The estimated components are extracted and the contributions made by each feature on four PCs are visualized next:

```
tidied_pca <- tidy(pca_prep,3)

tidied_pca |>
  filter(component %in% paste0("PC", 1:4)) |>
  mutate(component = fct_inorder(component)) |>
  ggplot(aes(value, terms), fill = "grey" ) +
  geom_col(show.legend = FALSE) +
  facet_wrap(~component, nrow = 1) +
```

```
    labs(x = "Absolute value of contribution"
        ,y = NULL )
```

*Figure 8.3* displays the loadings (contributions) of the original time series features for the first four PCAs computed from the M3 monthly dataset. From this plot, you can visualize the positive/negative contribution of a feature with a PC. The longer the bar, the more the specific feature contributes to defining the PC.

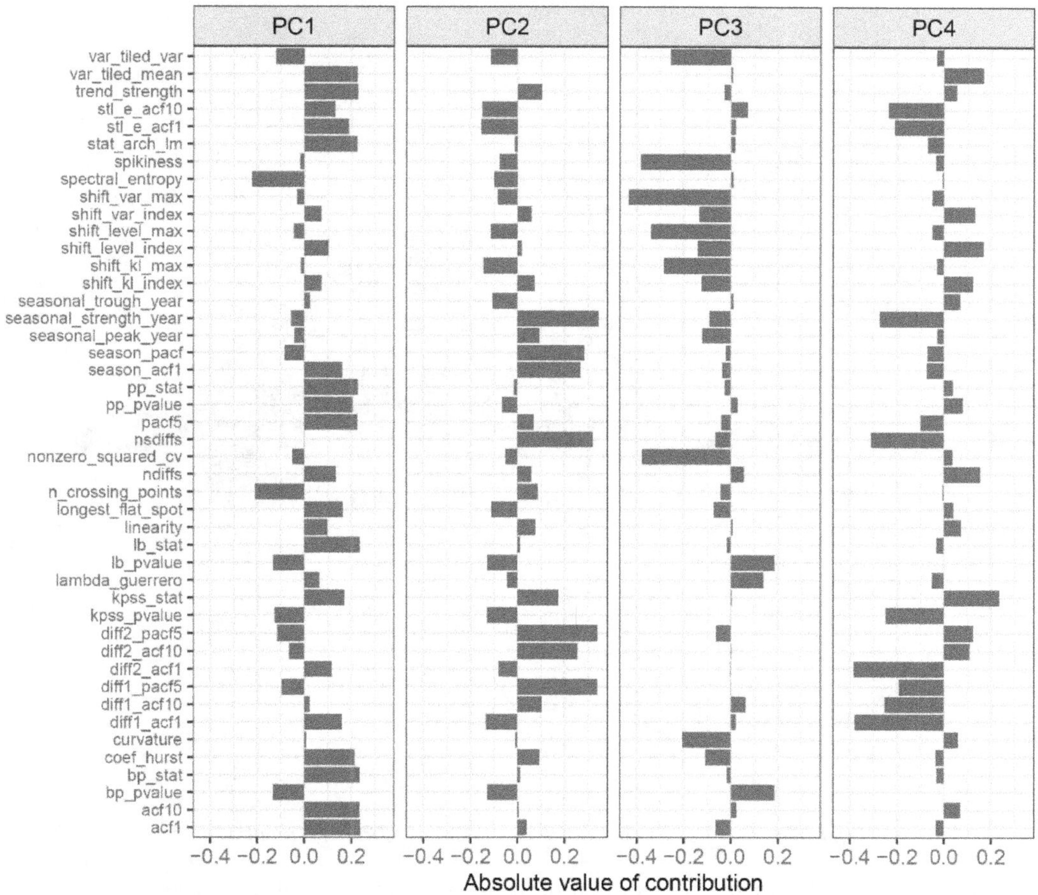


*Figure 8.3: Contribution of features on PCA of M3 monthly data*

The PC1 component seems to capture overall data complexity and trend/seasonality strength. It has strong positive loadings from `var_tiled_var`, `trend_strength`, and various ACF and PCAF features. It has negative loadings from stationarity `tests`, `kpss_stat` and `lb_stat` which makes sense as the stronger trend/seasonality means the data is less stationary. The PC1 component appears to differentiate between high trend/seasonal series and very stationary series. The PC2 component is dominated by seasonal features, PC3 seems to capture characteristics of the remainder terms and sudden variance shifts (`shift_var_max`), and PC4 is mostly dominated by the autocorrelation structure in the first differenced series.

Let's visualize each series' relative standing against the first two PCs using the following code:

```
juice(pca_prep) |>
  ggplot(aes(PC1, PC2, label = series_name)) +
  geom_point( alpha = 0.7, size = 2) +
  geom_text(check_overlap = TRUE, hjust = "inward") +
  labs(color = NULL)
```

You will see the generated plot in *Figure 8.4*:

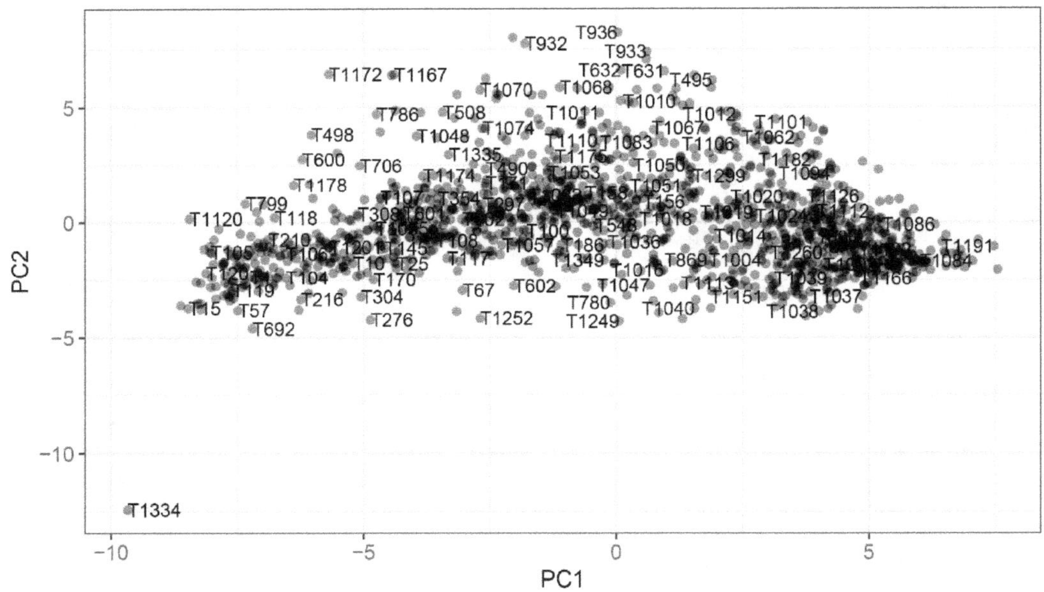


*Figure 8.4: First two PCs' space of features computed from M3 monthly data*

Most series in the M3 data exhibit a similar pattern with one clear outlier, series T1334, observed from *Figure 8.4*.

Even though PCA reduced data dimensionality by preserving complete variation information, it did not provide any clear separation based on the patterns of individual series.

Next, we apply another dimension reduction technique (UMAP) on the features, combined with a clustering technique (K-means), with the objective of clustering the monthly series in the *M3 Comp* dataset.

## 8.9.2 Clustering using time series features, UMAP, and K-means

UMAP (Leland et al., 2018) is a dimension reduction technique to simplify complex datasets by creating a graphical representation that shows how data points relate to each other. It compresses the information into a smaller space while keeping the most important connections and structures (both local and global) intact. This technique is particularly efficient at capturing non-linear relationships.

We use the `umap()` implementation from the `uwot` package. Before applying UMAP on the `monthly_features` data, the near zero variance features and the ID variables are removed. The parameters in the following code are provided as a guide, and you are encouraged to experiment. The `set.seed()` function is used for reproducibility.

```
nzv_var <- c("zero_run_mean","zero_start_prop","zero_end_prop")
id_var <- "series_name"

set.seed(121123)
umap_mod <- uwot::umap(
  X = monthly_features |>
    select(-all_of(c(id_var, nzv_var))),
  metric = "euclidean" ,
  scale = TRUE,
  approx_pow = TRUE,
  init = "spca",
  n_components = 2,
  min_dist = 0,
  n_trees = 25,
  nn_method = "annoy",
  n_threads = 12,
  ret_nn = TRUE ,
  ret_model = TRUE,
  verbose = FALSE
)
```

UMAP runs locally and saves the model output in a temporary folder. The model output can be saved in an RDS file and reloaded if needed:

```
temp_file <- fs::file_temp(pattern = "umap", ext = "rds")
umap_save_path <- here("data","data_chap8","umap_mod.rds")

# save
save_uwot(umap_mod, file = umap_save_path, unload = TRUE)

# reload
umap_mod <-load_uwot(file = umap_save_path)
```

The following code will store and visualize the embeddings from UMAP into a two-dimensional space:

```
umap_embed <- monthly_features |>
  select(series_name) |>
  dplyr::bind_cols(
    umap_mod$embedding |>
      as_tibble()
  ) |>
  rename(umap1 = V1, umap2 = V2)

umap_embed |>
  ggplot(aes(x = umap1, y = umap2)) +
  geom_point(size = 2, alpha = 0.8) +
  geom_text(aes(label = series_name),
            check_overlap = TRUE,
            hjust = "inward",
            size = 4)
```

The plot is presented in *Figure 8.5*:

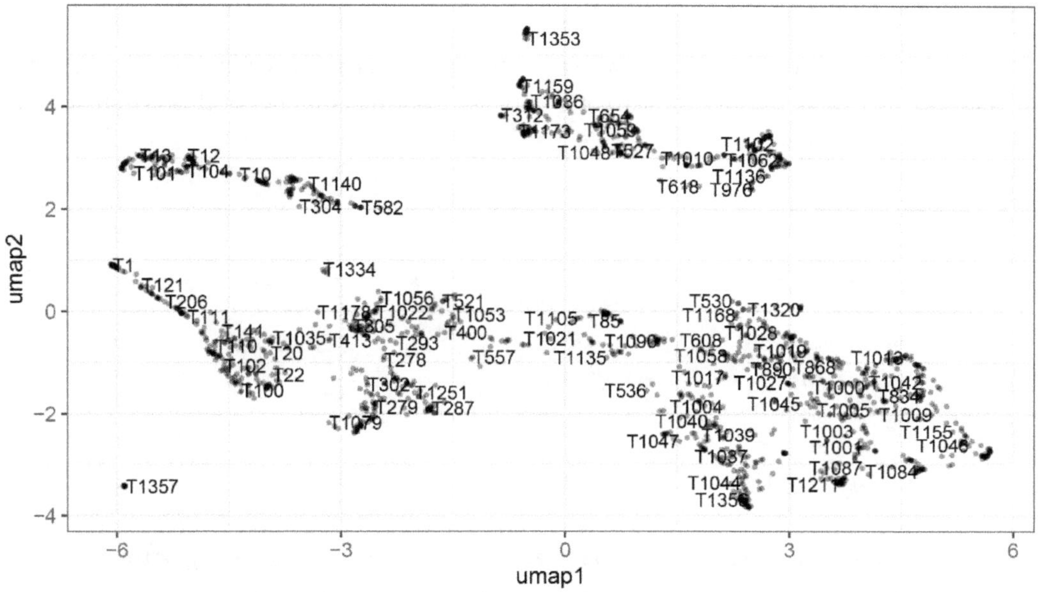


*Figure 8.5: UMAP embedding using features from M3 monthly data*

The UMAP embedding has successfully grouped all the 1,426 series into visible clusters. However, it is as yet unknown whether these groupings are optimal or if over-clustering has occurred. To identify the optimal number of clusters, we will repeatedly run K-means clustering on the UMAP embedding, testing between two to eight clusters.

There are many K-means implementations available in multiple R packages. Here, we use the `eclust()` function from the `factoextra` package:

```
kmean_list <- list()
for (j in 2:8){
  kmean_list[[j]] <- factoextra::eclust(
    umap_embed |>
      select(-series_name),
    FUNcluster = "kmeans",
    k = j,
    k.max = j,
    graph = FALSE,
    nboot = 0,
    verbose = FALSE
  )
```

```
}
names(kmean_list) <- 2:8
```

The optimal cluster is the one for which any further grouping provides minimal reduction in the total within sum of squares values. Let's extract the optimal number of clusters identified using the following code:

```
km_totwss <- tibble(
  clust_num = seq(2,8,1),
  totwss =  do.call(c, lapply(kmean_list, '[[', "tot.withinss")))

opt_cluster <- km_totwss |>
  mutate( first_diff = totwss - lag(totwss)) |>
  mutate(second_diff = first_diff - lag(first_diff)) |>
  drop_na() |>
  slice_max(second_diff)
```

The optimal number of clusters for this dataset is 4, as shown in the following output:

```
> km_totwss

# A tibble: 7 × 2
  clust_num totwss
      <dbl>  <dbl>
1         2 10454.
2         3  5115.
3         4  3465.
4         5  3083.
5         6  2492.
6         7  1803.
        8  1411.

> opt_cluster

# A tibble: 1 × 4
  clust_num totwss first_diff second_diff
      <dbl>  <dbl>      <dbl>       <dbl>
1         4  3465.     -1650.       3689.
```

A visualization such as a scree plot (the total within sum squares from each cluster against the number of clusters) helps detect the optimal cluster number. You can make a scree plot and a vertical line at 4 using the following code:

```
km_totwss |>
  ggplot(aes(clust_num, totwss)) +
  geom_line(linewidth = 0.9) +
  geom_point(size = 4) +
  scale_x_continuous(limits = c(2,8), breaks = seq(2,8,1)) +
  labs(x = "Number of clusters",
       y = "Total within sum of squares") +
  geom_vline(xintercept = opt_cluster$clust_num, linetype = 2)
```

*Figure 8.6* shows the scree plot:

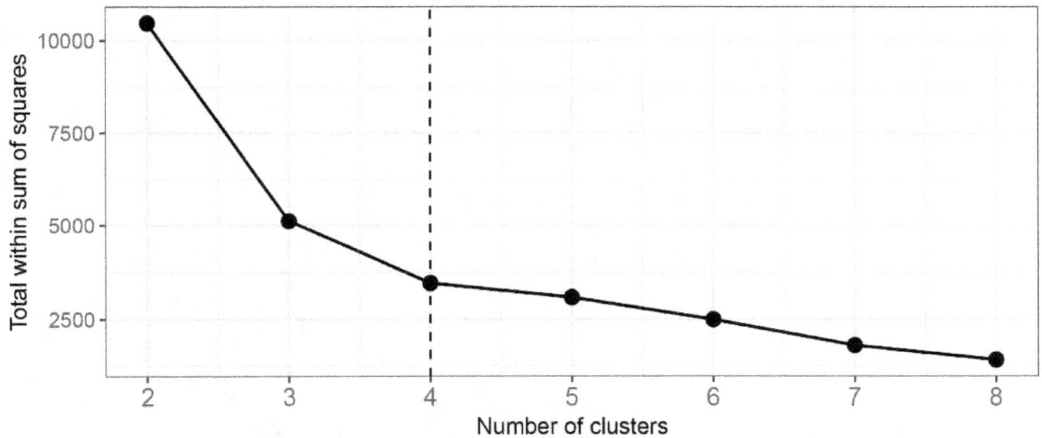


*Figure 8.6: Finding optimal number of clusters using Kmean on UMAP embedding of time series features from M3 monthly data*

The scree plot suggests four clusters are optimal, as an "elbow" appears at this point, indicating that further clustering provides only a minimal reduction in the total sum of squares.

Let's visualize the optimal clusters on the UMAP embedding space using the following code:

```
final_km <- kmean_list[[opt_cluster$clust_num]]
# add the cluster allocation
umap_embed <- umap_embed |>
  mutate(cluster_no = factor(final_km$cluster, levels = c(1:4)))

umap_embed |>
  ggplot(aes(x = umap1, y = umap2)) +
  geom_point(aes(colour = cluster_no, shape = cluster_no)
             ,size = 2
             ,alpha = 0.8) +
  geom_text(aes(label = series_name)
            ,check_overlap = TRUE
            ,hjust = "inward"
            ,size = 3) +
  labs(x = "UMAP Embedded dimension 1",
       y = "UMAP Embedded dimension 2") +
  theme(legend.position = "top")
```

The optimal cluster allocation is presented in *Figure 8.7*:

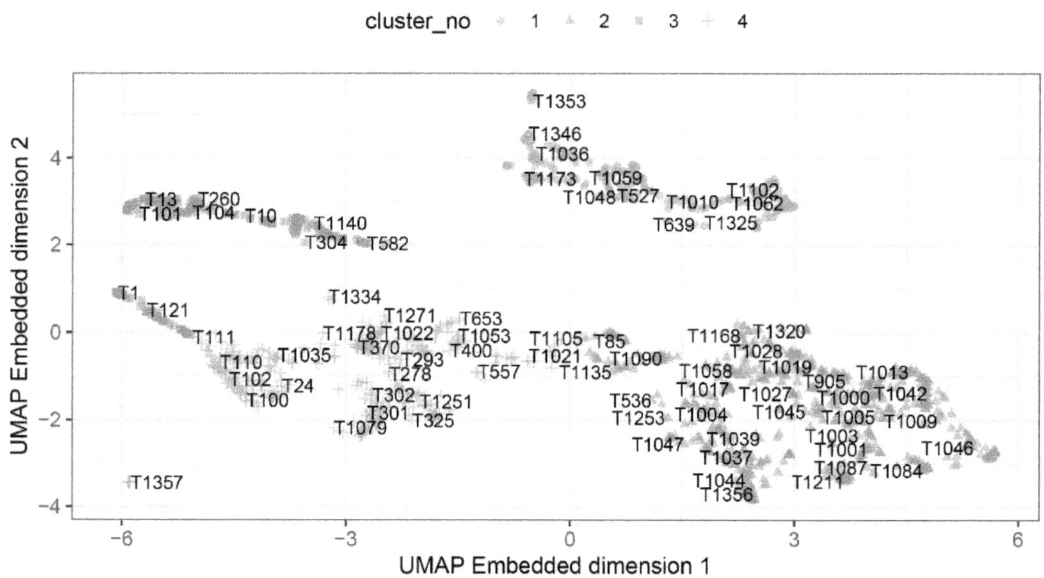


*Figure 8.7: Optimal clusters of UMAP embedding using time series features from M3 monthly data*

The series ($T1357$) identified by UMAP is different from the one identified as an outlier by PCA ($T1334$). The clusters are sparse on the UMAP dimensions. This indicates that within cluster variation is moderate to high. Cluster 3 has clearly two groups: even though their dimension 1 values are similar, their dimension 2 values are closer to cluster 4 elements. This might indicate artificial clustering, and excluding less informative features, such as test statistics providing similar outcomes, p-values, and time index, will improve the overall clustering.

Lastly, we will examine some series from different clusters to understand what characteristics caused them to group together. Let's plot four series, one from each cluster, for a quick identification:

```
# add the cluster allocation with the raw observation
m3_monthly_cluster <- m3_monthly |>
  left_join(umap_embed |>
            select(series_name, cluster_no),
            by = "series_name")

srname <- c("T527", "T1045","T29","T278")

m3_monthly_cluster |>
  filter(series_name %in% all_of(srname)) |>
  mutate(series = paste0(
    "Cluster no:", cluster_no, " & Series no:", series_name)) |>
  ggplot(aes(x = month, y = value)) +
  geom_line() +
  facet_wrap(~series, scales = "free_y",ncol = 1 )
```

*Figure 8.8* presents the time plots for the four selected series:

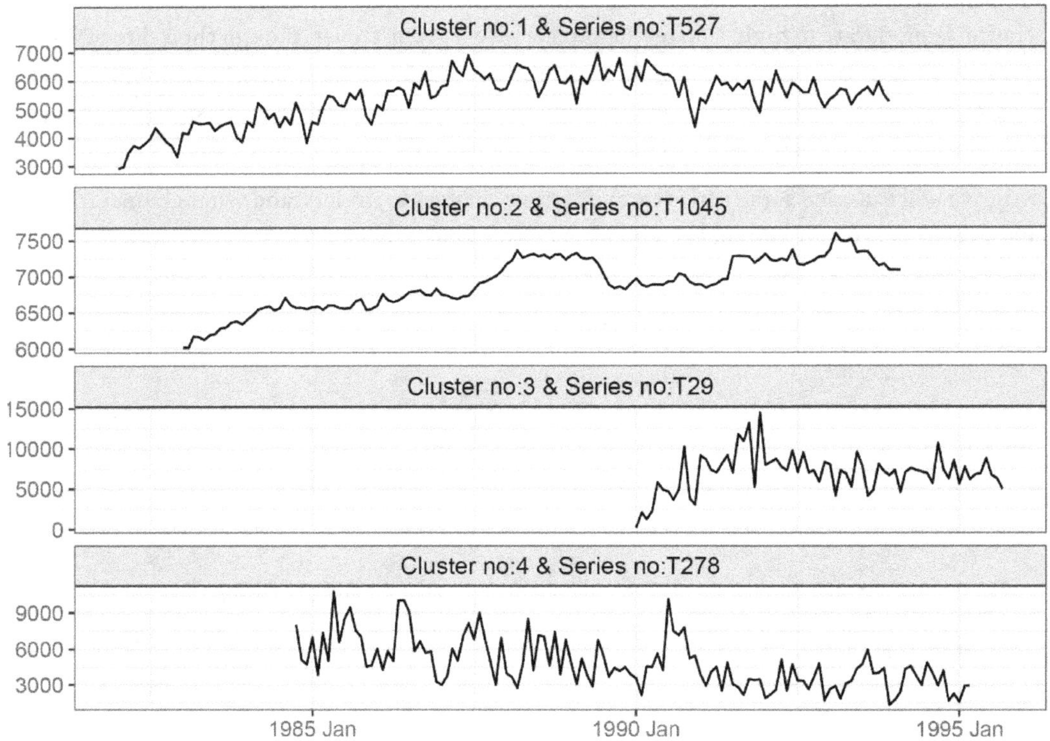


*Figure 8.8: Example of series from UMAP embedding clustering allocation*

Series T527 from cluster 1 shows some seasonality with an upward trend, while series T1045 from cluster 2 has an upward trend, weak seasonality, and periods of sudden upward and downward movements. Series T29 from cluster 3 does not show any particular pattern. However, there are clearly two distinct groups within cluster 3, so you may need to look in detail before taking an overall summary for cluster $3. Finally, series T278 from cluster 4 shows weak downward trend and seasonality, a reduction in volatility across time, and is a bit chaotic.

The distinct characteristics in the example series plotted in *Figure 8.8* suggest that the combination of UMAP and K-means has appropriately clustered the M3 monthly dataset using different features computed from the observed time series. Based on the characteristics of the series in each cluster, you can identify an appropriate analytics tool and forecasting models. Some features are highly correlated (statistical tests and their p-values). Performing the clustering again after excluding these multicollinear features might improve the results.

Even though both PCA and UMAP are used for dimension reduction, they produce different results when applied to a dataset. Understanding the methods themselves and aligning them with the objectives of your analysis will help you to decide on appropriate algorithms for your dataset.

# Summary

In this chapter, we have discussed time series features. There is a large collection of features that can be computed from a time series. Some features are based on the ACF, while some are purely based on statistical tests. They vary from simple descriptive statistics (mean) to information theory (Shannon entropy). The time series features provide valuable insights when you need to analyze a large collection of data on a regular basis.

The use of time series features was highlighted with two detailed examples: PCA and UMAP techniques were used for dimension reduction to reduce many time series features into a smaller feature space. Further, the optimal number of clusters was identified via K-means, which segmented the whole M3 monthly dataset into four distinct groups.

You also learned about a global forecasting data repository, *Monash Time Series Forecasting Repository*, from where you can download data to practice and learn. Working with real data requires patience and understanding from many different aspects before moving into modeling. Next, *Chapter 9* presents time series smoothing and filtering techniques, which are important tools for analyzing time series data.

# Further reading


- Kwiatkowski, D., Phillips, P. C. B., Schmidt, P., & Shin, Y. (1992). *Testing the null hypothesis of stationarity against the alternative of a unit root: How sure are we that economic time series have a unit root? Journal of Econometrics,* 54(1-3), 159–178.
- Leland, M., Healy, J., & Melville, J. (2018). *UMAP: Uniform manifold approximation and projection for dimension reduction.* arXiv Preprint arXiv:1802.03426.
- Makridakis, S. G., & Hibon, M. (2000). *The M3-competition: Results, conclusions and implications. International Journal of Forecasting,* 16.
- O'Hara-Wild, M., Hyndman, R., & Wang, E. (2023). *feasts: Feature extraction and statistics for time series* (Version 0.3.1). `https://CRAN.R-project.org/package=feasts`
- Wang, X., Smith, K., & Hyndman, R. (2006). *Characteristic-based clustering for time series data. Data Mining and Knowledge Discovery,* 13, 335–364.

# Join our community on Discord

Join our community's Discord space for discussions with the author and other readers:

https://discord.gg/G4YmB6DQvf

# 9

# Time Series Smoothing and Filtering

In time series, the presence of random fluctuations and noise that arise due to measurement error, could obscure the underlying data-generating process (DGP). The prominent time series components are not measurable until the noise is removed from the data. Time series smoothing and filtering techniques help us achieve this objective. They reduce random noise from the data and reveal the underlying structure and components ("signal").

In this chapter, you will learn techniques to either filter out noise or reduce it by smoothing. These techniques are very important to reveal the underlying time series characteristics from observed data.

The chapter is organized as follows:

- Section 9.1: Comparison of time series preprocessing techniques
- Section 9.2: Exponentially weighted moving average method (EWMA)
- Section 9.3: Time series filtering
- Section 9.4: Linear filtering with moving average (MA)
- Section 9.5: The Hodrick-Prescott filter
- Section 9.6: The Kalman Filter

Note

**Your purchase includes a free PDF copy + exclusive extras**

Your purchase includes a DRM-free PDF copy of this book, 7-day trial to the Packt+ library (no credit card required), and additional exclusive extras. See the *Free benefits with your book* section in the *Preface* to unlock them instantly and maximize your learning.

# Technical requirements

Load the following packages in your R workspace.

```
library(pacman)
p_load("conflicted"
       ,"datasets"
       ,"mFilter"
       ,"FKF"
       ,"qcc"
       ,"tidyverse"
       ,"tstibble"
       ,"zoo" )
```

The code file is available in the GitHub repo at `https://github.com/PacktPublishing/Modern-Time-Series-Analysis-with-R.git` for you to download and practice.

# 9.1 Comparison of time series preprocessing techniques

Time series decomposition techniques (e.g., STL) separate a time series into its components: trend, seasonal and random variation. Decomposition techniques do not remove any components, rather preserve them after separation. Time series smoothing removes random noise from the data to reveal underlying patterns. Time series filtering is a frequency-domain or time-domain operation that selectively attenuates or eliminates specific frequency components while preserving others.

All three techniques aim to extract meaningful information from a time series by reducing unmeasurable complexity (noise). They may seem to produce similar outcomes, but actually they have specific objectives, purposes and applications in time series analysis. *Table 9.1* summarizes some key aspects of these techniques.

| Aspect | Decomposition | Smoothing | Filtering |
|---|---|---|---|
| Primary goal | Separate time series into components (trend, seasonal, cyclical, irregular) | Reduce noise to reveal underlying patterns/trends | Remove or isolate specific frequency components |
| Purpose | Diagnostic | Descriptive | Prescriptive about what to remove |
| Output | Multiple component series (T, S, C, R) | Single smoothed series | Single filtered series or specific component |
| Approach | Structural breakdown into additive or multiplicative components | Local averaging | Frequency-selective transformation |
| Common methods | Classical decomposition, STL, X-12-ARIMA | Weighted Moving Average, exponential smoothing, LOESS, kernel smoothing | Linear filtering with Moving Average, Hodrick-Prescott filter, Kalman filter |
| Frequency selectivity | Component-based (separates by temporal pattern type) | Non-selective (diminish effect of all high frequencies) | Highly selective (targets specific frequency bands) |
| Reversibility | Yes, components can be recombined | No, information loss from averaging | Partial, depends on filter type |
| Preserved information | All identifiable structural components separately | Overall shape with reduced variability | Specified frequency components only |

| Aspect | Decomposition | Smoothing | Filtering |
|---|---|---|---|
| Causality | Non-causal; retrospective or post-processing only, need entire time series | Usually non-causal; a symmetric operation, requires both past (before) and future (after) data points; some methods are applicable in real-time scenario | Can be causal (uses only past and present data for real time application), or non-causal (uses future data points for symmetric filters) |
| Use case | Understanding structure; pre-processing for forecasting | Quality control; fast trend visualization; noise reduction for exploratory analysis | Signal processing; removing known interference; cycle extraction |
| Example | Analyzing the impact of seasons on sale | Annual sales review, identifying long-term climate trends, analyzing past marketing campaign | High-frequency trading algorithms, extract business cycle from GDP, biomedical signal processing (ECG analysis), image and motion processing |

*Table 9.1: Comparison of time series preprocessing techniques*

In a time series context, the terms smoothing and filtering are often used interchangeably. While filter broadly refers, in the frequency domain, to analyzing signal-to-noise ratio, smoothing refers only to reducing noise from the data to reveal underlying trends. Methods based on moving average are used for both filtering and smoothing. Advanced filtering techniques, such as the Kalman filter, are commonly used to filter out random noise from complex, often non-linear time series.

In *Chapter 7*, we discussed common decomposition and moving-average based smoothing methods. In this chapter, we will discuss the exponentially weighted moving average smoothing technique. For filtering, we will cover linear filtering with moving average, the Hodrick-Prescott filter, and the Kalman filter. Among them, the Kalman filter is the most advanced technique. We will apply all the mentioned techniques to a single series, so that you can compare the outputs from each method, analyze how they work, and understand what they achieve.

Next, the exponentially weighted moving average smoothing technique will be discussed which is widely used in quality control.

# 9.2 Exponentially weighted moving average method (EWMA)

The moving average smoothing discussed in *Section 7.6.1* and *7.6.2*, showed how to reduce random noise by taking averages over neighboring observations. In that scenario, all observations were given equal weight, irrespective of how close or far they are from the observation being smoothed. The moving average method has a few limitations, such as loss of information at the start and end of a series, and the optimal number of observations to average is more of a subjective choice.

Moving average requires both past ($y_{t-n}$,.., $y_{t-1}$) and future ($y_{t+1}$,...,$y_{t+n}$) data points relative to the time index, $t$. This information is not always available—especially if you would like to analyze the underlying pattern by reducing noise as soon as a period has passed. The **exponentially weighted moving average** (**EWMA**) method is applicable in such a scenario.

EWMA effectively removes random noise from a time series and smooths out fluctuations to reveal patterns. EWMA assigns less weight to past values, with more recent values receiving higher weights. EWMA has a single parameter, $\alpha$, the rate of exponentially decreasing weights. The formula for EWMA at time $t$ is given by:

$$\text{EWMA}_t = \alpha y_t + (1-\alpha)\text{EWMA}_{t-1} \qquad (9.1)$$

The EWMA at time $t$ depends on the present value, $y_t$ and the EWMA from previous steps. At the beginning of the process, $\text{EWMA}_1$ is set to the first observation of the series.

The smoothing parameter takes a value between $0$ and $1$, $(0 < \alpha \leq 1)$. A larger $\alpha$ gives more weight to the recent observation, making the method more responsive to recent changes. When $\alpha$ is small, it gives more weight to past observations, resulting in a smoother curve that reacts more slowly to the change. For illustration, we will use the `Nile` data from the `datasets` package which provides the annual flow of the river Nile during 1871-1970. Transform the `Nile` data in a `tsibble` object and visualize it using the following code:

```
nile_dt <- Nile |>
  as_tsibble() |>
  rename(year = index, flow = value)
```

```
nile_dt |>
  ggplot(aes(year, flow)) +
  geom_line() +
  scale_y_continuous(
    limits = c(400, 1400),
    breaks = seq(400, 1400, by=400)
  )
```

The time plot in *Figure 9.1* shows the annual river flow has more noise compared to any other pattern, such as trend:

*Figure 9.1: Annual flow of the river Nile (1871-1970)*

Now, we'll apply EWMA on the Nile data with three different smoothing parameters, $0.2$, $0.4$, and $0.7$ as shown in the following code:

```
smooth_list <- list()
alpha_vec <- c(0.2, 0.4, 0.7)
for (i in seq_along(alpha_vec))
{
  smooth_list[[i]] <- ewmaSmooth(
    x = nile_dt$year,
    y  =nile_dt$flow,
    lambda = alpha_vec[i]
  )$y
}
names(smooth_list) <- paste0("alpha_",alpha_vec)

# join the smoothed series
```

```
nile_ewma <- nile_dt |>
  bind_cols(
    smooth_list |> bind_cols()
  )
```

*Figure 9.2* presents the original series (in grey) with three EWMA smoothed series (in black):

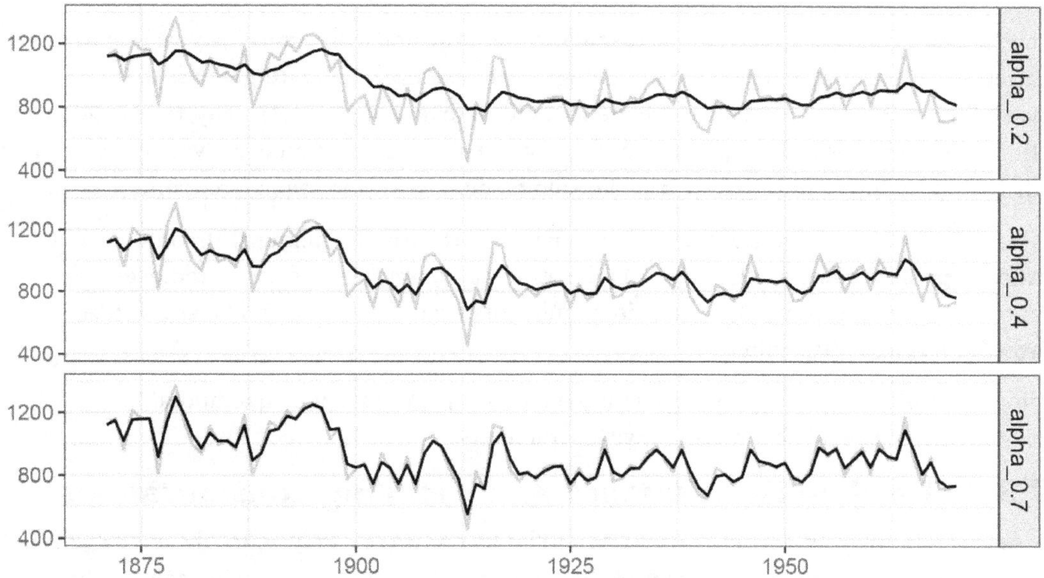


*Figure 9.2: Comparison of the effect of smoothing parameters on EWMA*

The comparison in *Figure 9.2* shows that the most randomness was removed (highest smoothing) with $\alpha = 0.2$, because it gave little weight to the most recent observation and more weights were distributed to the past values. When $\alpha = 0.7$, the majority weight was allocated to the most recent observation, hence the amount of smoothing was low- not much random noise has been filtered out. The $\alpha = 0.4$ seems a more reasonable smoothing parameter.

EWMA is helpful to identify the underlying pattern by removing the noise without any information loss. For its capabilities of handling high-dimensional data, EWMA is commonly used to generate algorithmic trading signals, and for volatility modeling. Process control and quality control are two prominent fields where EWMA is commonly used to measure the efficiency of a process or product.

Next, we'll investigate time series filtering techniques in detail.

# 9.3 Time series filtering

In general, the word filter means to pass a liquid, gas, light, or sound through a device to remove unwanted material. In the time series context, the filtering involves passing the observed series through a mathematical operation to remove or emphasize particular components. Time series filters are commonly used to isolate patterns at a specific frequency.

Time series filtering is applied in the **frequency domain**, and the prime objective is to isolate patterns that occur at various frequency lengths. For example, separating trends that occur over a 5-10 year period from short-frequency cyclical (over 2-4 years). Usually, time series filtering is mainly used for explaining and analyzing observed phenomena. Even though some advanced filtering techniques, such as the Kalman filter, are also used in forecasting models.

Time series smoothing concentrates on removing short-term fluctuations, also known as a **finite impulse response** (**FIR**) filter. On the other hand, time series filtering removes long-term fluctuations, technically an infinite number of observations and known as **infinite impulse response** (**IIR**) filter.

Time series filtering is a sophisticated mathematical operation in which temporal dependencies are explicitly modeled with parameters.

## 9.3.1 The Nyquist frequency and sampling considerations

In time series analysis, data is collected at regular intervals (i.e., daily, monthly, or quarterly). These intervals are known as the **sampling frequency**. In frequency domain analysis, the information that can be extracted from a time series is governed by the **Nyquist frequency**, which is half the sampling rate,

$$f_{\text{nyquist}} = \frac{f_{\text{sampling}}}{2}.$$

For any discrete time series, if you wish to see any variation/pattern at a particular frequency cycle, then your data collection (sampling) frequency must be at least twice that cycle.

For example, daily time series cannot capture the intra-day cycle; at least two days of observations are needed to detect such a cycle. An annual series ($f = 1$) could not provide enough information if there are quarterly patterns ($f = 4$) in the data. The annual series has a slow sampling rate, so the quarterly trend would be completely missed. You must sample at a higher frequency (e.g., monthly with $f = 12$) to detect any pattern that occurs at a lower frequency ($f = 4$).

This requirement eliminates **aliasing**, a well-known concern in signal processing. Aliasing is the distortion in the signals that occurs when a signal is sampled at a rate lower than twice its highest frequency component.

To ensure accurate representation, it is essential to choose a sampling frequency that is sufficiently high to capture the periodic behavior present in the business cycles or economic indicators being studied. These behaviors are defined by their frequency or cycle length:

- Lower frequencies typically refer to longer-term movements, such as trends or 2-5 year business cycles
- Higher frequencies refer to shorter, more rapid movements, such as annual cycles found in quarterly and monthly data, which we call as seasonality

The time series filtering techniques filter out a specific range of frequencies within a time series while attenuating others. They can be characterized as follows:

- **Low-pass filter**: A filter that dampens higher frequencies in the data and allows only lower frequencies to pass through. Ideal for separate trends.
- **High-pass filter**: A filter that dampens low frequencies and allows high frequencies to pass, which is used to separate short-and-long term business cycles.
- **Band-pass filter**: A filter that passes only the frequencies within a certain frequency band.

Most time series filters are based on the concept of moving averages: smooth out the white noise terms as much as possible. A variety of time series filtering techniques can be found in the literature with their own pros and cons. We include the following three that have wide application in time series analysis, especially business and economic time series contexts:

- Linear filtering with moving average: A finite impulse response filter that separates the underlying pattern by smoothing out past noise
- Hodrick-Prescott filter: A finite approximation of an infinite impulse response filter that separates trends from cycles and random noise
- Kalman filter: A general and versatile filter that considers observed series as a system of the true underlying data-generating process and could handle noise arising from multiple sources.

Let's start with linear filtering with moving average in the next section.

## 9.4 Linear filtering with moving average (MA)

In a **moving average** (**MA**) model, it is assumed that the current value of a time series is a linear combination of past white noise or error terms. The MA($q$) model considers the temporal dependency of a time series depends on the previous $q$ error terms.

$$y_t = \epsilon_t + \theta_1\epsilon_{t-1} + \theta_2\epsilon_{t-2} + \cdots + \theta_q\epsilon_{t-q} \qquad t = 1,2,\ldots,T \tag{9.2}$$

where, $\epsilon_t$ is the white noise term, and $\theta_1, \theta_2, \ldots, \theta_q$ are the MA coefficients.

We have already applied the MA technique to smooth out time series data. How is this MA model different from those? Technically, Equation (9.2) is the general version, and all other moving averages you have seen are special applications. For example, with $q = 3$ and $\theta_1 = \theta_2 = \theta_3 = 1$, it is equivalent to a 3-day moving average.

The `filter()` function in the `stats` package provides functionality to perform MA linear filtering. The `stats` package is a default package loaded with R. However, if `dplyr` (or `tidyverse`) is already loaded, then it is better to call via the namespace to avoid a conflict with the same name functions.

In the `stats::filter()` function, you need to pass a vector of length $q$ with values of the coefficients. For example, a vector of three $1/3$ is equivalent to applying a 3-MA with equal weights. The argument, `method=convolution`, assures moving average filtering, and `sides` controls whether to use past observations (`sides=1`) or values from both sides (`sides=2`). In the following code, two linear filters with MA are applied to the Nile data:

- MA(2): where $q = 2$ with weight $0.6$ and $0.4$, the smaller weight is allocated to the error further from the present value
- MA(3): three coefficients with equal weight of 1/3 for each of them.

```
ma_filter <- nile_dt |>
  mutate(
    "MA(2)" = stats::filter(
      flow,
      filter = c(0.6,0.4),
      method="convolution", sides = 1
    ),
    "MA(3)" = stats::filter(
      flow,
      filter = rep(1/3,3),
      method = "convolution", sides = 1
```

```
    )
  )
```

Let's plot the original and the filtered series for visual inspection. *Figure 9.3* presents the original (in grey) and the series obtained from $MA(2)$ and $MA(3)$ filtering (in black):

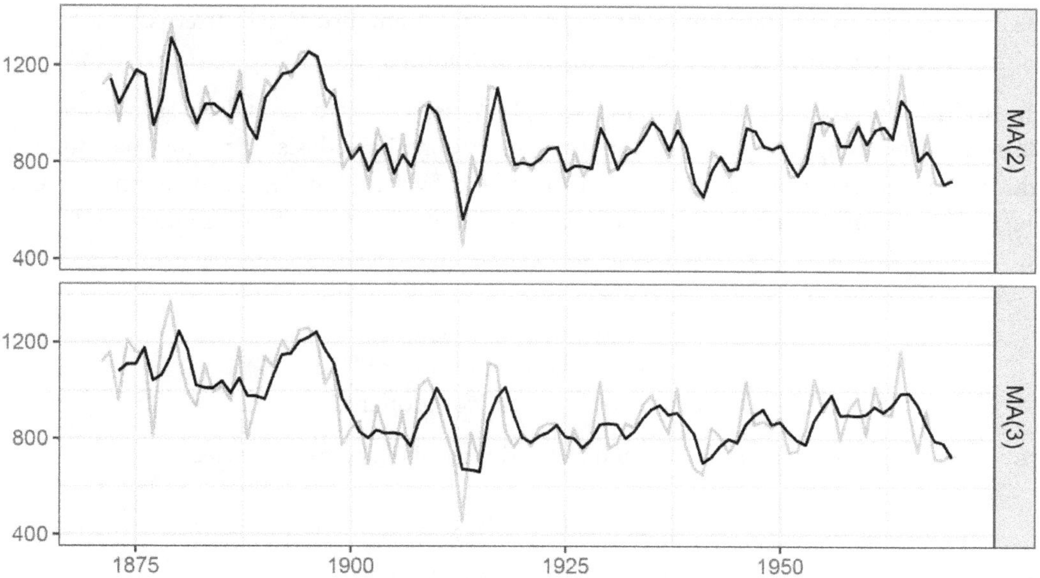


*Figure 9.3: Linear filtering using MA(2) and MA(3) on river Nile's annual flow*

*Figure 9.3* shows the $MA(3)$ filter has smoothed out noise to a higher degree compared to the $MA(2)$ filter. The MA coefficients represent the direct impact of previous error terms on the present value. They indicate the strength and direction of the short-term influence.

In business applications, the distinction between moving average smoothing and moving average filtering largely depends on the timing of the analysis: offline retrospective analysis versus real-time ongoing operational monitoring. In terms of data availability, filtering requires data that arrives sequentially, and decisions must be made immediately (e.g., a stock trading algorithm). For smoothing, you have the full dataset available to observe all outcomes (e.g., annual sales review).

Previously, you used STL decomposition to separate a time series into its components. In the next section, you will learn about the Hodrick-Prescott filter, which is used to separate trends and cycles from a time series.

## 9.5 The Hodrick-Prescott filter

Among the time series components, trends and cycles are often entangled with each other. You need longer periods to observe the true occurrence of these two components. Business cycles typically operate in 2-8 year periods, which create fluctuations around the long-term trend. On the other hand, when trends are very strong, they may dominate cyclical patterns, making it harder to detect without specific decomposition techniques.

The Hodrick-Prescott (HP) filter (Hodrick et al., 1997) separates trends and cycles from an observed series by applying the concept of frequency domain analysis. It is a high-pass filter, which means it preserves high-frequency components by damping low-frequency movement. The HP filter accomplishes this by minimizing the fluctuations from the estimated trend.

Considering

$$\begin{aligned} y_t&: \text{ is the observed series} \\ b_t&: \text{ is the trend component} \\ c_t&: \text{ is the cyclical component,} \end{aligned}$$

we can write $y_t$ as a combination of trend ($b_t$) and cyclical ($c_t$) components as,

$$y_t = b_t + c_t.$$

The HP filter optimizes *Equation (9.3)* with respect to $b_t$:

$$\min \sum_{t=1}^{T} \{(y_t - b_t)^2 + \lambda[(b_{t+1} - b_t) - (b_t - b_{t-1})]^2\}. \qquad (9.3)$$

where, $\lambda$ is the smoothing parameter.

The deviation from the trend, $z_t = y_t - b_t$, is commonly referred to as the business cycle component.

The HP filter provides an estimate of the trend by minimizing the squared difference between the original series and the trend (first term), and the curvature of the trend (the second term), while using $\lambda$ as a smoothing parameter to control the trade-off between these two terms. The higher the value of $\lambda$, the smoother the estimated trend.

We will use the implementation of the HP filter from the `mFilter` package on the U.S. unemployment rate data (`unemp`). The `mFilter()` function automatically selects the optimal $\lambda$ value for a given series. Users could supply their own value too.

In the following code, the HP filter is applied on the `unemp` dataset and estimated trend and cycle component are extracted and organized:

```
data(unemp)
unemp_hp <- mFilter(unemp, filter = "HP")

# organize
hp_filter <- unemp |>
  as_tsibble() |>
  mutate(trend = as.numeric(unemp_hp$trend),
         cycle = as.numeric(unemp_hp$cycle)) |>
  rename(quarter = index,
         unemployment = value)
```

For this data, the optimal smoothing parameter estimated was 1,600 as shown in the following output:

```
unemp_hp$lambda

# [1] 1600
```

Let's plot the trend and cycle estimated by the HP filter, along with the original series to examine.

In *Figure 9.4*, the original series is grey and the results from HP filter is presented in black:

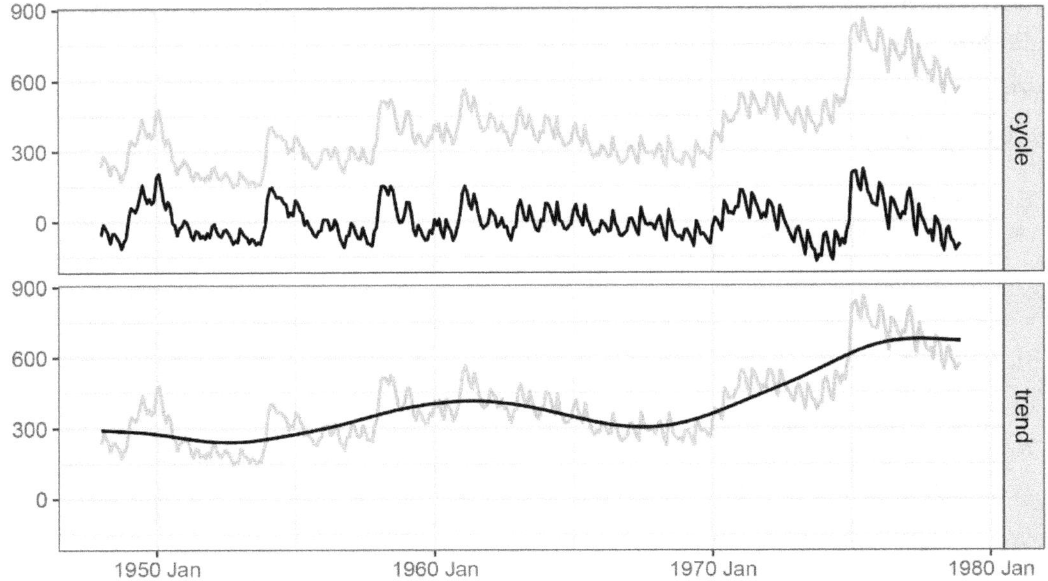


*Figure 9.4: Results from the Hodrick-Prescott filter on the U.S unemployment rate*

In the bottom panel in *Figure 9.4*, the estimated trend smoothed out the sharp, short-term spikes to extract a long-term trend that represents the low-frequency movements in the data. The trend captures the gradual, underlying direction of the unemployment rate, which increased over the observed period. The estimated cyclical pattern is the bottom line in the top panel. The cyclical component represents the higher-frequency, short-to-medium term fluctuations related to the business cycle. It oscillates around zero (i.e., mean-reverting), indicating the original data regularly moved above and below its long-term trend.

The identification of the business cycle is one of the primary goals in many macroeconomic, finance, and business analyses. That's where HP filtering helps.

So far, we have assumed that a time series is obtained from a single observable data generating process (DGP). What if the underlying DGP is actually unobservable; what we could measure is observations generated from a state of a dynamic system mixed with a series of noise. The Kalman filter was developed to handle such situation. It is a recursive algorithm for estimating the internal state of a linear dynamic system from a series of noisy measurements. It is also used as a filtering technique and can be used for forecasting.

In the next section, we'll discuss this advanced time series filtering technique, the Kalman filter.

# 9.6 The Kalman filter

In supply chain operations, the true customer demand for a product is "hidden" or "latent" state, because it cannot be measured accurately in real time. It is impossible to measure the exact demand until the sales are finalized over a period. On the other hand, the historical sales are measurable, but could be misleading to accurately predict product demand, as sales are often mixed with noise due to random fluctuations, promotional spikes, stock runouts and so on. In this scenario, what we observed (realized sales) is a combination of two different processes, unobservable demand and observed sales. Both simultaneously change over time and are mixed with their own noise.

The Kalman filter is a method for making accurate estimates about hidden demand using past sales. First, it predicts where the demand is likely to be based on the previous sales figures (sales forecasting). Then, in the second step, it combines this forecast with the actual new measurement (realized sales data) using a weighted average (the Kalman gain) to estimate the true underlying demand. By filtering out short-term noise from both the demand and sales systems, the Kalman filter provides a clear, continuously updated estimate of genuine market needs, which is essential for making better inventory and product decisions.

Kalman (1960) proposed this filter to estimate the actual system at discrete time steps using information hidden behind the noise. To be specific, the algorithm was first used to track the position, velocity, and orientation of the Apollo spacecraft guided by the **Apollo Guidance Computer** (**AGC**) in the presence of noise and uncertainty from the sensors. Kalman et al. (1961) extends the algorithm to be applicable to the state space framework of a linear model in a continuous time step.

So, what is the state space framework of a time series? The state space framework is a general and versatile representation of time series data. It represents the underlying dynamics of a system using a set of state variables and a set of measurement equations. The state variables represent the unobserved states of the system, while the measurement equations relate to the state variables of the observed data. Time series components, such as level, trend, and seasonality, can all be considered as state variables.

Suppose at time $t$,

$x_t$ is a vector of latent (unobserved) state variables

$y_t$ is observed, which represents the true but hidden DGP $x_t$

The movement of $x_t$ from period $(t-1)$ to period $t$ can be represented by the *state equation* or *transition equation*,

$$x_t = F_t x_{t-1} + u_t \qquad (9.4)$$

where,

$F_t$ is the transition matrix, which is applied to $x_{t-1}$ and gives the direction of movements from a time period $(t-1)$ to the next $t$, and

$u_t$ is the process error, which is assumed to be normally distributed with zero mean and covariance matrix $U_t$, i.e., $u_t \sim N(0, U_t)$.

At time $t$, the movement of $y_t$ can be represented by the *observation equation* or *measurement equation* as follows:

$$y_t = G_t x_t + v_t \qquad (9.5)$$

where,

- $G_t$ is the observation matrix that maps the relationship of the true state space into the observed space, and
- $v_t$ is the observation error made at time $t$, which is assumed to be normally distributed with zero mean and covariance matrix $V_t$, i.e, $v_t \sim N(0, V_t)$

The focus is to estimate the true state of $x_t$ using the information observed through $y_t$. The estimation quality of true state is assessed by the two measurements:

- $\hat{x}_{t|t}$: The estimate of the true state at time $t$ taking all the information up to time $t$, and
- $P_{t|t} = COV(x_t - \hat{x}_{t|t})$: The error covariance matrix, which is a measure of the accuracy of the state estimate.

The Kalman filter operates in two main phases: prediction and update in which the values of $\hat{x}_{t|t}$ and $P_{t|t}$ are obtained by applying recursive equations. However, we need an initial step to start the process.

- Step 0: Initialization phase:
    - State vector ($x_{0|0}$): Assign numerical values to initialize the state vector representing the current state of the system. This is the best estimate of the true state at the initial time.
    - State covariance matrix ($P_{0|0}$): Assign numerical values to initialize the covariance matrix representing the uncertainty in the state estimate. This matrix reflects the uncertainty or error in the initial state estimate.
- Step 1: Prediction phase:

$$\text{Predict state:} \quad \hat{x}_{t|t-1} = F_t \hat{x}_{t-1|t-1}, \quad (9.6)$$

$$\text{Prediction error covariance:} \quad P_{t|t-1} = F_t P_{t-1|t-1} F_t' + U_t. \quad (9.7)$$

- Step 2: Update phase:

$$\text{Measurement residual:} \quad w_t = y_t - G_t \hat{x}_{t|t-1}, \quad (9.8)$$

$$\text{Residual covariance:} \quad w_t = G_t P_{t|t-1} G_t' + V_t, \quad (9.9)$$

$$\text{Optimal Kalman gain:} \quad K_t = P_{t|t-1} G_t' W_t^{-1}, \quad (9.10)$$

$$\text{Updated state estimate:} \quad \hat{x}_{t|t} = \hat{x}_{t|t-1} + K_t w_t, \quad (9.11)$$

$$\text{Updated error covariance:} \quad P_{t|t} = P_{t|t-1} - K_t W_t K_t'. \quad (9.12)$$

The Kalman filter works under the following assumptions:

$$\begin{aligned} \mathrm{E}[x_t - \hat{x}_{t|t}] &= 0, \\ \mathrm{E}[x_t - \hat{x}_{t|t-1}] &= 0, \\ \mathrm{E}[w_t] &= 0, \\ W_t &= \mathrm{cov}(w_t), \\ P_{t|t-1} &= \mathrm{cov}(x_t - \hat{x}_{t|t-1}) \quad \textit{and} \end{aligned}$$

the initial step and the noise vectors at each step $\{x_0, u_1, \dots, u_t, v_1, \dots, v_t\}$ are mutually independent.

The Kalman gain, $K_t$, assures that for the linear state space framework, the algorithm provides the minimum one-step-ahead mean square error estimation without re-fitting the whole stochastic process every time a new observation is available. The formula of the Kalman gain in (9.10) suggests that if the measurement noise is high, then the Kalman gain will be close to 0, and the influence of the new data point $y_t$ will be small. If the measurement noise is small, the filtered value $x_{t|t}$ will be adjusted more towards $y_t$. This is important information to consider when tuning the Kalman filter algorithm for specific applications.

The state variable vector $x_t$ could be one-dimensional or multi-dimensional, depending on the unobserved components or states (level, trend, seasonal) that are associated with the underlying DGP. The number of parameters in the Kalman filter varies depending on how many states you have in your system; hence, this technique incorporates an optimization routine that can be extended based on the dynamics of the system.

Kalman filtering is a vast topic. There is an array of literature that covers different notations and representations of the Kalman filter, considering one to multiple states, estimation methods under different noise characteristics (Gaussian and non-Gaussian), how to tune parameters, how to choose initial state and covariance matrices, and their impact on convergence, to name a few. No wonder there are multiple R packages that offer Kalman filtering in various settings. See Tusell (2011) for a review of R packages on the Kalman filter.

We will use the FKF (Luethi et al., 2022) package in the following example. For simplicity, let's consider a local-level model for the annual flow of the river Nile (the Nile data).

Using the FKF package's notation, the system of equation is reduced to:

$$\begin{aligned} \text{transition equation:} l_{t+1} &= l_t + \eta_t, \quad && \eta_t \sim N(0, HH_t) \quad (9.13) \\ \text{measurement equation:} y_t &= l_t + \epsilon_t, \quad && \epsilon_t \sim N(0, GG_t) \quad (9.14) \end{aligned}$$

The initial parameters are considered as

$a_0$: a vector of the initial value of the state variable
$P_0$: a matrix giving the variane of$a_0$
$d_t$: a matrix giving the intercept of the transition equation
$c_t$: a matrix giving the intercept of the measurement equation
$T_t$: a matrix giving the factor of the transition equation
$Z_t$: a matrix giving the factor of the measurement equation
$HH_t$: an array giving the variance of the innovations in the transition equation
$GG_t$: an array giving the variance of the noises in the measurement equation

There are no intercept terms in the model in *Equations (9.13)* and *(9.14)*. Hence, $d_t$ and $c_t$ are both 0. The optim function runs the optimization algorithm and the fkf function runs the Kalman filter. The following code shows you how to perform those operations:

```
y <- Nile

## Set constant parameters:
dt <- ct <- matrix(0)
Zt <- Tt <- matrix(1)
a0 <- y[1] # Estimation of the first year flow
P0 <- matrix(100) # Variance of 'a0'

## Estimate parameters:
fit_fkf <- optim(
  c(HHt = var(y, na.rm = TRUE) * .5,
    GGt = var(y, na.rm = TRUE) * .5),
  fn = function(par, ...)
    -fkf(HHt = matrix(par[1]),
         GGt = matrix(par[2]), ...)$logLik,
  yt = rbind(y),
  a0 = a0, P0 = P0,
  dt = dt, ct = ct,
  Zt = Zt, Tt = Tt
```

```
)

## Filter Nile data with estimated parameters:
fkf_obj <- fkf(
  a0, P0, dt, ct, Tt, Zt,
  HHt = matrix(fit_fkf$par[1]),
  GGt = matrix(fit_fkf$par[2]),
  yt = rbind(y)
)
```

Let's check the estimated variances in the measurement equation in the following output:

```
fit_fkf$par

#       HHt        GGt
# 1300.777 15247.773
```

Now, extract the filtered series, the Kalman gain from the estimated model object, and organize as follows:

```
kf_filter <- nile_dt |>
  mutate("KF filtered" = as.numeric(fkf_obj$att)
         ,"Kalman gain" = as.numeric(fkf_obj$Kt))
```

The first 10 and the 10 observations are printed in the following output:

```
> head(kf_filter, 5)

# A tsibble: 5 x 4 [1Y]
   year  flow `KF filtered` `Kalman gain`
  <dbl> <dbl>         <dbl>         <dbl>
1  1871  1120          1120       0.00652
2  1872  1160         1123.       0.0841
3  1873   963         1100.       0.145
4  1874  1210         1121.       0.187
5  1875  1160         1129.       0.214
```

```
> tail(kf_filter, 5)

# A tsibble: 5 x 4 [1Y]
   year  flow `KF filtered` `Kalman gain`
  <dbl> <dbl>         <dbl>         <dbl>
1  1966   746          907.         0.253
2  1967   919          910.         0.253
3  1968   718          862.         0.253
4  1969   714          824.         0.253
5  1970   740          803.         0.253
```

In the beginning of the process, the filtered series starts with an initial value, the same as the observed data, and gets updated at every step along with the Kalman gain. When the system estimation is stabilized, the Kalman gain does not change, but the filtered values get updated. *Figure 9.5* presents the original (in grey) and the filtered series (in black) for visual inspection:

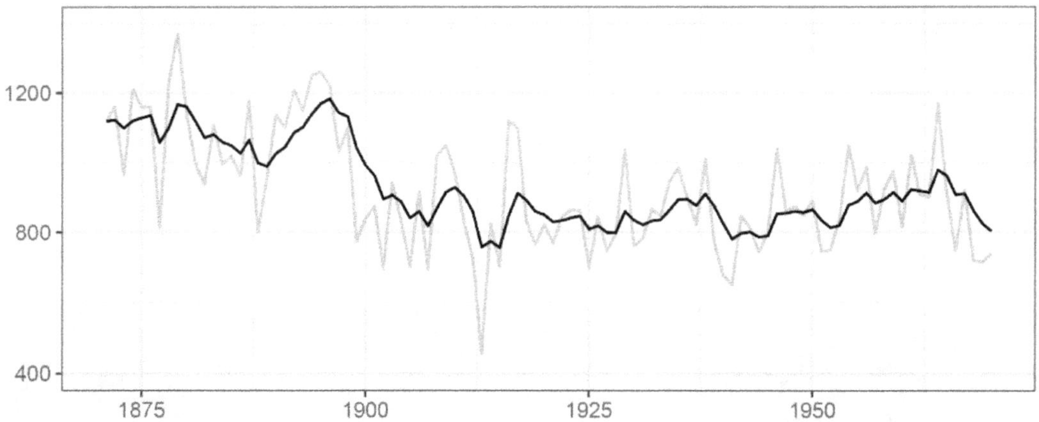


*Figure 9.5: Application of the Kalman filter using a local-level model*

A lot of the time, in a business context, we cannot measure the "actual states"; rather, what we observe is the output of the actual dynamics mixed with various measurement noise. In such situations, the Kalman filter shines by removing noise from multiple sources and identifying the true states of the system. The model is extensible to generate forecasts as well. Some prominent uses of the Kalman filter are signal processing and decomposition in communication networks, sensor data integration and **Internet of Things** (**IoT**), portfolio optimization, sales forecasting, and so on.

# Summary

In this chapter, you learned about the prominent smoothing and filtering techniques of time series. We provided a brief comparison of the concepts of time series decomposition, smoothing, and filtering techniques. This summary will help you identify the best techniques that are applicable to your problem.

We discussed how smoothing and filtering techniques are different in terms of their applications, either in time domain or frequency domain analysis. We also discussed three important time series linear filtering techniques that are widely applicable in many different industries.

Some models discussed in this chapter, such as $MA(q)$ and the Kalman filter, are extensible as forecasting models. In the next part of the book, you will be walking into the world of forecasting models, which range from applying simple heuristic models for univariate series to complex models for multiple series. We will start with the basics of forecasting, workflows, core packages, and some simple forecasting methods in *Chapter 10*.

# Further reading


- Hodrick, R., & Prescott, E. (1997). *Postwar U.S Business cycles: An empirical investigation. The Journal of Money, Credit and Banking*, 29(1).
- Kalman, R. E. (1960). *A new approach to linear filtering and prediction problems. Journal of Business and Economic Statistics*, 82, 34–45.
- Kalman, R. E., & Bucy, R. S. (1961). *New results in linear filtering and prediction theory. Journal of Basic Engineering*, 83, 95–108.
- Luethi, D., Erb, P., Otziger, S., McDonald, D., & Smith, P. (2022). *FKF: Fast kalman filter.* `https://CRAN.R-project.org/package=FKF`
- Tusell, F. (2011). *Kalman filtering in R. Journal of Statistical Software*, 39(2).

## Get this book's PDF version and more

Scan the QR code (or go to `packtpub.com/unlock`). Search for this book by name, confirm the edition, and then follow the steps on the page.

UNLOCK NOW

*Note: Keep your invoice handy. Purchases made directly from Packt don't require an invoice.*

# Part 4

# The Crystal Ball: Forecasting

Forecasting, the core use of a time series, is covered in this part. The chapters discuss a wide variety of forecasting techniques that range from simple methods to specialized algorithms. The contents are suitable for beginners to advanced professionals. We believe the forecasting methodologies covered in this part will be sufficient to solve the majority of business forecasting problems.

*Part 4* comprises the following chapters:

- *Chapter 10, Basics of Forecasting*
- *Chapter 11, Exponential Smoothing*
- *Chapter 12, ARIMA Forecasting Models*
- *Chapter 13, Advanced Computational Methods for Forecasting*
- *Chapter 14, Forecasting Models for Multiple Time Series*

# 10

# Basics of Forecasting

Forecasting models are specifically designed for time series data, where the order of observations matters. The inherent time dependency should be maintained in time series modeling. The predictive models that are commonly used on cross-sectional data can be used for forecasting when this time dependency in the data is respected and addressed appropriately. In statistical models, the focus is on how well the existing relationship between variables can be explained, which is captured in estimated parameters. The statistical tests conducted on these parameters assess the validity of modeling assumptions and offer confidence regarding the persistence of the identified relationship.

In contrast, machine learning or deep learning models do not rely on specific modeling assumptions. Instead, these models aim to make precise predictions without the requirement to explain the underlying relationship between variables. Some ML models, such as tree-based models, do capture some form of feature interactions and relationships.

The chapter is organized as follows:

- Section 10.1: Notations
- Section 10.2: Forecasting workflow
- Section 10.3: fable for forecasting
- Section 10.4: Naïve forecasting method
- Section 10.5: Seasonal Naïve method
- Section 10.6: Forecasting using average
- Section 10.7: Time Series Linear Model (TSLM)
- Section 10.8: Evaluating point forecast accuracy
- Section 10.9: Time series cross-validation

Note

**Your purchase includes a free PDF copy + exclusive extras**

Your purchase includes a DRM-free PDF copy of this book, 7-day trial to the Packt+ library (no credit card required), and additional exclusive extras. See the *Free benefits with your book* section in the *Preface* to unlock them instantly and maximize your learning.

# Technical requirements

For this chapter, the following packages are used.

```
library(pacman)
p_load("conflicted"
       ,"datasets"
       ,"fable"
       ,"fabletools"
       ,"feasts"
       ,"here"
       ,"patchwork"
       ,"rsample"
       ,"tidyverse"
       ,"tsibble"
       ,"tsibbledata" )

conflicts_prefer(
  fabletools::accuracy,
  fabletools::forecast,
  dplyr::filter )
```

The code file is available in the GitHub repo `https://github.com/PacktPublishing/Modern-Time-Series-Analysis-with-R.git` for you to download and practice.

Some common notations that are used throughout the book are presented in the next section as a reference.

## 10.1 Notations

The following standard notations are used in the context of forecasting in the remaining parts of the book. Some notations you have seen in the previous chapters already, however a revision would be beneficial.

- $t$: The $t$-th time period which takes values $1,2,\dots,n$.
- $y_t$: Observation at time $t$. $y$ denotes the full observed series.
- $x_{1,t}, x_{2,t}, \dots, x_{k,t}$: A set of explanatory variables $1,2,\dots,K$ measured at time point $t$ that may be correlated with a forecast series.
- $T$: The overall trend component of a time series. This can sometimes represent the combined trend-cycle ($T$ and $C$).
- $S$: The overall seasonal component of a time series.
- $E$: The unpredictable Irregular (or error) component of a time series.
- $l_t$: Denotes the level (average) of the observed series $y$ at time $t$.
- $b_t$: Denotes the growth at time $t$. This growth term is what measures the trend component within the models.
- $s_t$: Denotes the seasonal component at period $t$.
- $m$: Period of seasonality or the number of seasons in a year. For monthly data $m = 12$, for quarterly data $m = 4$, and so on.
- $h$: Forecasting horizon that takes values $1,2,\dots,H$.
- $\hat{y}_{n+h|n}$ or $\hat{y}_{n+h}$: The $h$-th period out-of-sample forecasts using all observations up to time $n$. Sometimes, how many past observations are used to forecast computation is written explicitly using the conditional symbol $|$. In the training data, the one-step ahead forecasts are known as fitted values.
- $e_t = y_t - \hat{y}_t$: Residuals in a time series model. This is calculated as the difference between observed and corresponding fitted/forecast values and what is left over after fitting a model. If a time series is transformed before fitting a model, then the difference between the transformed data and fitted values of the transformed data is known as innovation residuals. For example, if a model is fit on the transformed data, $w_t = \log(y_t)$, then $w_t - \hat{w}_t$ is the *innovation residuals.*
- $\epsilon_t$: Denotes the modeling error (or unexplained variation) at time $t$.

Effective forecasting requires a structured approach. The following section reviews the key steps in a forecasting problem-solving process.

## 10.2 Forecasting workflow

A centralized workflow ensures consistency in modeling approaches and methodologies used in a forecasting problem. This leads to uniformity in results and easier comparison between different models when applied to a dataset.

When developing solutions to business forecasting problems, you should adopt a workflow from the beginning of a project to streamline processes that cover all aspects of forecasting. By adopting a forecasting workflow, you will be able to collaborate better, increase efficiency, maintain consistency, and deploy models from proof-of-concept to production in a short time. A minimal forecasting workflow includes the following steps. You can add further granularity into it depending on the problem at hand.

1. Tidy data
2. Visualize
3. Build forecasting models
4. Check model performance
    - Residual diagnostics
    - Evaluate model fit
5. Generate forecasts
6. Measure forecast accuracy
7. Forecast generalizability via time series cross-validation

You learned how to create tidy time series for modeling through data cleaning and transforming, and specialized visualization in *Chapter 5*.

In this chapter, steps 3 to 7 in the preceding workflow are demonstrated repeatedly as we discuss three simple forecasting methods. This is to nudge you into the habit of adopting best practices while building forecasting models. The model fit evaluation is discussed in later chapters, when advanced statistical forecasting models are presented.

Efficiency and success are ensured when appropriate tools are used for the task at hand. Let's have a brief look at the functions from two R packages that will support you running end-to-end forecasting analysis.

# 10.3 fable for forecasting

A `fable` is a collection of stories that teaches moral lessons, and the term moral pertains to the principles of right and wrong behavior. When weaving forecasting storytelling from time series data, it's essential to adhere to the correct structure by adopting a forecasting workflow. Similar to a fable, the R packages `fable` and `fabletools` are a collection of functions developed upon shared forecasting morals and workflows.

The `fable` (**Forecasting Models for Tidy Time Series**) (O'Hara-Wild et al., 2023) and `fabletools` (Core Tools for Packages in the fable Framework) (O'Hara-Wild et al., 2021) packages provide a centralized process and interface for univariate and multivariate time series forecasting models. The `fable` package includes functions to fit models, whereas `fabletools` provide an integrated workflow for time series modeling and analysis.

The functions from these two packages follow a format consistent with the `tidyverse` workflow. They provide tools for the estimation of forecasting models, comparison, combination, visualization and forecast generation within a consistent framework.

The individual model functions from `fable` will be discussed in dedicated sections. Here is a list of some overarching functions from the `fabletools` package that you will frequently use in forecast modeling.

- `model()`: Fit a model on a single series. This can be extended to multiple models on a single series to multiple models on multiple series.
- `coef()` or `tidy()`: Extract the coefficients from fitted model(s).
- `glance()`: Provides one row summary of each fitted model, including residual variance, information criteria, and in-sample forecast errors.
- `report()`: Returns detailed information for a single model fit.
- `fitted()`: Returns in-sample fitted values.
- `residuals()`: Returns residuals from a fitted model.
- `augment()`: Provides the original series along with both fitted values and residuals.
- `accuracy()`: Returns common accuracy measures for a fitted model.
- `forecast()`: Generates both point forecasts and forecast distribution directly from the model fit.
- `hilo()`: Extracts multiple forecast intervals at once from a forecast distribution.

- `generate()`: Produces simulated paths from a model.
- `equation()`: Displays a model's equation.

Some functions are supported by particular models and are more suited in a production workflow, such as the following:

- `refit()`: Updates a model to new data
- `stream()`: Supplies new data to extend a model fit by bypassing the whole process of re-fitting a model
- `interpolate()`: Interpolates missing values in a series
- `components()`: Extracts various components from a fitted model

Most of these functions are used throughout this book. The forecasting workflow is easier to grasp when presented using simple methods. We will begin with the naïve forecasting method in the next section.

# 10.4 Naïve forecasting method

The naïve method uses the most recent observed value as a forecast for the upcoming periods. This approach allocates all the weight to the current observation in the series. Often referred to as random walk forecasts, this method essentially replicates the last observation as a forecast for the next period. This is the simplest method for a non-seasonal series when you have limited to no knowledge about the underlying data-generating process.

The equation of the naïve method is as follows:

$$\hat{y}_{n+h|n} = y_n \qquad h = 1,2,\ldots,H.$$

For illustration, we will use the WWWusage data, which provides the number of users connected to the internet through a server by minutes. First, tidy the data into `tsibble` format, which returns the number of observations in the index column and the internet usage in the value column. Print the data in the R console.

```
netuse <- as_tsibble(WWWusage)
netuse
```

There are $100$ observations in the data. The first $10$ observations are printed in the following output:

```
# A tsibble: 100 x 2 [1]
   index value
   <dbl> <dbl>
```

```
1     1    88
2     2    84
3     3    85
4     4    85
5     5    84
# i 95 more rows
```

Any modeling framework involves some explanatory analysis, and data visualization is commonly used for that purpose. Let's plot the WWWusage series using the following code:

```
autoplot(netuse) +
  labs(x = "Minutes", y = "Connections")
```

*Figure 10.1* shows the time plot that you can visually examine to identify any time series characteristics:

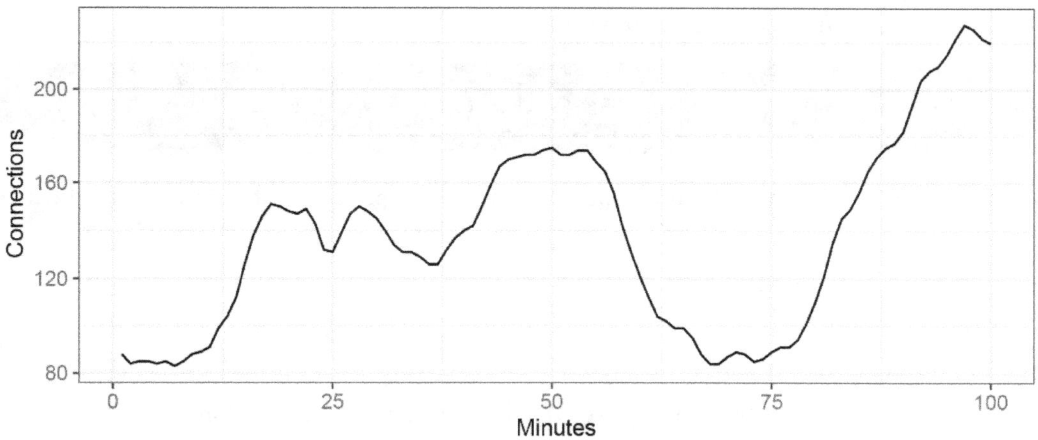


*Figure 10.1: WWWusage: number of users connected to the internet by minutes*

There is no apparent trend or seasonality in the WWWusage series. The number of internet users joined the server was not consistent, with strong irregularities and an influx of users after the 75-th minute. The naïve method provides strong baseline forecasts for data where no significant trend or seasonality is present, or for data that exhibits a random walk pattern.

It is standard practice to partition the original data into training and test sets, which are used for model development and performance evaluation, respectively. For time series, this can be done in two ways:

- A percentage of the original series is reserved for the test set: the `initial_time_split()` function from the `rsample` package can be used for such splitting
- Split the training and test set at a particular time point: the `filter()` or `filter_index()` functions achieve that.

Let's set aside 95% of observations from the internet usage data for training and the remaining 5% for the test set using the following code:

```
netuse_split <- initial_time_split(netuse, prop = 0.95)
netuse_split
```

The output returns data splitting information:

```
# <Training/Testing/Total>
# <95/5/100>
```

The `training()` and `testing()` functions extract the corresponding dataset from the split, as shown here:

```
netuse_train <- training(netuse_split)
netuse_train
```

The preceding code prints the training set, which has the first 95 observations:

```
# A tsibble: 95 x 2 [1]
   index value
   <dbl> <dbl>
 1     1    88
 2     2    84
 3     3    85
 4     4    85
 5     5    84
 6     6    85
 7     7    83
 8     8    85
 9     9    88
```

```
10     10    89
# i 85 more rows
```

Similarly, you can extract the test set and print it on the console as shown here:

```
netuse_test <- testing(netuse_split)
netuse_test
```

The test set has the last five observations as shown in the following output:

```
# A tsibble: 5 x 2 [1]
  index value
  <dbl> <dbl>
1    96   222
2    97   228
3    98   226
4    99   222
5   100   220
```

Each model has a dedicated function for specification. When those specification functions are passed into the `model()` function, the model is fit to the data and parameters are estimated.

To fit a naïve model on the internet usage data, use the `NAIVE()` function from the `fable` package. You just need to pass the target column into the `NAIVE()` function shown in the following code:

```
nv_fit <- netuse_train |>
  model(naive = NAIVE(value))

nv_fit
```

The output returns a `mable` (**M**odel T**able**), as in the following output:

```
# A mable: 1 x 1
    naive
  <model>
1 <NAIVE>
```

The naïve method has no parameters, hence the `coef()` or `tidy()` will return empty `tibbles`. You could use `glance()` to extract information from the model object:

```
glance(nv_fit)
```

This returns the following output that shows only one estimated parameter, the residual variance $\sigma^2$:

```
# A tibble: 1 × 2
  .model sigma2
  <chr>   <dbl>
1 naive    32.8
```

The `augment()` function extracts the observed series, fitted values, and residuals from a model object.

```
nv_aug <- augment(nv_fit)
nv_aug
```

You should see the following output:

```
# A tsibble: 95 x 6 [1]
# Key:       .model [1]
   .model index value .fitted .resid .innov
   <chr>  <dbl> <dbl>   <dbl>  <dbl>  <dbl>
 1 naive      1    88      NA     NA     NA
 2 naive      2    84      88     -4     -4
 3 naive      3    85      84      1      1
 4 naive      4    85      85      0      0
 5 naive      5    84      85     -1     -1
 6 naive      6    85      84      1      1
 7 naive      7    83      85     -2     -2
 8 naive      8    85      83      2      2
 9 naive      9    88      85      3      3
10 naive     10    89      88      1      1
# i 85 more rows
```

The Naïve method uses the value at time $t$ as the forecast for the next period, $t + 1$. Hence, the observations in the `.fitted` column are the actual observation shifted by one lag. As no data transformation is performed, the residuals and the innovation residuals are the same.

During model diagnostics, two assumptions must be checked for a good fit. They are (a) the normality assumption in residuals and (b) no autocorrelation in innovation residuals (should be a white noise series). The normality assumption is verified by a histogram plot of the residuals. Whereas an ACF plot confirms the existence of autocorrelation.

We can obtain standard residual diagnostic plots, including histogram and ACF plots, using the `gg_tsresiduals()` function from the `feasts` package on a `mable` object containing only one model as shown in the following code:

```
nv_fit |>  gg_tsresiduals()
```

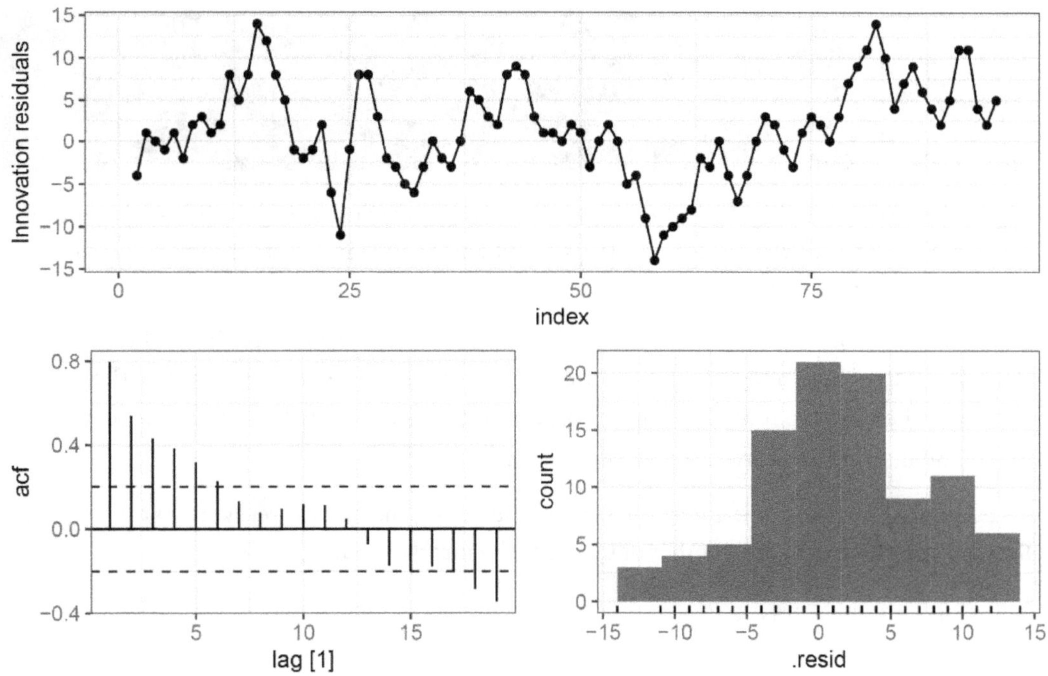


*Figure 10.2: Residual diagnostics of naïve method fitted on WWWusage*

The bottom-right histogram in *Figure 10.2* shows that the residuals are approximately normally distributed with a mean of 0 standard deviation of about 5. However, it is not a white-noise series (time plot) and has strong autocorrelation, as there are significant spikes in the ACF. This indicates that the naïve model is not enough to capture the data dimensions, and another model type could better fit the data.

Instead of visualization, statistical tests are commonly used to check the hypothesis that the autocorrelations in the residuals are similar to a white-noise series. These tests make the model diagnostics an objective and automated process. Either the Box-Pierce test or the Ljung-

Box test from the feasts package is used. These tests assume a null hypothesis that the autocorrelation in the residuals is similar to a white-noise series, and a p-value smaller than the common significance level, 0.05, will reject the null hypothesis. We can compute both the test statistics and corresponding p-values using the following code:

```
nv_aug |>
  features(.innov,list(box_pierce,ljung_box),lag = 10)
```

The output returns information in a `tibble` format with shorthand names, `bp` (Box-Pierce) and `lb` (Ljung-Box):

```
# A tibble: 1 × 5
  .model bp_stat bp_pvalue lb_stat lb_pvalue
  <chr>    <dbl>     <dbl>   <dbl>     <dbl>
1 naive     138.         0    144.         0
```

Both the tests reject the null hypothesis (p-values < 0.05) that provides enough evidence that the residual is not a white-noise series. Usually, when a model diagnostic confirms a poor fit, you should opt for a better model at this stage.

After model diagnostics, the next step is to generate forecasts for future periods. The `forecast()` function generates out-of-sample forecasts for the periods specified by the horizon, $h$. The function returns a `fable` (Forecast Table) object which has point forecasts (`.mean`) and distributions around each mean value.

Let's compute forecasts for the internet usage series using the naïve forecast method. The following code forecasts the number of users for the next 5 minutes:

```
nv_fcst <- forecast(nv_fit, h = 5)
nv_fcst
```

The function returns a fable object like the following, where `.mean` is the point forecasts and the value column provides the distribution around them as shown in the following output:

```
# A fable: 5 x 4 [1]
# Key:     .model [1]
  .model index       value .mean
  <chr>  <dbl>      <dist> <dbl>
1 naive     96  N(215, 34)   215
2 naive     97  N(215, 69)   215
3 naive     98 N(215, 103)   215
4 naive     99 N(215, 137)   215
5 naive    100 N(215, 171)   215
```

In the preceding output you may have noticed that the forecasts are flat, 215 for the next 5 minutes. The mean of the forecast distribution is also the same, however, the variance of that distribution increased at every step. It reflects the phenomenon that the further we forecast from the current time period, the less confident we become.

Now evaluate the forecast performance of the naïve method in both the training and the test set. The `accuracy()` function simplifies the evaluation process. The same function can be used for both in-sample and out-of-sample forecast accuracy measures. The forecast accuracy metrics are discussed in detail in *Section 10.8*.

For in-sample forecast accuracy measures, pass the model object into the `accuracy()` function as shown here:

```
accuracy(nv_fit, measures = point_accuracy_measures)
```

To compute out-of-sample accuracy measures, you need to pass the forecast object and the full observed series (training and test set combined) into the `accuracy()` function, as shown in the following code:

```
nv_fcst |>
  accuracy(netuse, measures = point_accuracy_measures)
```

The code provides the following output:

```
# A tibble: 1 × 10
  .model .type    ME  RMSE   MAE   MPE  MAPE  MASE RMSSE  ACF1
  <chr>  <chr> <dbl> <dbl> <dbl> <dbl> <dbl> <dbl> <dbl> <dbl>
1 naive  Test    8.6  9.09   8.6  3.83  3.83  1.89  1.55 0.126
```

The function is smart enough to detect what information is provided and the metrics are computed for that period accordingly. Notice `.type = Training` vs `.type = Test` in the preceding two outputs.

In general, a smaller value in the forecast accuracy metric indicates more accurate forecasts. As standalone metrics, they are informative enough to evaluate forecasting performance on a single model in the training and test set. In addition, these accuracy metrics provide directional evidence in model selection when multiple models are compared.

For performance evaluation, a common visual aid is used to plot out-of-sample forecasts along with prediction intervals. The `autoplot()` function on the forecast object produces the plot of point forecasts and prediction intervals around it. By default, two prediction intervals, 80% and 95% are plotted. A user can specify a single interval level or turn them off completely if needed.

There are two ways you can use the plotting functionality—plot the forecasts with prediction intervals or plot the training data along with forecasts and intervals. Both ways are shown in the following code:

```
p1 <- autoplot(nv_fcst) +
  labs(x = "minutes", y = "connections")

p2 <- nv_fcst |>
  autoplot(netuse_train, level = 95, size = 0.9) +
  geom_vline(xintercept = 96, color = "grey",
             linetype = "dashed", linewidth= 0.8) +
  labs(x = "minutes", y = "connections")

(p1/ p2)
```

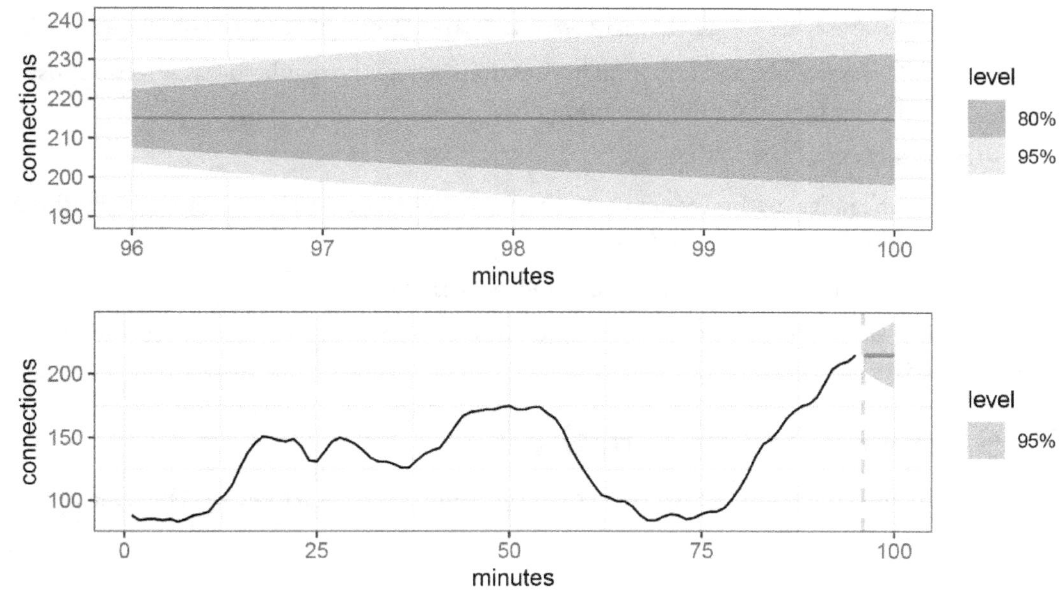


*Figure 10.3: Visual evaluation of the naïve method on the WWWusage data*

These forecasting plots are visual aids to detect any issues in forecasting and model fitting. In the top panel (*Figure 10.3*), the mean forecasts are plotted with two intervals, 80% (the darker shade) and 95% (the lighter shade). As time progressed, the estimated variance of the distributions around mean forecasts has increased. Hence, the intervals widen. The plot in the bottom panel shows the naïve forecasts actually did a good job on the shorter horizons.

So far, you have implemented the naïve method within a comprehensive forecasting framework. This framework enables you to perform data splitting, model fitting, model diagnostics, and evaluation, using both error metrics and visual analysis.

Let's see some key advantages and limitations of the naïve method:

- Advantages:
    - Even though it's overly simplistic, the naïve method provides good forecasts for many economic and financial time series.

- This method is useful as a baseline for forecast accuracy performance. Naïve forecasts are used in **Mean Absolute Scaled Error** (**MASE**) as a baseline model.
- The naïve method is helpful in situations where the regular data collection cost is very high and time-consuming.

- Limitations:
    - The naïve method is not usable beyond one-step ahead forecasts, as the method uses a flat forecasting function.
    - It's sensitive to extreme values and noise in the data.

The variation of the naïve method for seasonal data is seasonal naïve, which is discussed in the following section.

# 10.5 Seasonal Naïve method

Seasonal naïve uses the last observed value from the same season as a forecast. The forecast equation is:

$$\hat{y}_{n+h|n} \quad = \quad y_{n+h-m(k+1)} \qquad h = 1,2,\ldots,H.$$

where, $m$ is the seasonal period, and $k$ is the number of complete years in the forecast period prior to time $n + h$.

For example, for quarterly data, seasonal naïve will use the observed value in $Q_1$ in the last year as the forecast for the next year's $Q_1$.

In the following example, a seasonal naïve method is fitted on the quarterly electricity production in Australia during 2000 and 2008, and generates forecasts for 2009. We will use `filter_index()` to split data into training and test sets, which is convenient for data with a date-time-based index.

```
elctr <- aus_production |>
  select(Quarter, Electricity) |>
  filter_index("2000-Q1" ~ "2009-Q4" )

elctr_train <- elctr |>
  filter_index("2000-Q1" ~ "2008-Q4" )

elctr_test <- elctr |>
  filter_index("2009-Q1" ~ "2009-Q4" )
```

Let's look at the full series, including the training and test period, in *Figure 10.4*:

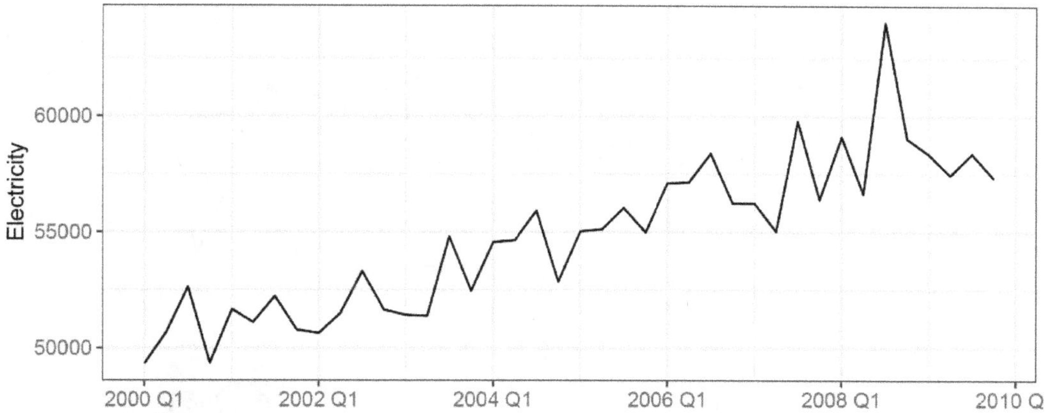


*Figure 10.4: Quarterly electricity production, Australia, 2000-2008*

The training data has an upward trend and strong irregularities. Even though the data is collected quarterly, there is no significant seasonal component detected visually.

Fit the seasonal naïve model using the dedicated function, `SNAIVE()`, and pass it to the `model()` function as shown here.

```
snv_fit <- elctr_train |>
  model(snaive = SNAIVE(Electricity))
glance(snv_fit)
```

Like the naïve method, seasonal naïve only provides the standard deviation of the observed data.

```
# A tibble: 1 × 2
  .model    sigma2
  <chr>      <dbl>
1 snaive 1861959.
```

To evaluate the model fit, let's generate the residual diagnostics plots presented in *Figure 10.5*:

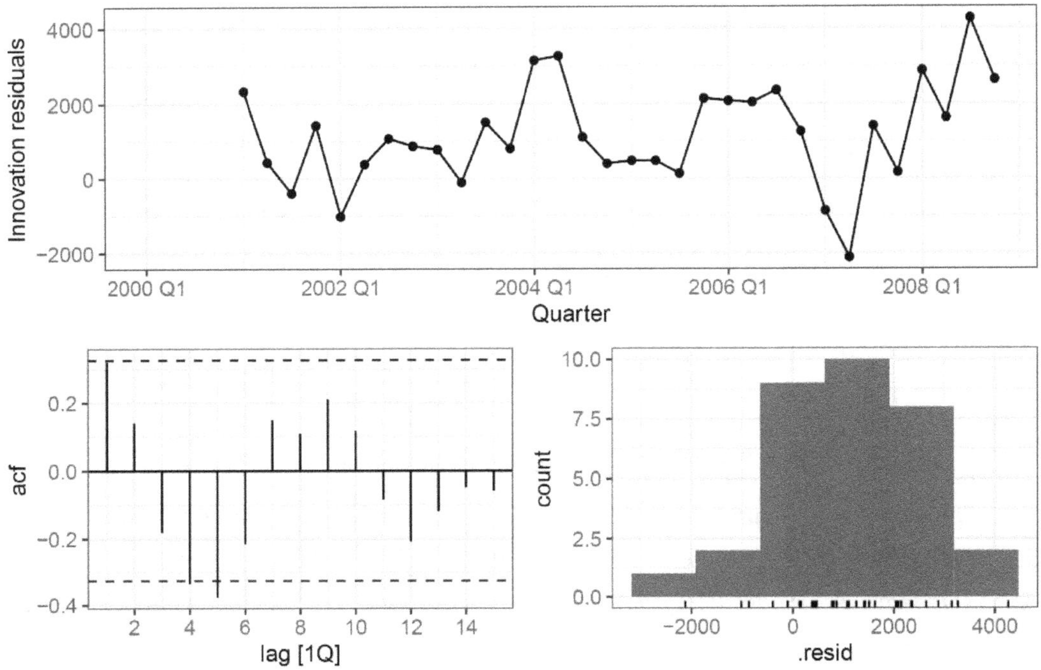


*Figure 10.5: Residual diagnostics for the seasonal naïve method on AUS electricity production*

The large errors in the residual time plot in *Figure 10.5* show that the seasonal naïve method could not model the observed series properly. The ACF has a small number of weak spikes very close (lag 1 and 4) or outside (lag 5) of the significant bands. The histogram appears roughly Gaussian with a non-zero mean and high variance. According to the overall fit, seasonal naïve can be a possible model for this series, however, other models could produce more accurate forecasts. Let's confirm whether the residuals is a white noise series by conducting statistical tests as shown in the following code:

```
augment(snv_fit) |>
  features(.innov, list(box_pierce,ljung_box ),lag = 10)
```

The test statistics and corresponding p-values are provided in the following output:

```
# A tibble: 1 × 5
  .model bp_stat bp_pvalue lb_stat lb_pvalue
  <chr>    <dbl>     <dbl>   <dbl>     <dbl>
1 snaive    17.6    0.0623    21.9    0.0155
```

The Box–Pierce p-value (0.0623) is borderline, and the Ljung–Box p-value (0.0155) rejects the null at the 5% level. This mixed result suggests there may still be some autocorrelation present in residuals.

Nonetheless, let's generate forecasts for the next 4 quarters with the seasonal naïve method:

```
snv_fcst <- forecast(snv_fit, h = 4)
```

This time, compute accuracy metrics for both the training and test period and combine them as shown in the following code:

```
bind_rows(accuracy(snv_fit) ,
          accuracy(snv_fcst, elctr))
```

The results are presented in the following output:

```
# A tibble: 2 × 10
  .model .type        ME  RMSE   MAE   MPE  MAPE  MASE RMSSE   ACF1
  <chr>  <chr>     <dbl> <dbl> <dbl> <dbl> <dbl> <dbl> <dbl>  <dbl>
1 snaive Training  1154. 1771. 1436.  2.04  2.56  1     1     0.321
2 snaive Test     -1830. 3013. 2236. -3.14  3.85  1.56  1.70 -0.339
```

For Seasonal naïve model, MASE will be always 1 as this is used as the benchmark method. All other accuracy metrics have large values indicating the forecasting model could be worth trying.

For a visual evaluation, you can plot the observed and fitted values in the training set, and observe and forecast values in the test set. You may organize the data before plotting as shown in the following code:

```
snv_comb <- elctr |>
  mutate(source = "observed") |>
  as_tibble() |>
  bind_rows(
    augment(snv_fit) |>
      select(Quarter,Electricity = .fitted) |>
      as_tibble() |>
      mutate(source = "snaive")
  ) |>
  bind_rows(
    as_tibble(snv_fcst) |>
      select(Quarter, Electricity =.mean) |>
```

```
        mutate(source = "snaive")
    ) |>
    mutate(source = factor(source, levels = c("observed","snaive")))
```

*Figure 10.6* presents the observed series (grey) and fitted and forecast values in black. The dashed vertical line separating the training and test period:

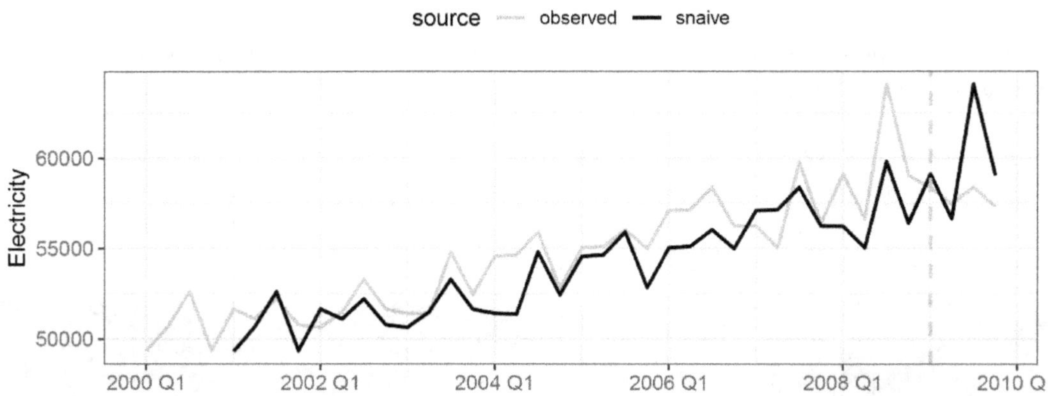


*Figure 10.6: Observed, fitted and forecasts from seasonal naïve of electricity production*

As shown in *Figure 10.6*, the seasonal naïve model captured the upward trend in the training set even though the seasonal variation was weak. The actuals plummeted unexpectedly in the test set, whereas seasonal naïve used the corresponding quarters from preceding years as forecasts. That's why the model produced larger errors for this series. You may have noticed the irregularities measured with the seasonal component, which could mask the forecasting performance of the seasonal naïve. A better approach could be to first estimate seasonality component by reducing the irregularities before applying any forecasting model.

In this section, the seasonal naïve forecasting method was discussed, following the forecasting framework. Even though simple, this method has some key advantages and limitations, summarized as follows:

- Advantages:
    - Seasonal naïve works well with data with consistent seasonal patterns

    - The method also provides good forecasts for data with weak seasonality, even when a complex model is not sufficient for capturing data dynamics
    - This method provides base forecasts in the Mean Absolute Scaled Error (MASE), which provides a justification against using a complex model in a forecasting exercise.
- Limitations:
    - Forecasts are heavily influenced by the recent seasonal observations
    - It's not suitable to produce forecasts for longer horizons

Next, we discuss another simple forecasting method, the average method of forecasts.

# 10.6 Forecasting using average

The average forecasting method involves using the mean of the observed series as the out-of-sample forecasts. Similar to the naïve method, this results in a flat forecasting function where all future values are the same. However, unlike the naïve method, the average method assigns equal weight to all observations in the series. This is also known as the mean method.

The mathematical equation of the average method is as follows.

$$\hat{y}_{n+h|n} \quad = \quad \frac{1}{n}\sum_{t=1}^{n} y_t \qquad h = 1,2,\dots,H.$$

The `MEAN()` function specifies the average forecasting method. Following the forecasting workflow shown previously, you can fit the model on training data, perform diagnostics checks, generate forecasts for the test set, and calculate accuracy. The average method estimates a single parameter: the mean of the data. You have used the `glance()` function earlier on a model object. This time we will use `tidy()` on a model object as shown in the following code:

```
avg_fit <- netuse_train |>
  model(mean = MEAN(value))

tidy(avg_fit)
```

The `tidy()` function returns the following output, which provides estimated parameters and results of the significance test on them:

```
# A tibble: 1 × 6
  .model term  estimate std.error statistic  p.value
  <chr>  <chr>    <dbl>     <dbl>     <dbl>    <dbl>
1 mean   mean      133.      3.65      36.3 4.14e-57
```

The mean method has only one estimate, the average level of the series (133). The very small p-value provides enough evidence that it is significantly different from 0.

Now, check the white noise assumption in the residuals by performing statistical tests:

```
augment(avg_fit) |>
  features(.innov, list(box_pierce, ljung_box), lag=10)
```

The following output shows the test results:

```
# A tibble: 1 × 5
  .model bp_stat bp_pvalue lb_stat lb_pvalue
  <chr>    <dbl>     <dbl>   <dbl>     <dbl>
1 mean      353.         0    373.         0
```

The residuals from the average model did not show a white-noise structure as the p-values from statistical tests are very small; hence, the null hypothesis of the white noise series is rejected.

Let's evaluate the model's performance of the average forecasts in the training and test sets. You can combine the two accuracy metric tables into one, as shown in the following code:

```
avg_fcst <- forecast( avg_fit, h = 5)
avg_acc <- accuracy(avg_fit) |>
  bind_rows(accuracy(avg_fcst, netuse))

avg_acc
```

The result is presented in the following output:

```
# A tibble: 2 × 10
  .model .type        ME  RMSE   MAE   MPE  MAPE  MASE RMSSE  ACF1
  <chr>  <chr>     <dbl> <dbl> <dbl> <dbl> <dbl> <dbl> <dbl> <dbl>
```

```
1 mean   Train… 1.35e-14  35.4  29.9 -7.80  25.1  6.58  6.04 0.950
2 mean   Test    9.11e+ 1  91.1  91.1 40.7   40.7 20.0  15.6  0.126
```

The average method showed better performance on the training set. It has large forecast errors in the test sets across all of the accuracy metrics.

Two forecasting methods, **naïve** and **average**, are applied to the same training data. Since the test set is the same, we can evaluate their forecasting performance by comparing respective accuracy metrics.

According to the MASE, naïve has performed better compared to the average method (1.894 versus 20). The same is true when RMSEs are compared; naïve (5.85) forecast errors are smaller compared to average method (91.1).

Let's visualize forecasts from these two methods along with the test set. Like before, you need to organize the data before plotting it. *Figure 10.7* presents two plots, the full training and test periods on left, and only the test set on the right to facilitate a closer inspection:

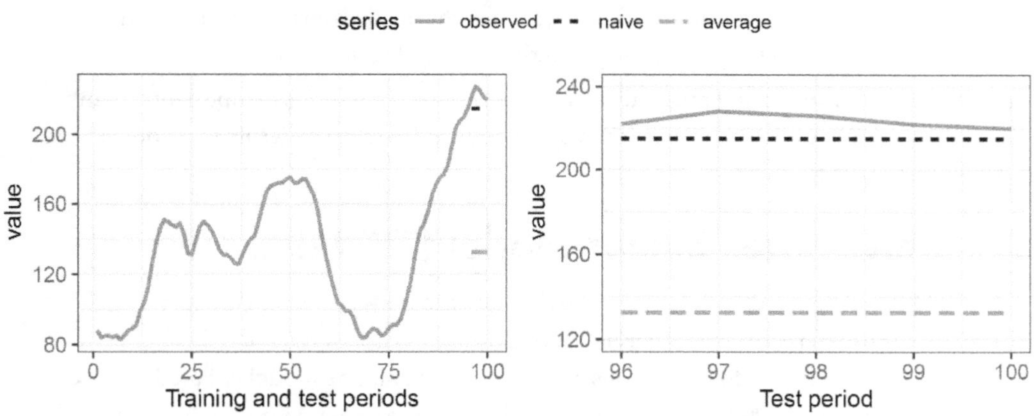


*Figure 10.7: Forecast comparison of naïve and average methods on WWWusage*

The naïve method produced better forecasts as it captured the recent history well. On the other hand, the average level in the data remained at 133 despite the significant movement across the observed period. Visualizing forecasts from multiple potential models with original data provides insights into their suitability in forecasting scenario. You can assess where a model has suffered to produce accurate forecasts and that can potentially lead to use alternative forecasting models for a given data.

The advantages and limitations of the average forecasting method is summarized as follows:

- Advantages:
    - The average method is used for initial forecasts when data is limited for fitting any meaningful model, or in a situation where a crude average is sufficient in the context of the forecasting problem
    - The average method smooths out short-term fluctuations and noise in a time series.
    - The average method can be used as a baseline to assess the performance of sophisticated forecasting models
- Limitations:
    - Being too simple, the average method fails to capture any pattern and uncertainty in the data
    - It is highly affected by outliers in the data
    - The average method assigns equal weight to all observations, and is not responsive to recent changes in the observed series.

None of the simple forecasting methods discussed use any time series components or any explanatory variables. In the next section, we will look at the time series linear model in the context of forecasting.

# 10.7 Time Series Linear Model (TSLM)

In a linear regression model, we examine the relationship between a dependent variable ($y$) with one or more explanatory variables ($x$'s). Usually, in a regression model, both the dependent and the independent variables are cross-sectional. The model is called a **Time Series Linear Model** (**TSLM**) when a time series is used as a dependent variable. The independent variables may or may not be a time series, but they must be measured at the same time period as $y$. A TSLM model is represented by *Equation (10.1)*.

$$y_t \quad = \quad \beta_0 + \beta_1 x_{1,t} + \beta_2 x_{2,t} + \cdots + \beta_k x_{k,t} + \epsilon_t \qquad t = 1,2,\dots,n \qquad (10.1)$$

where, $x_{i,t}$ are the explanatory variables. Time series components, such as trend and seasonal dummies can be used as explanatory variables.

$\beta_0$ is the intercept term, which measures the average level of the dependent variable in the absence of any explanatory variable,

$\beta_1, \dots, \beta_k$ are the parameters that measure the impact of corresponding explanatory variable on the dependent variable,

$\epsilon_t$ is the model's error term, which follows these assumptions:

- They have zero mean
- Not autocorrelated
- Unrelated to the predictor variables

The zero mean assumption of the error term assures that the forecasts are not systematically biased. Forecasts become inefficient if there is autocorrelation among residuals because, in that scenario, there was information in the data that was not captured by the model and, as a result, it ends up as residuals. The third assumption makes sure that no systematic information is left out by the model.

Along with these base assumptions, it is also assumed that the errors are normally distributed, so that prediction intervals can be easily calculated. *Equation (10.1)* does not have any lagged term of the forecast variable, $y$, or any of the predictors $x$. Adding lag terms would violate the no autocorrelation assumption.

As a first step of using regression model in forecasting, visualize the electricity production in a scatter plot with a straight line to check the direction of the trend shown in the following code:

```
elctr_train |>
  ggplot(aes(x = Quarter, y = Electricity)) +
  geom_point() +
  geom_smooth(method = "lm", se = FALSE) +
  labs(x = "", y = "")
```

The plot is presented in *Figure 10.8*:

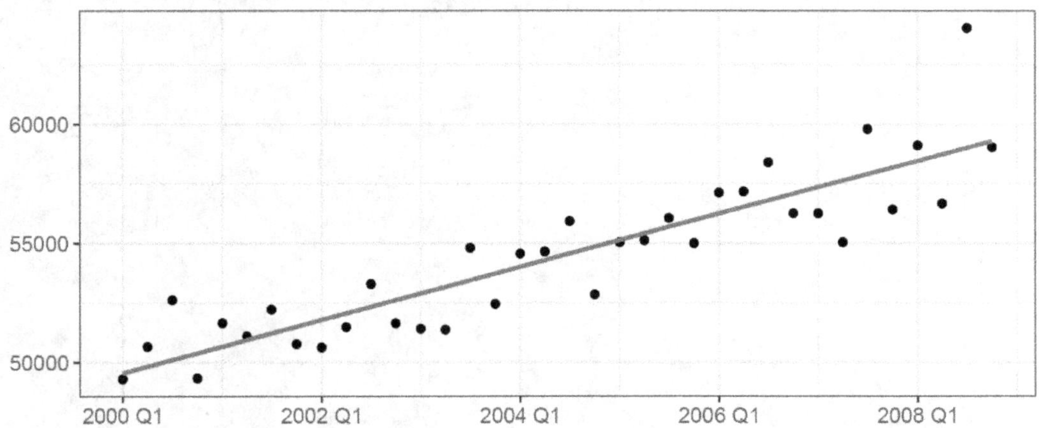


*Figure 10.8: Quarterly electricity production during 2000-2008*

The trend line in the scatter plot of electricity production against quarters has an upward movement. The observations above or below the trend, such as third-quarter values, are usually higher, and the fourth-quarter values are lower than the trend line, indicating seasonal variations.

We can fit a linear model of the electricity production on the estimated trend and seasonality components. We don't need to estimate these two components separately, but rather pass them in the `TSLM()` model specification as explanatory variables. In standard regression modeling, these features must be computed separately before passing them into a model formula. In time series regression, the `trend()` and `season()` functions perform similar operations inside the `TSLM()` function. The `trend()` function adds a linear predictor $x_{1,t} = t$. The `season()` function adds seasonal dummy variables. The total seasonal dummies will be one less than the number of seasons detected in the observed data. For the quarterly electricity series, there will be three seasonal dummies.

Let's fit a linear regression model on the training data using trend and seasonal components as explanatory variables. The following code shows how to perform this operation:

```
lmcomp_fit <- elctr_train |>
  model(tscomp = TSLM(Electricity ~ trend() + season()))

report(lmcomp_fit)
```

The `report()` function on the model fit produces the following below output:

```
Series: Electricity
Model: TSLM

Residuals:
     Min       1Q   Median       3Q      Max
-2005.32  -792.10    40.98   737.06  3250.69

Coefficients:
              Estimate Std. Error t value Pr(>|t|)
(Intercept)   49159.13     472.29 104.088  < 2e-16 ***
trend()         278.89      17.65  15.797 2.23e-16 ***
season()year2  -484.44     516.02  -0.939  0.35508
season()year3  1896.11     516.92   3.668  0.00091 ***
season()year4  -984.22     518.43  -1.898  0.06698 .
---
Signif. codes:  0 '***' 0.001 '**' 0.01 '*' 0.05 '.' 0.1 ' ' 1
```

```
Residual standard error: 1094 on 31 degrees of freedom
Multiple R-squared: 0.9025, Adjusted R-squared: 0.8899
F-statistic: 71.75 on 4 and 31 DF, p-value: 3.1907e-15
```

The estimated coefficients suggest the average electricity production was 49,000 gigawatt hours (GWh), which increased every quarter by 279 GWh. The model assumes the first quarter as the base, and the coefficients for the remaining three quarters are relative to this base. On average, the second quarter had production of 484 GWh lower than the first quarter, the third quarter had production of 1,896 GWh higher than the first quarter, and the fourth quarter had production of 984 GWh lower than the first quarter.

The p-values provide evidence that the intercept, trend, and the third quarter coefficients are significant, but the second and fourth quarter estimates are not significant. The overall fit of the model is good, as 91% variability of the data is explained by trend and seasonal dummies. The bias-corrected adjusted R-squared (88%) also suggests a good overall fit. The adjusted R-squared should be considered for linear model fit evaluation rather than R-squared, because the latter always increases as the number of predictors increases.

Let's check the residual diagnostic plots for TSLM fit presented in *Figure 10.9*:

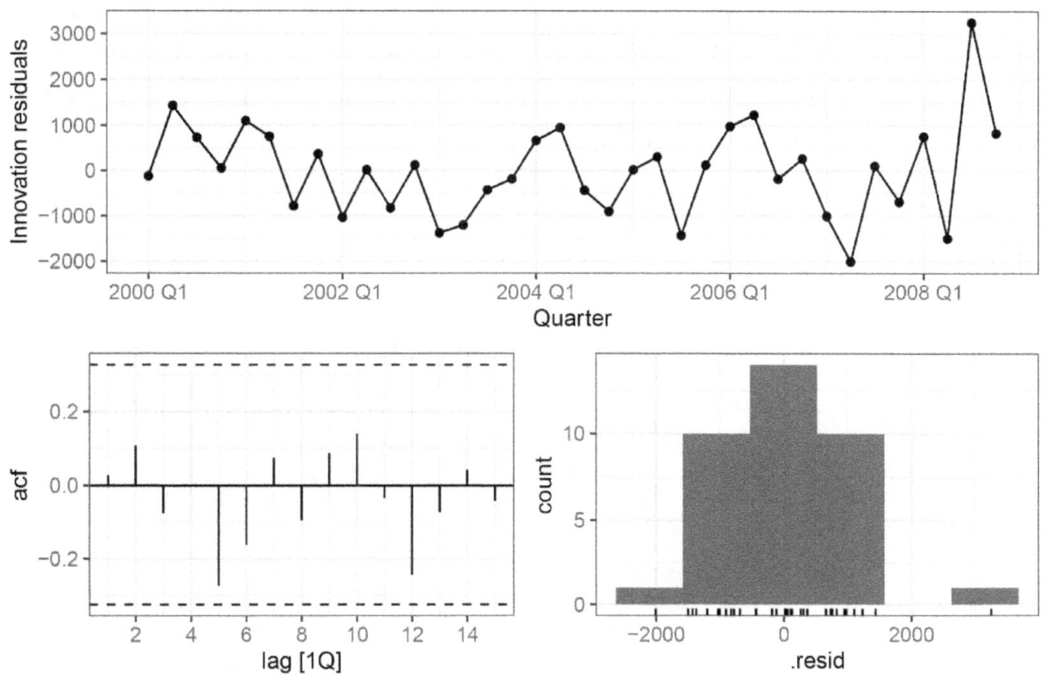


*Figure 10.9: Residual diagnostics of TSLM on electricity production*

The ACF in *Figure 10.9* shows no significant spikes, and the distribution is roughly Gaussian. The line chart indicates that some pattern might be left in the residuals. The p-values from the white-noise statistical tests will confirm that, computed in the following code snippet.

```
augment(lmcomp_fit) |>
  features(.innov, list(box_pierce, ljung_box), lag = 10)
```

The white-noise tests produce the following output.

```
# A tibble: 1 × 5
  .model bp_stat bp_pvalue lb_stat lb_pvalue
  <chr>    <dbl>     <dbl>   <dbl>     <dbl>
1 tscomp    5.81     0.831    7.36     0.691
```

Both the tests have higher p-values, Box-Pierce $(0.831)$ and Ljung-Box $(0.691)$. There is not enough evidence to reject the null hypothesis at a $5\%$ significance level. What it means is that the residuals have an ACF structure similar to a white-noise series.

For linear regression models, you can plot residuals against the fitted value to check: (a) whether the spread of the residuals is roughly constant across all fitted values (homoscedasticity assumption), (b) the relationship between predictors and the response variable is linear (linearity assumption), and (c) not correlated with any predictor variable.

You can also identify the impact of outlier or influential data points, if they exist, from residual versus fitted values plot presented in *Figure 10.10*:

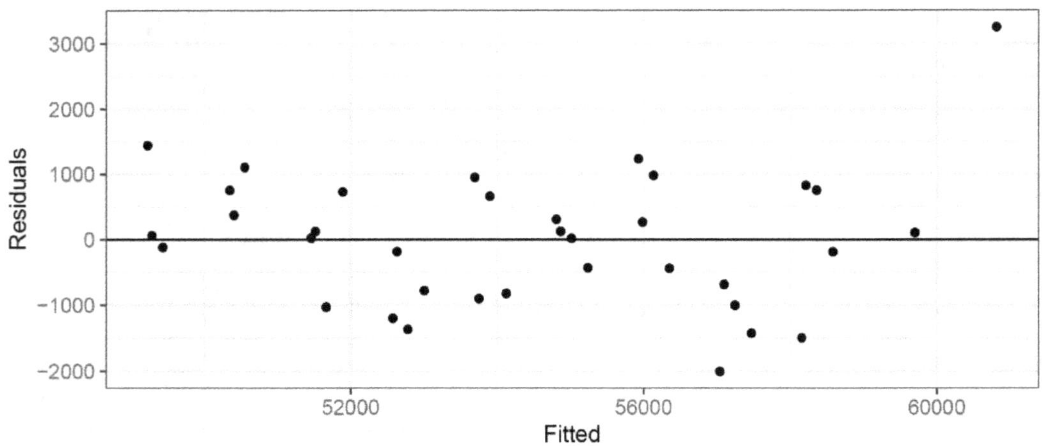


*Figure 10.10: Residuals diagnostic of TSLM with time series components*

The overall model looks good, since there is no pattern detected in the residuals when plotted against fitted values. However, the model could be improved as there are more positive residuals (under forecasts) than negative residuals (over forecasts). It suggests possible bias in certain parts of the time series. Also, the presence of outliers could influence forecasting performance.

Now, generate forecasts for the test set (the next 4 steps) and obtain in-sample and out-of-sample accuracy measures as shown in the following code:

```
lmcomp_fcst <- lmcomp_fit |>
  forecast(h = 4)

lmcomp_acc <- accuracy(lmcomp_fit) |>
  bind_rows(accuracy(lmcomp_fcst, elctr))

lmcomp_acc
```

You can see the results provided by the preceding code in the following output:

```
# A tibble: 2 × 10
  .model .type            ME  RMSE   MAE      MPE  MAPE  MASE RMSSE    ACF1
  <chr>  <chr>         <dbl> <dbl> <dbl>    <dbl> <dbl> <dbl> <dbl>   <dbl>
1 tscomp Training  8.08e-13 1015.  783. -0.0299  1.42 0.545 0.573  0.0280
2 tscomp Test     -2.11e+ 3 2290. 2111. -3.64    3.64 1.47  1.29  -0.0947
```

A training MASE smaller than 1 indicates that the model is a better choice that will produce more accurate forecasts than the seasonal naïve model. For the electricity production series, both the training (0.545 versus 1) and test (1.47 versus 1.56) MASE are smaller than the seasonal naïve.

The time plot of the training set and the forecasts for the test set are presented in *Figure 10.11*:

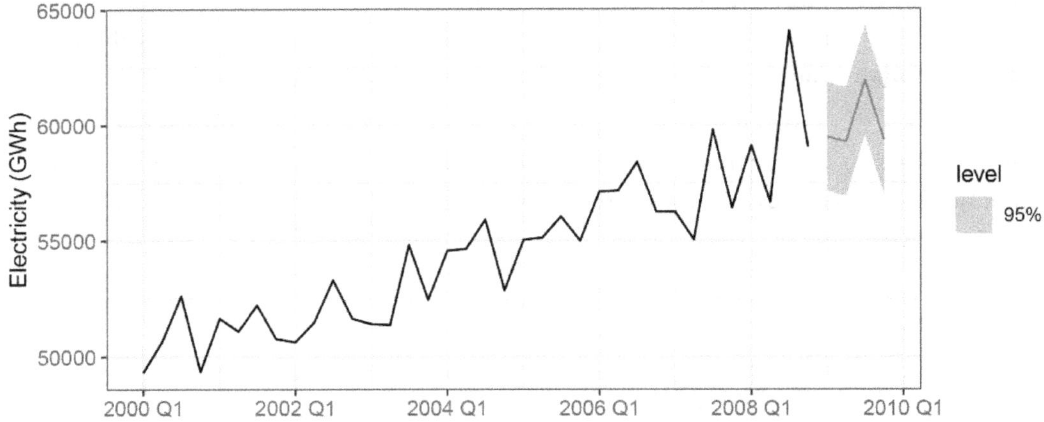


*Figure 10.11: Forecasts of electricity production from the TSLM model*

The TSLM model captures the upward trend and some seasonality without being influenced by the recent changes in the historical data. Once the forecasting performance is promising in the test set, you can refit the model with the full data and generate forecasts beyond 2009.

One advantage of using time series components as predictors is that the out-of-sample forecasts from the linear regression are valid, as the trend predictor will be extrapolated, and the seasonal dummies stay the same. However, this is not always the case for TSLM, as the forecasts generated outside of the predictors' range may not be valid, or the linearity assumption may not hold. In such a situation, the out-of-sample forecasts need to be used with caution.

The TSLM forecasting model can be used to conduct a what-if analysis to make strategic business decisions. Let's see an example of such an application in the next section.

## 10.7.1 Scenario analysis

The Texas housing data (`txhousing`) has monthly listings, number of sales, sales volume, and inventory for 46 cities in Texas, USA from January 2000 to July 2015 (186 months). We will fit a TSLM model to forecast sales using the number of listings, inventory, and months for Dallas after tidying data into a `tsibble` using year and month as the index.

```
dallas <- txhousing |>
  filter(city == "Dallas") |>
  mutate(year_month = yearmonth("2000 Jan") + 0:186) |>
  as_tsibble(index = year_month)
```

The sales are regressed on listings, inventory, and monthly dummies created via the `season()` function. The benefits of creating dummy variables via this function will become clearer when we use this model for scenario analysis and forecasting. Fit the model as shown in the following code and examine the fitted model's information:

```
lm_fit <- dallas |>
  model(lm = TSLM(sales ~ listings + inventory + season()))

report(lm_fit)
```

For the TSLM model, the report produces detailed output like the following:

```
Series: sales
Model: TSLM

Residuals:
     Min       1Q   Median       3Q      Max
-1150.95  -218.82    10.19   220.36  1110.83

Coefficients:
                  Estimate Std. Error t value Pr(>|t|)
(Intercept)      3.422e+03  1.216e+02  28.143  < 2e-16 ***
listings         1.661e-01  8.513e-03  19.509  < 2e-16 ***
inventory       -8.279e+02  3.773e+01 -21.944  < 2e-16 ***
season()year2    6.205e+02  1.253e+02   4.953 1.74e-06 ***
season()year3    1.769e+03  1.254e+02  14.110  < 2e-16 ***
season()year4    1.857e+03  1.255e+02  14.796  < 2e-16 ***
season()year5    2.417e+03  1.256e+02  19.244  < 2e-16 ***
season()year6    2.674e+03  1.257e+02  21.280  < 2e-16 ***
season()year7    2.451e+03  1.258e+02  19.486  < 2e-16 ***
season()year8    2.294e+03  1.302e+02  17.629  < 2e-16 ***
season()year9    1.430e+03  1.278e+02  11.193  < 2e-16 ***
season()year10   1.284e+03  1.276e+02  10.062  < 2e-16 ***
season()year11   8.124e+02  1.274e+02   6.376 1.62e-09 ***
season()year12   9.790e+02  1.274e+02   7.687 1.11e-12 ***
---
Signif. codes:  0 '***' 0.001 '**' 0.01 '*' 0.05 '.' 0.1 ' ' 1

Residual standard error: 354.3 on 172 degrees of freedom
```

```
Multiple R-squared: 0.8905, Adjusted R-squared: 0.8822
F-statistic: 107.6 on 13 and 172 DF, p-value: < 2.22e-16
```

The model provides a good fit (adjusted R-squared = 88%), and each coefficients are significant with very low p-values. The average level of sales was around $3422 which increases as the number of listings increases (positive coefficient) but decreases as the inventory increases (negative coefficient). Taking January as the base, sales were significantly higher in the other months. It is important to note that listings and inventory may be correlated in housing markets. This model could suffer from multicollinearity, and such a situation needs to be addressed in business applications on real data.

Let's visualize the fitted values and actual sales in a single plot presented in *Figure 10.12*:

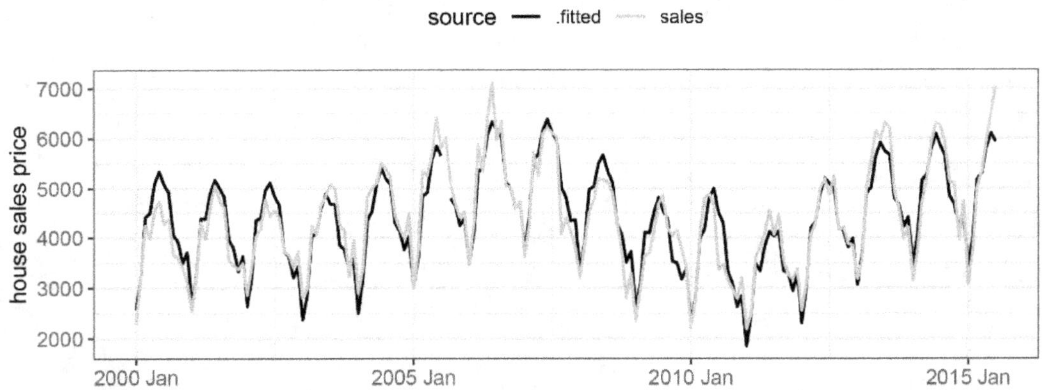


*Figure 10.12: Actual and fitted values from TSLM on Dallas housing sales*

The fitted TSLM model on the Dallas housing data has good in-sample fit. It captured the seasonalities and peak-trough movement in the trend-cycle by the variables used in the fit. We can use the model for what-if analysis, also known as scenario analysis.

Suppose a real estate company wanted to know what the sales would be for the next five months if the listings changed while keeping inventory fixed at a certain level. Using the `scenario()` and `new_data()` functions you can create combinations of data points to forecast. Essentially, this is the equivalent of creating new datasets under two different conditions. Since the seasonal dummies were created directly from the time variable (index) of the tidy `tsibble`, there is no need to create those in the new datasets again. The future values for the next five months are created assuming two number of listings, 30,000 and 15,000, while keeping inventory fixed at 5 as shown in the following code:

```
future_dt <- scenarios(
  `High listings` = new_data(dallas, 5) |>
    mutate(listings = 30000, inventory = 5),
  `Low listings` = new_data(dallas, 5) |>
    mutate(listings = 15000, inventory = 5)
)
```

Technically, `new_data`(dallas,5) extends the time index for the next 5 months. Now we will generate forecasts from the estimated model on the new datasets.

```
scr_fcst <- forecast(lm_fit, new_data = future_dt)
scr_fcst
```

The forecast object provides the following results:

```
# A fable: 10 x 7 [1M]
# Key:     .scenario, .model [2]
   .scenario     .model year_month           sales .mean listings inventory
   <chr>         <chr>       <mth>          <dist> <dbl>    <dbl>     <dbl>
 1 High listings lm       2015 Aug N(6560,140137) 6560.    30000         5
 2 High listings lm       2015 Sep N(5696,139269) 5696.    30000         5
 3 High listings lm       2015 Oct N(5549,139298) 5549.    30000         5
 4 High listings lm       2015 Nov N(5078,139209) 5078.    30000         5
 5 High listings lm       2015 Dec N(5244,139462) 5244.    30000         5
 6 Low listings  lm       2015 Aug N(4068,137343) 4068.    15000         5
 7 Low listings  lm       2015 Sep N(3204,136923) 3204.    15000         5
 8 Low listings  lm       2015 Oct N(3058,136861) 3058.    15000         5
 9 Low listings  lm       2015 Nov N(2586,136901) 2586.    15000         5
10 Low listings  lm       2015 Dec N(2753,136789) 2753.    15000         5
```

A plot presented in *Figure 10.13* could provide visual aids in the comparison of different scenarios:

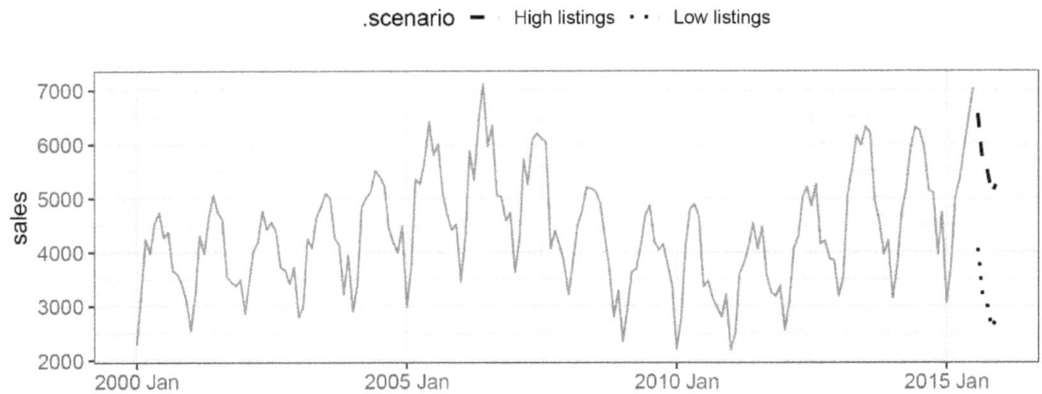


*Figure 10.13: Scenario analysis using TSLM on Dallas housing sales*

The scenario forecasts in *Figure 10.13* show that irrespective of listings numbers, future sales will trend in a lower direction when the inventory is kept constant at the average level.

Conducting a what-if analysis is a crucial step before making any significant and strategic business decisions. Scenario-based forecasting offers a practical method to test hypothetical situations, aiding businesses in making informed and strategic choices.

Let's summarize the advantages and limitations of the TSLM model:

- Advantages:
    - The time series linear regression model has wide applications in fields, such as economic studies, finance, marketing, advertising, and sales forecasting.
    - The TSLM model provides interpretable insights into the relationship between the forecast variable and predictors. The estimated coefficients, along with the statistical tests, provide an impact and variable importance on generated forecasts.
    - The model can capture the underlying behavior better when the forecast variable exhibits a strong linear relationship with predictors.
    - The TSLM model facilitates scenario-based forecasting, which is crucial to make important business decisions.
- Limitations:
    - A key drawback of TSLM is that it requires accessible predictor data to generate forecasts beyond the observed period. For instance, if you've built a time series

regression model of the unemployment rate based on variables, such as interest rates, productivity, and average salary using quarterly data, forecasting the unemployment rate for the next quarter outside the observed data period necessitates having future values of these predictors. Often, these predictor values are unknown in advance, limiting the usefulness of time series regression models for out-of-sample forecasts.

- The non-linear relationship between the forecast variable and predictors is not captured by the linear model. However, transforming the variables before modeling could help.

So far, you have built and tested four forecasting models using a tidy workflow. The residual diagnostic plots are used for model evaluation. For forecasting performance, a set of accuracy measures are evaluated.

Let's have a deeper look into forecast accuracy metrics, how they are calculated, and how they are interpreted. Each of these forecast accuracy measures has its own pros and cons. A brief summary is presented on the forecast accuracy metrics in the next section.

## 10.8 Evaluating point forecast accuracy

Accuracy measures provide a quantitative way to evaluate the performance of forecasting models. They assess how well a model is capturing the underlying pattern and trends in the data. However, a well-fitted model does not guarantee accurate forecasts. That's why it is important to compare both the in-sample and out-of-sample accuracy of a single model to identify whether the model is over-fitted (i.e., fits well in the training data but has poor performance in the test set).

If you have competing models for a dataset, then you can objectively compare the performance of each model using forecast accuracy measures at different horizons and choose the most effective one.

The point forecast metrics are also called errors, as they are usually computed on the one-step ahead forecast errors, calculated as follows:

$$e_t = y_t - \hat{y}_{t+1}$$

For the training set, $t = 2, \ldots, n$, and for the test set, $t$ is replaced by $h$, which varies $h = 1, \ldots, H$.

The formulas for computing accuracy metrics are the same in both the training and the test set. For simplicity, we will use the subscript $t$ irrespective of the dataset used.

There are many forecast accuracy measures available in the literature. However, we will look deeper into five of them as they are widely used in business. Based on how they are computed, they are categorized into three categories.

- Scaled-dependent errors:
  - Mean Absolute Error (MAE): A magnitude-driven measure that provides equal weight to all errors.

$$\text{MAE} = \frac{\sum_{t=2}^{n} |e_t|}{n-1}$$

MAE is easy to interpret and less sensitive to outliers. The metric has the same scale as the data, allowing it to compare multiple forecasting models for a single series.

- Root Mean Squared Error(RMSE): A metric that puts a heavy penalty on larger errors.

$$\text{RMSE} = \sqrt{\frac{\sum_{t=2}^{n} e_t^2}{n-1}}$$

RMSE is on the same scale as the data and is used to compare competing models for a series. It is well-suited for situations where a small number of large errors have greater business consequences.

The main drawback of the scale-dependent measures is that they are not suitable for comparing the forecasting performance of a model across multiple series with different units.

- Percentage errors:
  - Mean Absolute Percentage Errors (MAPE): This is a scale-free measure that first transforms forecast errors as a percentage of the observed value, and then the mean is computed over the percentages.

$$\text{MAPE} = \sum_{t=2}^{n} \frac{(|100e_t/y_t|)}{n-1}$$

Being a unit-free metric, MAPE is widely used to compare forecast performance across many datasets. However, this cannot be used when $y_t = 0$, which is a common scenario in demand forecasting.

- Scaled errors:

Scaled errors tackle issues in scaled metrics and the 0 denominator problem in the percentage error metric by using a scaling factor.

$$\text{non-seasonal series: } q_j = \frac{e_j}{\frac{1}{n-1}\sum_{t=2}^{n} | y_t - y_{t-1}}$$

$$\text{seasonal series: } q_j = \frac{e_j}{\frac{1}{n-m}\sum_{t=m+1}^{n} |y_t - y_{t-m}|}$$

where, $m$ is the number of seasonal periods.

The quantity $q_j$ is scale-free as both the numerator and denominator are computed at the same scale as the original data. The denominator in $q_j$ is nothing but the MAE of naïve forecasts in the training data. Using the scaling factor $q_j$, the following two metrics are proposed by Hyndman et al. (2006).

- Mean Absolute Scaled Error (MASE): The average of the absolute forecast errors scaled by the MAE in the training dataset.

$$\text{MASE} = \text{mean}(|q_j|)$$

- Root Mean Squared Scaled Error (RMSSE): The same as RMS,E where the forecast errors are scaled using the MAE of naïve forecasts.

$$\text{RMSSE} = \sqrt{\text{mean}(q_j^2)}$$

The primary benefit of scaled errors is the ability to compare the forecasting performance of numerous models across datasets.

In the previous sections, forecast accuracy measures were compared for a single forecasting method on a single series. Let's see an example in which forecasts are obtained from multiple methods on a single series and out-of-sample performance is compared. Three methods naïve, average (mean), and random walk with drift (RW) were fitted on the training set on WWWusage data. For each model, forecasts are computed for the next 5 horizons, and then point accuracy measures are computed in both the training and test sets.

```
nv_mult_fit <- netuse_train |>
  model(naive = NAIVE(value),
        mean = MEAN(value),
        drift = RW(value ~ drift()))

nv_mult_fcst <- nv_mult_fit |>
  forecast(h = 5)

nv_mult_acc <- bind_rows(accuracy(nv_mult_fit),
                         accuracy(nv_mult_fcst, netuse))
```

```
nv_mult_acc
```

Let's see the forecast accuracy metrics for the three models summarized in one table:

```
# A tibble: 6 × 10
  .model .type          ME  RMSE   MAE    MPE  MAPE   MASE  RMSSE  ACF1
  <chr>  <chr>       <dbl> <dbl> <dbl>  <dbl> <dbl>  <dbl>  <dbl> <dbl>
1 naive  Training 1.35e+ 0  5.85  4.54  0.849  3.48  1      1     0.797
2 mean   Training 1.35e-14 35.4  29.9  -7.80  25.1   6.58   6.04  0.950
3 drift  Training 6.20e-15  5.70  4.42 -0.245  3.44  0.974  0.973 0.797
4 drift  Test     4.55e+ 0  6.19  5.25  2.01   2.33  1.16   1.06  0.358
5 mean   Test     9.11e+ 1 91.1  91.1  40.7   40.7  20.0   15.6   0.126
6 naive  Test     8.6 e+ 0  9.09  8.6   3.83   3.83  1.89   1.55  0.126
```

The random walk with drift method had the smallest errors in all metrics in both the training and test data, suggesting that this model is the best among the alternatives. *Figure 10.14* shows a plot of the forecast values from the three methods along with the observed data to visually inspect them:

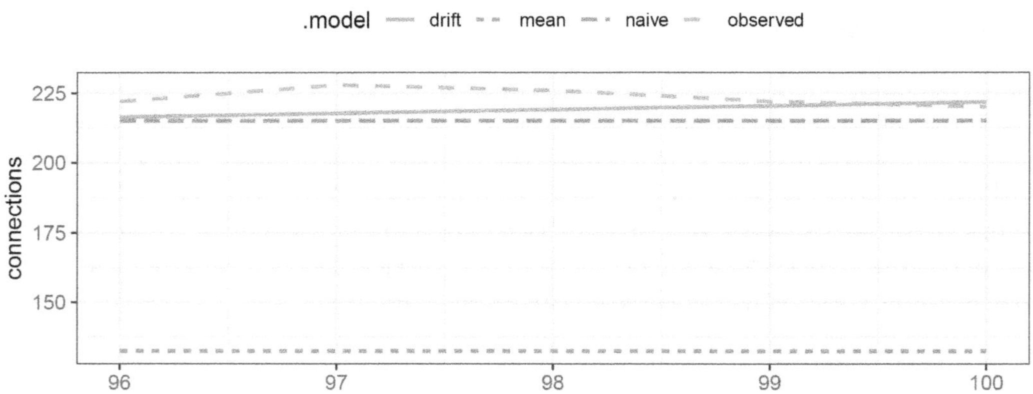


*Figure 10.14: Comparing the forecasting performance of naïve, average and random walk models on WWWusage data*

As expected, the RW method produces forecasts that are close to the observed values, and the mean and naïve methods had flat forecasts. You may choose the RW method as optimal for this series. However, be mindful that the drift term will continue to increase whereas observations in the test set changed direction.

Our chosen model may not produce accurate forecasts if the final model relies on the assessment of a single test set. Performing time series cross-validation is the answer to this problem. In cross-validation, forecasts are evaluated on multiple test sets. In the next section, you will see how to perform time series cross-validation and assess forecasting performance.

## 10.9 Time series cross-validation

Cross-validation is a standard method of performance comparison in any modeling framework. In cross-validation, observations are randomly allocated in cross-sectional data and corresponding models. However, for time series forecasting, preserving the time index is crucial during cross-validation due to temporal-dependency. A rolling window-based approach is used for time series cross-validation by expanding observations in the training set, allocating successive observations in the corresponding test set.

Using the `stretch_tsibble()` function, you can create multiple training sets for time series data. The options:

- `.init`: Requires that the number of observations belongs to the original training set
- `.step`: Specifies the increment in observations between successive training sets.

We will cross-validate the performance of the Random Walk with the drift model (RW) on WWWusage data. Even though this model performed better on one occasion, it is important to check how this model's forecasting performance is impacted when the data window changes.

To facilitate cross-validation, we first need to generate multiple training and test datasets, then fit the model on each training set, and generate a forecast for the corresponding test set. Finally, compute accuracy metrics are averaged over all test windows to evaluate the model's forecasting performance.

The WWWusage series has 100 observations. The first training set will have 60 observations and will increase by 5 observations in the successive training sets. The `stretch_tsibble()` function facilitates this operation in a single call, as shown in the following code snippet:

```
netuse_cv <- netuse |>
  stretch_tsibble(.init = 60, .step = 5) |>
  relocate(.id)
```

Now we have 9 training sets of expanding windows. You can see the number of observations included in each training set by organizing them in a nested object as follows:

```
netuse_cv |>
  group_by(.id) |>
  nest()
```

The preceding code will produce the following output, in which you can confirm the number of observations in each set:

```
# A tibble: 9 × 2
# Groups:   .id [9]
    .id data
  <int> <list>
1     1 <tbl_ts [60 × 2]>
2     2 <tbl_ts [65 × 2]>
3     3 <tbl_ts [70 × 2]>
4     4 <tbl_ts [75 × 2]>
5     5 <tbl_ts [80 × 2]>
6     6 <tbl_ts [85 × 2]>
7     7 <tbl_ts [90 × 2]>
8     8 <tbl_ts [95 × 2]>
9     9 <tbl_ts [100 × 2]>
```

For illustrative purposes, we use the random walk with a drift model. A model will be fitted on each training set, and 5-step-ahead forecasts will be generated. The whole operation is combined in the following code snippet. The final forecast object is transformed into a fable object and organized for simplified computation as shown in the following code:

```
nv_cv_fcst <- netuse_cv |>
  model(drift = RW(value ~ drift())) |>
  forecast(h = 5) |>
  group_by(.id) |>
  mutate(h = row_number()) |>
  ungroup() |>
  as_fable(response = "value", distribution = value)

nv_cv_fcst
```

You will see the following output:

```
# A fable: 45 x 6 [1]
# Key:     .id, .model [9]
     .id .model index        value .mean     h
   <int> <chr>  <dbl>       <dist> <dbl> <int>
 1     1 drift     61  N(122, 33)  122.      1
 2     1 drift     62  N(122, 66)  122.      2
 3     1 drift     63 N(123, 101)  123.      3
```

```
 4      1 drift     64 N(123, 137) 123.       4
 5      1 drift     65 N(124, 174) 124.       5
 6      2 drift     66   N(99, 33)  99.2      1
 7      2 drift     67   N(99, 66)  99.3      2
 8      2 drift     68 N(100, 101)  99.5      3
 9      2 drift     69 N(100, 137)  99.7      4
10      2 drift     70 N(100, 174)  99.9      5
# i 35 more rows
```

We have point forecasts (`.mean` column) from the 9 fitted models. The `accuracy()` function will compute the metrics for each horizon, and provide average performance over all test sets in a single line of code shows as follows:

```
nv_cv_acc <- nv_cv_fcst |>
  accuracy(netuse, by = c("h", ".model"))

nv_cv_acc
```

The code produces a summary table as output:

```
# A tibble: 5 × 11
      h .model .type    ME  RMSE   MAE   MPE  MAPE  MASE RMSSE  ACF1
  <int> <chr>  <chr> <dbl> <dbl> <dbl> <dbl> <dbl> <dbl> <dbl> <dbl>
1     1 drift  Test   3.10  7.33  6.53  1.63  4.86  1.44  1.26 0.521
2     2 drift  Test   5.19 14.7  12.6   1.65  9.23  2.78  2.54 0.498
3     3 drift  Test   5.92 18.4  15.5   1.41 11.7   3.41  3.17 0.502
4     4 drift  Test   6.51 20.3  16.7   1.88 12.9   3.70  3.50 0.487
5     5 drift  Test   9.73 23.6  19.6   4.29 14.4   4.33  4.07 0.458
```

The cross-validated RMSE and MASE are reliable measures to select a final model when multiple alternative forecasting models' performance is evaluated on a single series. From the cross-validated forecasting performance, we see that RW forecasting performance degraded as the forecast horizon increases irrespective of metrics.

The WWWusage data has a significant structural break near the end of the series, which affected the overall forecasting performance. This highlights the model's inability to adapt quickly to new, unobserved patterns based solely on older data.

The training sets in the cross-validation are shorter compared to the full dataset. It is important to note that shorter training sets can lead to higher variance in performance because the model has less data to learn long-term trends and complex patterns. The initial cross-validation metrics might overestimate the error rate compared to a model trained on a full dataset.

Time series cross-validation provides a robust way to evaluate performance across different time periods, especially how the model handles new data, closely simulating a real-world forecasting exercise.

# Summary

This chapter was a steppingstone in your time series forecasting journey. You applied the standard forecasting workflow; (visual) explanatory analysis on the observed data, splitting data into training and test sets, fitting forecasting models, evaluating the model fit with diagnostic checks, generating forecasts and evaluating forecasting performance using the forecast accuracy measures.

You applied three of the simplest forecasting methods: naïve, seasonal naïve and average from the detailed examples provided in this chapter. You also used **Time Series Linear Model** (**TSLM**) for forecasting, and learned how to extract time series components and use them as explanatory variables. A what-if scenario analysis with TSLM showed you the versatile use of this model in business decision-making. The relative advantages and limitations of each model will help you decide the appropriate application of each of them. The examples presented in this chapter showed you how to fit multiple models on a single series and generate accuracy metrics for them.

We discussed prominent forecast accuracy measures with their relative pros and cons. The chapter presented a detailed worked-out example of time series cross-validation, which is a robust technique to find an optimal model by generalizing forecasting performance.

In the next chapters, we will look at advanced statistical models that are specifically developed for time series forecasting. *Chapter 11* will discuss the exponential smoothing family of forecasting models, and *Chapter 12* will present ARIMA models.

# Further reading


- Hyndman, R. J., & Koehler, A. B. (2006). *Another look at measures of forecast accuracy. International Journal of Forecasting*, 22(4), 679–688.
- O'Hara-Wild, M., Hyndman, R., & Wang, E. (2021). *fabletools: Core tools for packages in the fable framework* (Version 0.3.2). `https://CRAN.R-project.org/package=fabletools`
- O'Hara-Wild, M., Hyndman, R., & Wang, E. (2023). *fable: Forecasting models for tidy time series* (Version 0.3.3).

# Join our community on Discord

Join our community's Discord space for discussions with the author and other readers:

`https://discord.gg/G4YmB6DQvf`

# 11

# Exponential Smoothing

Exponential smoothing is a simple, yet powerful time-series forecasting technique. The method was introduced in the late 1950s when computers were in their infancy. Still, this family of models effectively solves real-life forecasting problems, and is widely used in business decision-making in sales, inventory management, finance and so on.

In exponential smoothing, recent observations are more important to forecasts compared to values in the distant past. The weights attached to observations decay exponentially as the data points move further from the present, hence the name exponential smoothing. This model leverages trend and seasonality components of a time series along with various specifications for how the error components are incorporated.

In this chapter, we will first explore the fundamental theory behind exponential smoothing methods. It is important to gain a deep understanding of how the exponential smoothing model framework has evolved over time, so that you can choose the optimal model for your forecasting problem. There are plenty of coding examples and discussions presented in this chapter for you to learn from, which will allow you to generate accurate, actionable forecasts for decision-making.

The chapter is organized as follows:

- Section 11.1: Forecasting methods versus models
- Section 11.2: Exponential smoothing forecasting
- Section 11.3: Simple Exponential Smoothing (SES)
- Section 11.4: Holt's linear trend method
- Section 11.5: Holt-Winters' trend and seasonality method
- Section 11.6: Automated model selection using ETS()

# Technical requirements

Load the packages that are used in this chapter.

```
#install.packages("pacman")
library(pacman)
p_load("conflicted"
       ,"datasets"
       ,"fable"
       ,"fabletools"
       ,"Mcomp"
       ,"tidyverse"
       ,"tsibble"
       ,"tsibbledata" )

conflicts_prefer(
  fabletools::accuracy,
  fabletools::forecast,
  dplyr::filter )
```

The code file is available in the GitHub repo `https://github.com/PacktPublishing/Modern-Time-Series-Analysis-with-R.git` for you to download and practice.

# 11.1 Forecasting methods versus models

Initially, if exponential smoothing was used for forecasting, they were known as *methods* because the underlying statistical properties were not explicitly modeled. In 2002, the statistical models underlying each exponential method were introduced in the literature, known as *state space exponential smoothing models.*

The terms *forecasting methods* and *forecasting models* are often used interchangeably. The difference lies in the formulation of equations and the levels of detail.

Forecasting methods are simple, and involve minimal mathematical representation and computation, such as an average method or naïve method. On the other hand, forecasting models incorporate elaborate mathematical or statistical frameworks, which range from simple to complex computation.

Time series data is a combination of three main components: trend (or trend-cycle), seasonality and random fluctuations (errors). They could be combined in several ways in an observed series:

- Additive: $y = T + S + E$,
- Multiplicative: $y = T \times S \times E$,
- Additive trend and seasonality with multiplicative error: $y = (T + S) \times E$.

When time series characteristics were discussed in *Chapter 4*, we used $R$ to represent the random variations instead of $E$ as shown here. This convention is adopted from technical literature to distinguish decomposition of historical components from the development of active forecasting models. Theoretically, $R$ (Remainder) is preferred in decomposition contexts, where the random variation is treated as an identifiable component that can be extracted from an observed series using proper techniques, such as STL decomposition. In contrast, in forecasting and regression modeling, the error term $E$ (typically denoted as $\epsilon_t$) is used to represent the unexplained residual that is minimized through the estimation of model parameters.

The main differences between forecasting methods and forecasting models are based on two facts: the assumptions made about the error term ( $\epsilon_t$), and the estimation techniques. Let's elaborate on these differences.

Firstly, the forecasting method assumes no distributional characteristics of the error term in the forecasting process. In the case of the forecasting model, we explicitly assume that the error term follows a distribution, usually Gaussian with mean 0 and variance $\sigma^2$, i.e. $\epsilon_t \sim N(0, \sigma^2)$. Making explicit assumptions of error terms is the backbone of any statistical modeling. This difference is discussed in later sections.

Secondly, forecasting methods usually find optimal parameters by minimizing the sum of squared one-step forecasting error:

$$\sum_{t=1}^{n} \epsilon_t^2 = \sum_{t=1}^{n} \left(y_t - \hat{y}_{t|t-1}\right)^2 .$$

In contrast, parameters of forecasting models are estimated by maximizing the likelihood calculated from the data. The likelihood is the probability of the data arising from a specified model. The technique of estimating model parameters by maximizing the log-likelihood (equivalently, minimizing the negative log-likelihood) is known as **Maximum Likelihood Estimation** (**MLE**). When the error term in the model is assumed to follow a Gaussian distribution, the estimation technique is called the **Ordinary Least Squares Estimation** (**OLS**). The specific estimation technique of a likelihood depends on the assumed distribution of the error term and the parameters of a model. However, in most scenarios, the likelihood reduces to the **Sum of Squared Errors** (**SSE**) computed based on the assumed model and the observed data. For example, for a regression model, the parameters are estimated by maximizing the SSE:

$$\sum_{t=1}^{n} \epsilon_t^2 = \sum_{t=1}^{n} (y_t - \beta_0 - \beta_1 x_1 - \cdots - \beta_k x_k)^2$$

Forecasting models are much more complicated to estimate than regression models. Different software uses different estimation techniques and optimization algorithms. You may get slightly different answers from different software even though the same data and model specifications are fed into them.

Thirdly, purely procedural forecasting methods typically provide only point forecasts, while forecasting models provide both point forecasts and prediction intervals. However, the examples presented in *Chapter 10* showed prediction intervals for the forecasting methods, such as average, naïve, and seasonal naïve. The implementation of these methods computes prediction intervals by deriving the one-step ahead forecast standard deviation ($\widehat{\sigma_h}$) from the residual standard deviation ($\hat{\sigma}$) under the assumption of uncorrelated residuals. It is also possible to generate prediction intervals from bootstrapped residuals, a technique used for some forecasting methods and machine learning forecasting models in which explicit assumptions for error distribution are not required.

The distinction between forecasting methods and models can be illustrated using exponential smoothing families. Each member of the exponential smoothing family has two mathematical formula representations: a forecasting method and a corresponding forecasting model. The estimation technique also changes accordingly. We are going to explore the exponential smoothing family of forecasting in detail in the following sections.

## 11.2 Exponential smoothing forecasting

Between 1957 and 1960, exponential smoothing techniques were first introduced as forecasting methods. The formulas had no dedicated residual term and parameters were estimated by minimizing one-step-ahead forecast errors.

Hyndman et al. (2002) identified the statistical models underlying each exponential smoothing method. The models were presented in a state space framework with recursive equations that facilitate automatic forecasting for any horizons. The state space representation of models provides both point forecasts and prediction intervals.

The variations in exponential smoothing methods are predominantly based on which components are present and how they are combined in the data. We know that components can be combined in either an additive or multiplicative way.

A two-way classification of exponential smoothing methods was first proposed by Pegels (1969), considering the presence or absence of two components: trend ($T$) and seasonality ($S$). A pair of letters is used to define the presence of components, as well as how they are combined in the observed data. For example, $(A, A)$ stands for additive trend and additive seasonality. ($N$, $N$) indicates that neither of the components is observable in the data.

We have adapted the classification shown in *11.1*, from Hyndman et al. (2021), in which the methods based on multiplicative trend are excluded as they do not provide good forecasts in practice. Rather, variations on the additive damped trend ($A_d$) methods are considered. A "damping" is performed when there is a trend in the observed series that is less likely to persist for a longer period in the future.

There are nine possible combinations, and each of these exponential smoothing methods has its own name. However, the six methods in *Table 11.1* are the ones that are widely used in business forecasting:

| **Trend component** | **Seasonality: None ($N$)** | **Seasonality: Additive ($A$)** | **Seasonality: Multiplicative ($M$)** |
|---|---|---|---|
| None ($N$) | $(N, N)$: Simple exponential smoothing | $(N, A)$ | $(N, M)$ |
| Additive ($A$) | $(A, N)$: Holt's linear trend method | $(A, A)$: Additive Holt-Winters' method | $(A, M)$: Multiplicative Holt-Winters' method |
| Additive damped ($A_d$) | $(A_d, N)$: Additive damped trend method | $(A_d, A)$ | $(A_d, M)$: Holt-Winters' damped method |

*Table 11.1: A two-way classification of exponential smoothing methods*

If we assume the error term ($E$) is present in the data along with trend and seasonality, we could use three letters to concisely represent various scenarios. Taking $E$ for error, $T$ for trend, and $S$ for seasonality, the possible combinations are made up using the following sets:

- Error = $\{A, M\}$
- Trend = $\{N, A, A_d\}$
- Seasonality = $\{N, A, M\}$.

Each possible unique combination can be conveniently defined by ETS(.,.,.). For example, ETS$(A, N, N)$ represents the exponential smoothing model with additive errors and no trend or seasonality, whereas ETS$(M, A, N)$ represents the exponential smoothing model with multiplicative error, additive trend and no seasonality.

By making explicit assumptions about error distribution, we are transitioning from *exponential smoothing methods* to *exponential smoothing models*. The mathematical formulas are adapted from the state space representation proposed by Hyndman et al. (2002).

There are two automated implementations of exponential smoothing models in R:

- The `ETS()` function from the `fable` package is based on classifications included in Hyndman et al. (2008)
- The `ets()` function from the `forecast` package uses conventions from Hyndman et al. (2002) and Hyndman et al. (2008), which cover a wider variety of models, including multiple trend.

We use `ETS()` from the `fable` package in this book.

Let's take a deeper look at three prominent exponential smoothing methods and corresponding models with examples:

- Simple exponential smoothing (SES): Applicable to data with no trend and no seasonality,
- Holt's linear trend method (Holt, 1957): Applicable to data with trend and no seasonality, and
- Holt-Winters' method (Winters, 1960): Applies to data with trend and seasonality. The seasonality could be of any frequency but works better with smaller periods, such as, quarterly, or monthly.

We will discuss mathematical equations of the component form (without error) and corresponding state space models (with error component) for each of them. The discussions and examples will help you apply this knowledge to other models that are not covered in this chapter.

# 11.3 Simple Exponential Smoothing (SES)

SES (Brown, 1959) is the simplest exponential smoothing method. This forecasting method is applicable to a series that has no trend and no seasonal pattern. So, SES only provides an estimate of the level ($l$) in an observed data. While estimating the level, the associated smoothing parameter, $\alpha$, controls the rate at which the weights on past values decrease.

## 11.3.1 Component form of the (N, N) method

The component form has one smoothing equation to estimate level. The forecast equation uses the estimated level for all future values. The equations are as follows:

$$\text{Smoothing equation:} \quad l_t = \alpha y_t + (1-\alpha)l_{t-1} \quad (11.1)$$
$$\text{Forecast equation:} \quad \hat{y}_{t+h|t} = l_t \quad (11.2)$$

where, $0 \leq \alpha \leq 1$.

Parameter $\alpha$ in *Equation (11.1)* is estimated by minimizing one-step-ahead forecast errors. The forecast function is flat as the smoothed level in *Equation (11.2)* does not update as the forecast horizon, $h$, changes.

The state space representation of SES has two versions: ETS($A, N, N$) with additive errors and ETS($M, N, N$) with multiplicative errors. We discuss both of them in the following sub-sections.

## 11.3.2 SES with additive errors: ETS(A, N, N)

When the error component is additively combined with trend and seasonality components in the data, the one-step-ahead residuals can be obtained from the one-step-ahead forecast error:

$$\epsilon_t = y_t - \hat{y}_{t|t-1} = y_t - l_t$$

Now we make an explicit assumption about the residuals to follow a Gaussian distribution with mean $0$ and variance $\sigma^2$, $\epsilon \sim \text{NID}(0, \sigma^2)$.

Using the error distribution in a state space representation, the SES with additive errors has the following system of equations:

$$\text{Measurement equation:} \quad y_t = l_{t-1} + \epsilon_t \quad (11.3)$$
$$\text{State equation:} \quad l_t = l_{t-1} + \alpha\epsilon_t \quad (11.4)$$

where $\alpha$ is the smoothing parameter with condition, $0 \leq \alpha \leq 1$.

You may have noticed that a single error term is being used in both equations (11.3) and (11.4). For this reason, the system of equations is also known as a **single source of error** model or an **innovations state space model**.

The measurement *equation (11.3)* shows that the current observation ($y_t$) is linearly dependent on the previous level ($l_{t-1}$) and residual term, $\epsilon_t$. The state equation (11.4) shows how the current level changes from the previous period by smoothing the current error term using parameter $\alpha$, which controls the amount of smoothing. With a smaller $\alpha$ value, the level does not change with time, whereas a large $\alpha$ value ensures a drastic change.

Parameter $\alpha$ is estimated by minimizing the log-likelihood calculated from the observed series. The estimation process starts by taking the initial value of the parameter and finding the final estimate for which the log-likelihood is the maximum (or, equivalently, the negative log-likelihood is minimum). A user can also supply a value to initiate the estimation process.

For illustration, we will use the N0003 dataset from the M3 competition, available in the `M3comp` package. You can access any series by its number as shown in the following code:

```
M3$N0003
```

The output provides information about the series and training and test sets:

```
Series: Y3
Type of series: MICRO
Period of series: YEARLY
Series description: SALES ( CODE= ARC)

HISTORICAL data
Time Series:
Start = 1975
End = 1988
Frequency = 1
 [1] 1461.57 1692.50 2193.82 2459.68 3246.80 4748.86
 [7] 5559.46 5292.42 5029.40 4753.60 4344.60 2897.40
[13] 3256.40 3525.20

FUTURE data
Time Series:
Start = 1989
End = 1994
Frequency = 1
[1] 3070.2 3601.6 3407.4 3500.6 3437.8 3007.0
```

The N003 series has 14 observations in the training set and 6 in the test set. This might look like a very short series. However, it is common to forecast short annual series in many business scenarios. First, organize the series into tsibble format, as shown in the following code snippet. x denotes the training observations and xx denotes the test observations:

```
sesa_train <- as_tsibble(M3$N0003$x) |>
  rename(year = index, sales = value)

sesa_test <- as_tsibble( M3$N0003$xx) |>
  rename(year = index, sales = value)
```

Now plot the training series to visually examine it. The time plot in *Figure 11.1* presents the annual sales:

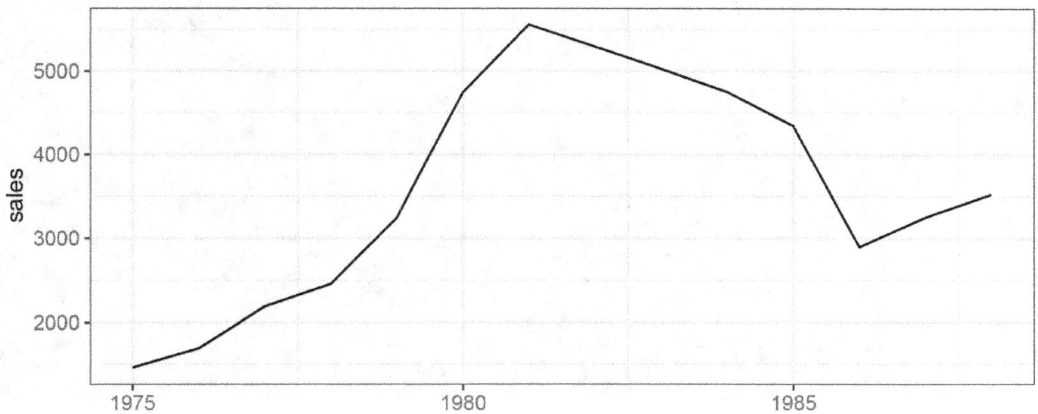


*Figure 11.1: Micro sales series N0003 (1975-1988) from M3 competition dataset*

There is no seasonality and no apparent trend in the annual sales in *Figure 11.1*. That's why the SES model with additive error, no trend, and no seasonality is an ideal choice for this data. A 5-7-year business cycle may be present, but for a short series like this, that will be hard to capture.

You can specify the respective components inside the `ETS()` function, as shown in the following code block. The function will fit the model, and the `report()` function prints the output in a tidy format:

```
sesa_fit <- sesa_train |>
  model(ses_additive = ETS(
    sales ~ error("A") + trend("N") + season("N")))

report(sesa_fit)
```

This code will give the following output:

```
Series: sales
Model: ETS(A,N,N)
  Smoothing parameters:
    alpha = 0.9998999

  Initial states:
     l[0]
 1461.183

  sigma^2:  549010

     AIC     AICc      BIC
225.8109 228.2109 227.7281
```

SES smoothed out most of the noise with a high smoothing parameter, $\hat{\alpha} = 0.99$. The data variance is 549,010. The information criteria, $AIC$, $AIC_c$, and $BIC$, provide information on overall model fit. A lower information criterion indicates a better fit. These metrics are widely used for comparing the relative quality of different statistical models for a given dataset.

For forecasting models, residuals are nothing but the one-step-ahead forecast errors. *Figure 11.2* presents the time plot of one-step-ahead forecasts errors, ACF and its distribution:

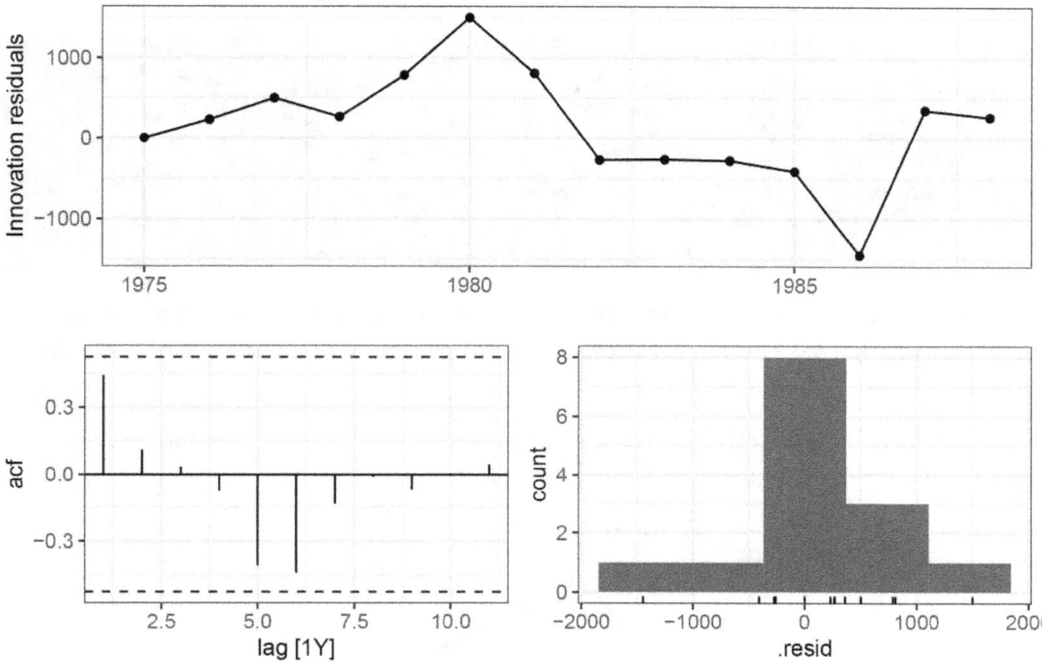


*Figure 11.2: Residual diagnostics from a SES model on the N0003 series*

There is no significant spike in the ACF plot; the histogram looks roughly Gaussian with zero mean. Apparently, the SES model fits quite well on the N0003 series.

Let's compute forecasts for the next six years and measure performance. In the following code, the forecast accuracy measures are computed in both the training and test sets:

```
sesa_fcst <- sesa_fit |>
  forecast(h = 6)

sesa_acc <-  bind_rows(
  accuracy(sesa_fit) ,
  accuracy(sesa_fcst,  bind_rows(sesa_train,sesa_test)))

sesa_acc
```

The accuracy measures will be printed in a tidy format, as follows:

```
# A tibble: 2 × 10
  .model       .type         ME  RMSE   MAE   MPE  MAPE  MASE
  <chr>        <chr>      <dbl> <dbl> <dbl> <dbl> <dbl> <dbl>
1 ses_additive Training  147.   686.  528.   4.36 15.1  0.929
2 ses_additive Test     -188.   290.  213.  -6.10  6.81 0.375
# i 2 more variables: RMSSE <dbl>, ACF1 <dbl>
```

The MASE and RMSE are lower in the test set compared to the training set. The test set MASE (0.375) tells us that SES showed better forecasting performance compared to a naïve model for this series.

Now we will combine the original series, the fitted values, and the forecasts, and plot them together for visual inspection presented in *Figure 11.3*:

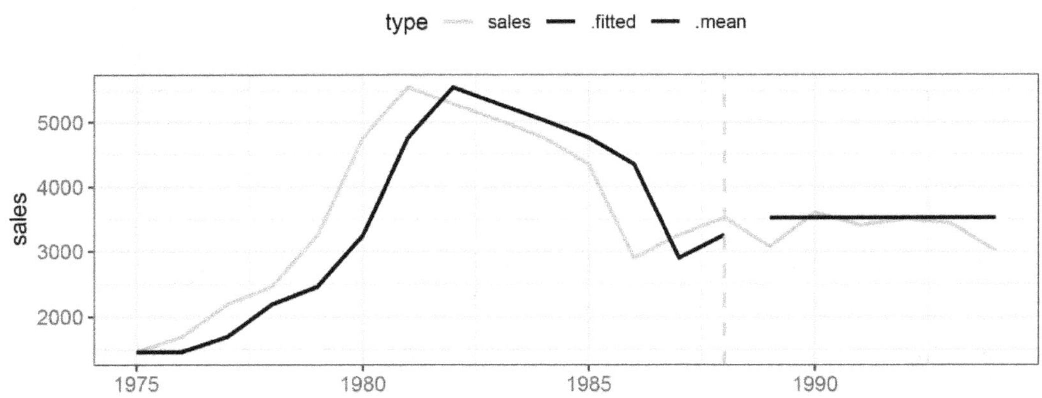


*Figure 11.3: Actual, fitted, and forecasts from an ETS(A,N,N) model on N0003 series*

In *Figure 11.3*, the fitted values (black solid line) seem to trail one step behind the original observations (grey solid line). This is because ETS(A,N,N) used a large smoothing value (0.99), which smoothed out errors from the most recent value. The forecast is a flat line, providing the estimated level of the observed series. In other words, SES uses the last observed value for $h = 1$ and the higher smoothing parameter kept almost the same information in the consecutive horizons.

This example demonstrated an application of the SES model on a series with additive errors. Next, we will discuss the model structure of SES with multiplicative errors.

## 11.3.3 SES with multiplicative errors: ETS(M, N, N)

The multiplicative error structure is observed when the variance of random noise changes as time increases. With multiplicative error, the one-step-ahead training residuals become the following:

$$\epsilon = \frac{y_t - \hat{y}_{t|t-1}}{\hat{y}_{t|t-1}}$$

where $\epsilon \sim \text{NID}(0, \sigma^2)$.

The corresponding state space form can be written as follows:

$$\begin{aligned} \text{Measurement equation:} \quad & y_t = l_{t-1}(1 + \epsilon_t) \\ \text{State equation:} \quad & l_t = l_{t-1}(1 + \alpha\epsilon_t) \end{aligned}$$

It is unlikely that a real-world business data would have no trend and no seasonality but have a multiplicative error structure. If you want to fit the multiplicative version of SES, you can do so by changing the letter representation in the model specification, as shown:

```
ETS(y ~ error("M") + trend("N") + season("N"))
```

We have discussed SES forecasting models for two error structures. The additive error version was illustrated with a detailed example. Let's see what some of the advantages and limitations of this model are:

- Advantages:
  - The SES method is suitable for short-term forecasting for series that have no trend and no seasonality. It is particularly effective when the underlying pattern is relatively smooth and does not exhibit complex fluctuations.
  - One of the main advantages of SES is its simplicity. It is easy to understand and implement with minimal computational resources. This makes it particularly efficient for forecasting large collections of individual series that exhibit no trend and no seasonal components.
  - SES allows for real-time forecasts as soon as a new data point is available. This feature makes it suitable for applications where the forecast needs to be continuously adjusted based on the latest information.

- Limitations:
    - SES is sensitive to the choice of initial values for the smoothing process. Different initial values could lead to different forecasts when subjective judgment is involved.
    - Assumptions of constant variance and lack of an explicit error model make SES a challenging option to assess the uncertainty around forecasts or to identify any systematic errors in the model.
    - SES is sensitive to outliers and usually underperforms for long forecast horizons.

Business processes are usually measured with some form of trend, even though they may not have seasonality. You should use Holt's linear trend exponential smoothing model for forecasting such data. The model is discussed in the next section.

# 11.4 Holt's linear trend method

Holt's linear trend method is widely used on data with a trend and no seasonality. The model assumes the trend increases or decreases in a linear fashion. The model is not suitable for a non-linear trend.

The method has two smoothing parameters: $\alpha$ for the level, $l_t$, and $\beta^*$ which is associated with the trend component, $b_t$.

## 11.4.1 Component form of Holt's linear trend method

The component form of this method is presented as follows:

$$\begin{aligned} \text{Forecast equation:} \quad & \hat{y}_{t+h|t} = l_t + hb_t \\ \text{Level equation:} \quad & l_t = \alpha y_t + (1-\alpha)(l_{t-1} + b_{t-1}) \\ \text{Trend equation:} \quad & b_t = \beta^*(l_t - l_{t-1}) + (1-\beta^*)b_{t-1} \end{aligned}$$

where * $l_t$ is the level of the series at time $t$ with $0 \leq \alpha \leq 1$ and * $b_t$ is the slope (trend) at time $t$ with condition $0 \leq \beta^* \leq 1$.

Like SES, Holt's linear trend model has two versions depending on how the error component is incorporated into the data. Let's use this model to forecast a series with additive errors.

## 11.4.2 Holt's linear trend model with additive errors: ETS(A, A, N)

For this model the one-step-ahead training residuals are as follows:

$$\epsilon_t = y_t - l_t - b_t$$

where $\epsilon \sim \text{NID}(0, \sigma^2)$.

The state space representation of Holt's linear trend model becomes the following:

$$\begin{aligned} y_t &= l_{t-1} + b_{t-1} + \epsilon_t \\ l_t &= l_{t-1} + b_{t-1} + \alpha\epsilon_t \\ b_t &= b_{t-1} + \alpha\beta^*\epsilon_t \end{aligned}$$

where $\beta = \alpha\beta^*$ is set in the third equation for simplicity.

For illustration, we will use the series no N0011 from the *M3comp* dataset, which is annual sales from 1975 to 1988. Let's extract the train and test set using the following code block:

```
hl_train <- as_tsibble(M3$N0011$x)
hl_test <- as_tsibble(M3$N0011$xx)
```

Let's examine the training set characteristics from the time plot presented in *Figure 11.4*:

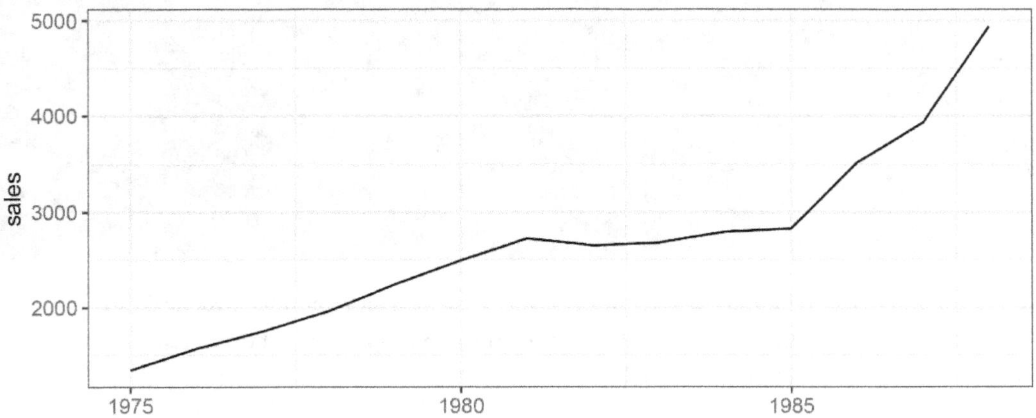


*Figure 11.4: Annual sales (code=GE); series N0011 from M3COMP dataset*

There was a strong upward trend in the annual sales with no seasonality. The error component is additive because the noise didn't fluctuate as time progressed. The trend followed a consistent linear change. These characteristics of this data make Holt's linear model a good choice for forecasting.

We fit Holt's linear model using the model specification, `ETS(A,A,N)` and use the fitted model to generate forecasts for the test set as shown in the following code:

```
hl_fit <- hl_train |>
  model(holt_additive = ETS(
    value ~ error("A") + trend("A") + season("N")))

hl_fcst <- hl_fit |>
  forecast(h = 6)

report(hl_fit)
```

The model provides estimated parameters, as well as the overall goodness of fit as presented in the following output:

```
Series: value
Model: ETS(A,A,N)
  Smoothing parameters:
    alpha = 0.7801431
    beta  = 0.7801431

  Initial states:
     l[0]     b[0]
 1195.501 135.7589

  sigma^2:  73647.98

     AIC     AICc      BIC
199.1349 206.6349 202.3302
```

Holt's linear model estimated moderate smoothing parameters for both level and trend: $\hat{\alpha} = 0.078$ and $\widehat{\beta^*} = 0.78$. That means the model places a higher weight on the most recent observation's error for level and trend.

To evaluate fitted model's performance, let's first inspect *Figure 11.5* that plots the full data (solid grey), the in-sample fitted values (dashed black), and out-of-sample forecast values (solid black):

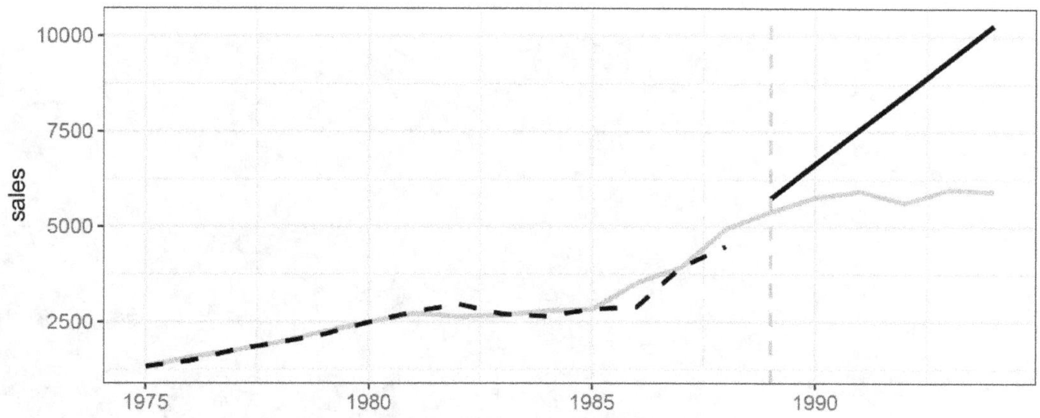


*Figure 11.5: Actual and forecasts values from Holt's linear additive error model on annual sales series (N0011 from M3Comp)*

As shown in *Figure 11.5*, Holt's linear model provides good in-sample fit for the N0011 series. However, the model failed to capture the plateaued effect of the trend as the forecasts increased in a linear fashion dominated by the trend and level estimated from the training data.

Next, we discuss Holt's linear model that assumes the error is multiplicative.

## 11.4.3 Holt's linear trend model with multiplicative errors: ETS(M, A, N)

Following the process of finding the one-step-ahead training residuals as the ratio of the components, the state space formula for the multiplicative Holt's linear trend method becomes the following:

$$\begin{aligned} y_t &= (l_{t-1} + b_{t-1})(1 + \epsilon_t) \\ l_t &= (l_{t-1} + b_{t-1})(1 + \alpha\epsilon_t) \\ b_t &= b_{t-1} + \beta(l_{t-1} + b_{t-1})\epsilon_t \end{aligned}$$

where $\epsilon_t = \frac{y_t - (l_{t-1} + b_{t-1})}{(l_{t-1} + b_{t-1})}$,

$\beta = \alpha\beta^*$ and $\epsilon \sim \text{NID}(0, \sigma^2)$.

Let's use the annual sodium silicate production (`N0195` series from M3Comp) data in the example. The following code shows how to fit a Holt's linear trend model with a multiplicative error after organizing training and test set:

The fitted model information is presented in the following output:

```
> report(hlm_fit)

Series: value
Model: ETS(M,A,N)
  Smoothing parameters:
    alpha = 0.9442556
    beta  = 0.0001000366

  Initial states:
     l[0]     b[0]
 4750.919 133.4805

  sigma^2:  0.0062

     AIC     AICc      BIC
666.9179 668.6322 675.4858
```

The model used a large smoothing parameter for the level (0.94) to smooth out the errors from the level to a greater degree. A small smoothing parameter for the trend (0.0001) indicates that the error measured with the trend required minimal smoothing.

We can visually evaluate forecasting performance of the fitted model from the plot presented in *Figure 11.6*:

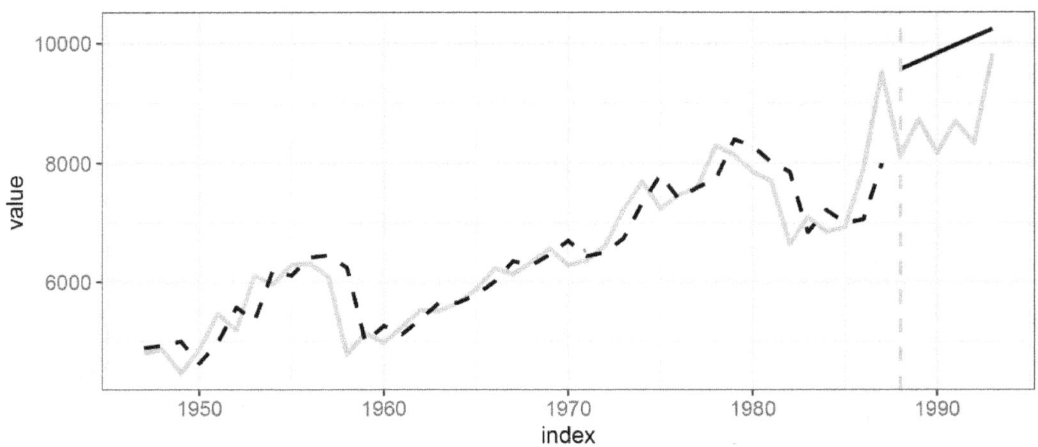


*Figure 11.6: Actual and forecast values from Holt's linear multiplicative error model on annual sodium silicate production (1947-1987)*

The training data (solid grey) has an upward trend with fluctuations (or, noise) around the trend increase as time progressed, indicating the suitability of Holt's linear model to handle the multiplicative error structure. The in-sample fit (dashed black) shows the good in-sample fit but the model was not good enough for forecasting. The forecast was dominated by the last observation and the associated large smoothing parameter for the level.

The benefits and limitations of Holt's linear model are summarized next:

- Advantages:
  - This method provides a separation between the level and trend components using two different smoothing parameters. This provides a way to measure the impact of each component on forecasts, as well as how to smooth them individually. This separation can be useful for analysis and decision-making.
  - The method is flexible and adaptable to changing trends over time.
- Limitations:
  - The method is sensitive to the choice of initial values. This is especially an issue when applied to short series.
  - The forecast equation allows either an increase (or, decrease) in trend indefinitely in the future. This may produce large forecast errors for longer forecast horizons.

When a forecast equation includes a one-directional estimated trend, that model is less likely to perform well for longer forecast horizons. The additive damped trend method has a mechanism to handle such scenarios. Let's explore that in the next section.

## 11.4.4 Additive damped trend method

Gardner et al. (1985) introduced a damping parameter to control the constant trend factor in Holt's linear trend method. The parameter $\phi$ $(0 \leq \phi \leq 1)$ dampens the constant trend to a flat line in the future. This damping parameter is incorporated in the set of equations in the component form as follows,

$$\begin{aligned} \text{Forecasting equation:} \quad & \hat{y}_{t+h|t} = l_t + (\phi + \phi^2 + \cdots | + \phi^h) b_t \\ \text{Level equation:} \quad & l_t = \alpha y_t + (1 - \alpha)(l_{t-1} + \phi b_{t-1}) \\ \text{Trend equation:} \quad & b_t = \beta^*(l_t - l_{t-1}) + (1 - \beta^*)\phi b_{t-1} \end{aligned}$$

Let's see how the damping parameter is incorporated in the state space representation of the additive damped trend model with additive error:

$$
\begin{aligned}
y_t &= l_{t-1} + \phi b_{t-1} + \epsilon_t \\
l_t &= l_{t-1} + \phi b_{t-1} + \alpha \epsilon_t \\
b_t &= \phi b_{t-1} + \beta \epsilon_t
\end{aligned}
$$

When $\phi = 1$, the method is identical to Holt's linear method. If $\phi < 0.8$ there is strong dampens impact on the short-term forecasts too. Empirical studies suggest restricting the value of $\phi$ in between $0.8$ to $0.98$.

The main advantage of Holt's damped trend method over Holt's linear trend method is the capability of bringing the trend component into a flat line in the distant future. In practice, this is a desirable property of a forecasting model. That's why the damped trend method works well in business contexts.

To illustrate the benefit of having a damping trend in forecasting model, we fit Holt's linear trend model, and damped trend with additive error model on the `N0011` series from the `M3Comp` dataset. We use cross-validation to find the best model by minimizing the one-step ahead forecast errors over the cross-validated test sets. You can fit the models, forecasts and compute accuracy metrics using the following code:

```
s11_cv <- hl_train |>
  stretch_tsibble(.init = 10) |>
  model(
    `Linear trend` = ETS(
      value ~ error("A") + trend("A") + season("N")),
    `Damped trend` = ETS(
      value ~ error("A") + trend("Ad") + season("N"))
  ) |>
  forecast(h = 1) |>
  accuracy(hl_train)

s11_cv
```

Check the output of the forecast accuracy metrics in the test data:

```
# A tibble: 2 × 10
  .model       .type     ME  RMSE   MAE   MPE  MAPE  MASE
  <chr>        <chr> <dbl> <dbl> <dbl> <dbl> <dbl> <dbl>
1 Damped trend Test   484.  594.  524.  11.5  12.8  1.82
2 Linear trend Test   539.  717.  602.  12.3  14.5  2.09
# i 2 more variables: RMSSE <dbl>, ACF1 <dbl>
```

The cross-validated forecasting performance shows that the damped trend model produced more accurate forecasts according to the RMSE, MAE, MAPE and MASE.

For visual inspection, *Figure 11.7* presents forecasts generated from the damped trend model and the linear trend model (without cross-validation):

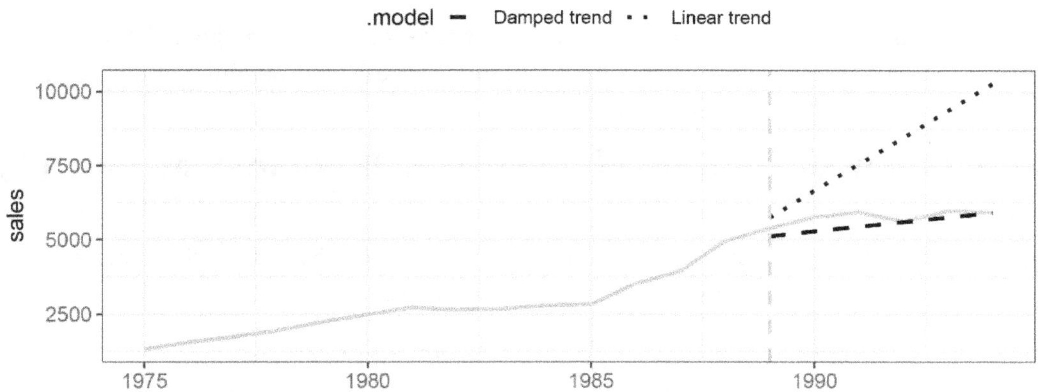


*Figure 11.7: Forecast performance between Holt's additive and damped additive models on series: N0011*

*Figure 11.7* shows that the damped trend model (dashed black) has a flat slope that produces better forecast compared to the Holt's linear trend model in the example data. This model works well in many common business scenarios where future growth is dampening. Unless there is strong indication that future growth will be linear, it is better to use the dampen trend instead of the linear trend model as far as accurate forecasting is concerned.

In the example, we let the `ETS()` function estimate the parameters automatically. You can also control the damping effect by passing a user defined value of phi within the trend (`"Ad", phi=?`) component if you have better knowledge about the future growth direction.

So far, we discussed SES and Holt's models that consider only trends and no seasonality. In each case, we presented the components form and corresponding state space representation to establish the relationship between forecasting methods and forecasting models. Next, we will look at that branch of exponential smoothing that handles both trend and seasonality in the observed series.

## 11.5 Holt-Winters' trend and seasonality method

The work of Holt (1957) and Winters (1960) allows us to capture seasonality in exponential smoothing and is also known as Holt-Winters exponential smoothing method. This extension incorporates three components: level ($l_t$), trend ($b_t$), and seasonal pattern ($s_t$), each with smoothing parameters $\alpha$, $\beta^*$, and $\gamma$, respectively. In component form, the equation for the seasonal component operates by smoothing the most recent observation (once its level and trend have been removed) with the corresponding seasonal observation from the prior year. As with previous models, the state space representation utilizes a single error term, $\epsilon_t$, associated with seasonal component; the smoothing parameter, $\gamma$, then determines the extent to which this error updates the seasonal estimates.

To simplify the discussion, we only present component forms for each method. The set of equations of corresponding state space representation for both additive and multiplicative errors are present in *Table 11.3* for reference.

### 11.5.1 Component form of additive Holt-Winters' method (A, A)

Additive Holt-Winters' method assumes additive trend and additive seasonality, represented by the combination $(A,A)$. The set of equations is as follows:

$$\begin{aligned}
\text{Forecasting equation:} &\quad \hat{y}_{t+h|t} = l_t + hb_t + s_{t+h-m(k+1)} \\
\text{Level equation:} &\quad l_t = \alpha(y_t - s_{t-m}) + (1-\alpha)(l_{t-1} + b_{t-1}) \\
\text{Trend equation:} &\quad b_t = \beta^*(l_t - l_{t-1}) + (1 - beta^*)b_{t-1} \\
\text{Seasonality equation:} &\quad s_t = \gamma(y_t - l_{t-1} - b_{t-1}) + (1-\gamma)s_{t-m}
\end{aligned}$$

### 11.5.2 Component form of multiplicative Holt-Winters' method (A, M)

The multiplicative Holt-Winters' method assumes additive trend and multiplicative seasonality with the following set of equations:

$$\begin{aligned}
\text{Forecasting equation:} &\quad \hat{y}_{t+h|t} = (l_t + hb_t)s_{t+h-m(k+1)} \\
\text{Level equation:} &\quad l_t = \alpha\frac{y_t}{s_{t-m}} + (1-\alpha)(l_{t-1} + b_{t-1}) \\
\text{Trend equation:} &\quad b_t = \beta^*(l_t - l_{t-1}) + (1 - beta^*)b_{t-1} \\
\text{Seasonality equation:} &\quad s_t = \gamma\frac{y_t}{l_{t-1} + b_{t-1}} + (1-\gamma)s_{t-m}
\end{aligned}$$

### 11.5.3 Component form of Holt-Winters' damped method ($A_d$, A)

The Holt-Winters' damped method assumes additive trend and seasonality and the damping parameter, $\phi$, attached to the trend only. The set of equations takes the following form:

$$\begin{aligned}
\text{Forecasting equation:} \quad & \hat{y}_{t+h|t} = [l_t + (\phi + \phi^2 + \cdots + \phi^h)b_t)]s_{t+h-m(k+1)} \\
\text{Level equation:} \quad & l_t = \alpha \frac{y_t}{s_{t-m}} + (1-\alpha)(l_{t-1} + \phi b_{t-1}) \\
\text{Trend equation:} \quad & b_t = \beta^*(l_t - l_{t-1}) + (1 - beta^*)\phi b_{t-1} \\
\text{Seasonality equation:} \quad & s_t = \gamma \frac{y_t}{l_{t-1} + \phi b_{t-1}} + (1-\gamma)s_{t-m}
\end{aligned}$$

In all the component forms, the parameters are bounded as follows:

- $\alpha$ is the smoothing parameter for level with $0 \leq \alpha \leq 1$
- $\beta^*$ is the smoothing parameter for trend with $0 \leq \beta^* \leq 1$
- $\gamma$ is the smoothing parameter for seasons with $0 \leq \gamma \leq 1 - \alpha$
- $k$ is the integer part of $\frac{h-1}{m}$

The set equations ensure the following:

- The estimates of the seasonal indices used for forecasting come from the final year of the sample, determined by $k$
- The level equation is the weighted average of the seasonally adjusted observations and the non-seasonal forecast for time $t$
- The trend equation gives the adjusted trend with the estimated level for time $t$
- The seasonal equation is the weighted index of the current seasonal index and the seasonal index of the same season in the prior year

You can review how each component updates by incorporating respective smoothing parameters when model framework makes explicit assumption of error, either additive or multiplicative, from the state space equations presented in *Table 11.3*. The same function, `ETS()`, is used to fit all the variations of Holt-Winters' model by passing the letter representing respective components when you fit these models on an observed data.

We will apply three variations of Holt-Winters' forecasting models, fitted on a seasonal data, such as quarterly Australian Gas production between 1956 and 2007. Using the following code, extract the gas production from the `aus_production` data and split into training and test set:

```
gas <- aus_production |>
  select(Quarter, Gas)|>
```

```
  filter_index("1956 Q1" ~ "2009 Q4")

gas_train <- gas |>
  filter_index("1956 Q1" ~ "2007 Q4")

gas_test <- gas |>
  filter_index("2008 Q1" ~ "2009 Q4")
```

Let's visually examine which components are present in the training data from the time plot presented in *Figure 11.8*:

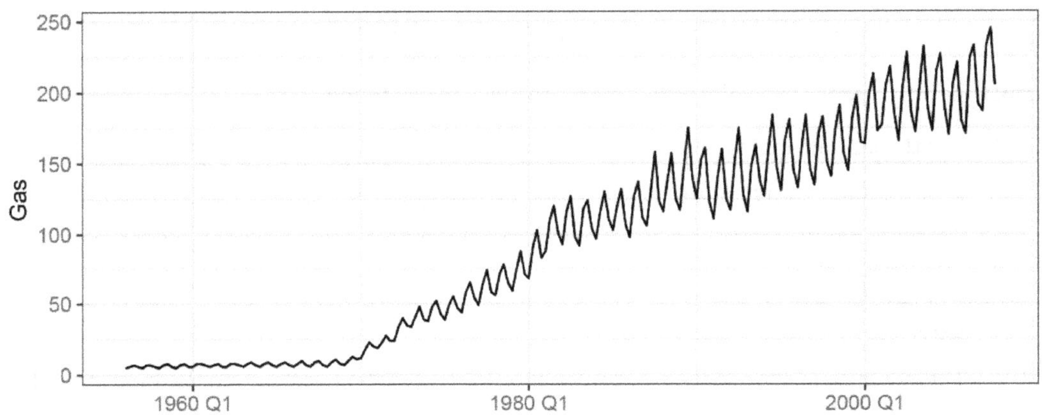


*Figure 11.8: Australian quarterly gas production (1956 Q1-2009 Q4)*

The quarterly gas production observed over 53 years has significant changes as time progressed. There is an upward trend that has a plateau effect in the last 10 years or so, multiplicative seasonality (the variations within a year changed over time). The error could be additive influenced by the strong multiplicative seasonality or could be multiplicative itself. For this data, the forecasting models worth trying are:

- ETS$(A, A, A)$: Model with additive error, trend, and seasonality
- ETS$(M, A, M)$: Model with multiplicative error, additive trend, and multiplicative seasonality
- ETS$(A, A_d, M)$: Model with additive error, damped additive trend, and multiplicative seasonality

Let's fit the three variations of Holt-Winter's models on the training set, generate forecasts for two years, and compute forecast accuracy metrics using the following code:

```
gas_fit <- gas_train |>
  model(
    `HW additive` = ETS(Gas ~ error("A") + trend("A") + season("A")),
    `HW multiplicative` = ETS(
      Gas ~ error("M") + trend("A") + season("M")),
    `HW damped` = ETS(Gas ~ error("A") + trend("Ad") + season("M"))
  )

gas_fcst <- gas_fit |>
  forecast(h = "2 years")

gas_acc <- accuracy(gas_fcst, gas)

gas_acc
```

The forecasting performance of the three models is presented in the following output:

```
# A tibble: 3 × 10
  .model       .type    ME  RMSE   MAE   MPE  MAPE  MASE
  <chr>        <chr> <dbl> <dbl> <dbl> <dbl> <dbl> <dbl>
1 HW additive Test  -4.52  5.44  4.52 -2.07  2.07 0.806
2 HW damped    Test  -2.56  3.88  3.05 -1.11  1.37 0.545
3 HW multipl… Test  -7.08  8.57  7.70 -3.09  3.41 1.38
# i 2 more variables: RMSSE <dbl>, ACF1 <dbl>
```

The out-of-sample accuracy metrics suggests that the Holt Winters' damped model showed the best forecasting performance by producing the smallest error.

*Figure 11.9* shows a plot of the last three years from the training set and the test set, along with forecasts from the three fitted models:

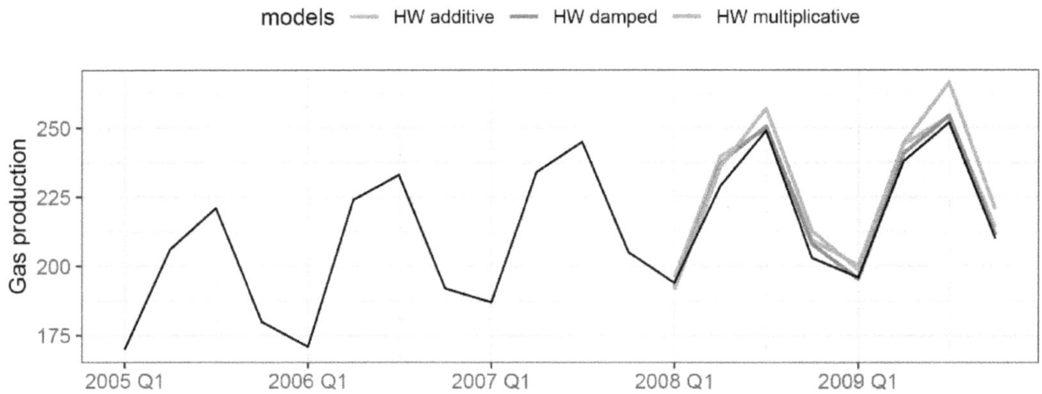


*Figure 11.9: Australian Gas production forecast using three variations of Holt-Winters' models*

According to the forecast accuracy measures and the visualization in *Figure 11.9*, the damped trend with multiplicative seasonality, ETS$(A, A_d, M)$, is the best model for this data.

Given a large pool of plausible models, how do you determine which Holt-Winters' model is suitable for your data? The summary presented in *Table 11.2* might help you to assess the components observed in the data and then choose an optimal forecast model for it:

| **Characteristics** | **Holt-Winter additive seasonality** | **Holt-Winter multiplicative seasonality** |
|---|---|---|
| How seasonality is incorporated | Constant across time | Changes proportionally with respect to the level of the series |
| How seasonal variation is expressed | In absolute terms on the scale of the observed series | In relative terms, that is, in percentages |
| How seasonal indexes are estimated within a year | Adds up to approximately zero | Approximately adds up to the total seasonal frequency, $m$ |

| Characteristics | Holt-Winter additive seasonality | Holt-Winter multiplicative seasonality |
|---|---|---|
| How seasonal adjustments are performed in the level equation | Subtracting the seasonal component | Dividing by the seasonal component |

*Table 11.2: Characteristics to consider while determining Holt-Winters' seasonal exponential smoothing methods*

Let's summarize the advantages and limitations of Holt-Winter's exponential smoothing models:

- Advantages:
  - The Holt-Winters' exponential smoothing method is designed to forecast series with trends and seasonality.
  - It can capture both the additive and multiplicative seasonality, which makes it a sought-after method in business for generating short-term forecasts.
  - The damped version allows the trend to accommodate any changes in growth in the future
- Limitations:
  - Like other exponential smoothing methods, the Holt-Winters' methods are also sensitive to initial value choices and outliers
  - They are not suitable for long-term forecasts
  - There are many smoothing parameters that need sufficiently large historical data for estimation

So far, we have{ XE "Holt-Winters' trend method:component form, damped method" } discussed nine exponential smoothing forecasting models that are widely used in business forecasting problems. These models are robust and usually produce accurate forecasts by capturing the dynamics present in an observed data. There are other versions of models available from the exponential family that we have not covered in this chapter. For advanced forecasters, the complete state space representation of all exponential smoothing models is presented in *Table 11.3*. We adapted this table from Hyndman et al. (2021).

| ADDITIVE ERROR MODELS | |
|---|---|
| **Trend** | **Seasonal** |

| ADDITIVE ERROR MODELS | | | |
|---|---|---|---|
| | **N** | **A** | **M** |
| **N** | $y_t = l_{t-1} + \epsilon_t$<br>$l_t = l_{t-1} + \alpha\epsilon_t$ | $y_t = l_{t-1} + s_{t-m} + \epsilon_t$<br>$l_t = l_{t-1} + \alpha\epsilon_t$<br>$s_t = s_{t-m} + \gamma\epsilon_t$ | $y_t = l_{t-1}s_{t-m} + \epsilon_t$<br>$l_t = l_{t-1} + \alpha\epsilon_t/s_{t-m}$<br>$s_t = s_{t-m} + \gamma\epsilon_t/l_{t-1}$ |
| **A** | $y_t = l_{t-1} + b_{t-1} + \epsilon_t$<br>$l_t = l_{t-1} + b_{t-1} + \alpha\epsilon_t$<br>$b_t = b_{t-1} + \beta\epsilon_t$ | $y_t = l_{t-1} + b_{t-1} + s_{t-m} + \epsilon_t$<br>$l_t = l_{t-1} + b_{t-1} + \alpha\epsilon_t$<br>$b_t = b_{t-1} + \beta\epsilon_t$<br>$s_t = s_{t-m} + \gamma\epsilon_t$ | $y_t = (l_{t-1} + b_{t-1})s_{t-m} + \epsilon_t$<br>$l_t = l_{t-1} + b_{t-1} + \alpha\epsilon_t/s_{t-m}$<br>$b_t = b_{t-1} + \beta\epsilon_t/s_{t-m}$<br>$s_t = s_{t-m} + \gamma\epsilon_t/(l_{t-1} + b_{t-1})$ |
| **A$_d$** | $y_t = l_{t-1} + \phi b_{t-1} + \epsilon_t$<br>$l_t = l_{t-1} + \phi b_{t-1} + \alpha\epsilon_t$<br>$b_t = \phi b_{t-1} + \beta\epsilon_t$ | $y_t = l_{t-1} + \phi b_{t-1} + s_{t-m} + \epsilon_t$<br>$l_t = l_{t-1} + \phi b_{t-1} + \alpha\epsilon_t$<br>$b_t = \phi b_{t-1} + \beta\epsilon_t$<br>$s_t = s_{t-m} + \gamma\epsilon_t$ | $y_t = (l_{t-1} + \phi b_{t-1})s_{t-m} + \epsilon_t$<br>$l_t = l_{t-1} + \phi b_{t-1} + \alpha\epsilon_t/s_{t-m}$<br>$b_t = \phi b_{t-1} + \beta\epsilon_t/s_{t-m}$<br>$s_t = s_{t-m} + \gamma\epsilon_t/(l_{t-1} + \phi b_{t-1})$ |

| MULTIPLICATIVE ERROR MODELS | | | |
|---|---|---|---|
| **Trend** | **Seasonal** | | |
| | **N** | **A** | **M** |
| **N** | $y_t = l_{t-1}(1 + \epsilon_t)$<br>$l_t = l_{t-1}(1 + \alpha\epsilon_t)$ | $y_t = (l_{t-1} + s_{t-m})(1 + \epsilon_t)$<br>$l_t = l_{t-1} + \alpha(l_{t-1} + s_{t-m})\epsilon_t$<br>$s_t = s_{t-m} + \gamma(l_{t-1} + s_{t-m})\epsilon_t$ | $y_t = l_{t-1}s_{t-m}(1 + \epsilon_t)$<br>$l_t = l_{t-1}(1 + \alpha\epsilon_t)$<br>$s_t = s_{t-m}(1 + \gamma\epsilon_t)$ |

| MULTIPLICATIVE ERROR MODELS | | | |
|---|---|---|---|
| **A** | $y_t = (l_{t-1} + b_{t-1})(1 + \epsilon_t)$<br>$l_t = (l_{t-1} + b_{t-1})(1 + \alpha\epsilon_t)$<br>$b_t = b_{t-1} + \beta(l_{t-1} + b_{t-1})\epsilon_t$ | $y_t = (l_{t-1} + b_{t-1} + s_{t-m})(1 + \epsilon_t)$<br>$l_t = l_{t-1} + b_{t-1} + \alpha(l_{t-1} + b_{t-1} + s_{t-m})\epsilon_t$<br>$b_t = b_{t-1} + \beta(l_{t-1} + b_{t-1} + s_{t-m})\epsilon_t$<br>$s_t = s_{t-m} + \gamma(l_{t-1} + b_{t-1} + s_{t-m})\epsilon_t$ | $y_t = (l_{t-1} + b_{t-1})s_{t-m}(1 + \epsilon_t)$<br>$l_t = (l_{t-1} + b_{t-1})(1 + \alpha\epsilon_t)$<br>$b_t = b_{t-1} + \beta(l_{t-1} + b_{t-1})\epsilon_t$<br>$s_t = s_{t-m}(1 + \gamma\epsilon_t)$ |
| **A$_d$** | $y_t = (l_{t-1} + \phi b_{t-1})(1 + \epsilon_t)$<br>$l_t = (l_{t-1} + \phi b_{t-1})(1 + \alpha\epsilon_t)$<br>$b_t = \phi b_{t-1} + \beta(l_{t-1} + \phi b_{t-1})\epsilon_t$ | $y_t = (l_{t-1} + \phi b_{t-1} + s_{t-m})(1 + \epsilon_t)$<br>$l_t = l_{t-1} + \phi b_{t-1} + \alpha(l_{t-1} + \phi b_{t-1} + s_{t-m})\epsilon_t$<br>$b_t = \phi b_{t-1} + \beta(l_{t-1} + \phi b_{t-1} + s_{t-m})\epsilon_t$<br>$s_t = s_{t-m} + \gamma(l_{t-1} + \phi b_{t-1} + s_{t-m})\epsilon_t$ | $y_t = (l_{t-1} + \phi b_{t-1})s_{t-m}(1 + \epsilon_t)$<br>$l_t = (l_{t-1} + \phi b_{t-1})(1 + \alpha\epsilon_t)$<br>$b_t = \phi b_{t-1} + \beta(l_{t-1} + \phi b_{t-1})\epsilon_t$<br>$s_t = s_{t-m}(1 + \gamma\epsilon_t)$ |

*Table 11.3: State space representation for each of the models in the ETS framework*

From the discussion and examples presented, you may have understood that the optimal model choice largely depends on what components are observed and how they are combined in the data. Fitting a model is the simplest task once an optimal model is identified by assessing the characteristics of observed data. Practically, it is not desirable to assess each series individually to determine the best forecasting model. In a modern business scenario, you might have to forecast thousands of series on a regular basis. So, how could you accomplish such a mammoth task in a short time?

You can leverage the automated model selection capability of the `ETS()` function. The function evaluates the characteristics in a time series and suggests an optimal model for that series. Let's discuss the automated model selection of the exponential smoothing family in the following section.

## 11.6 Automated model selection using ETS()

There are three core parts of a forecasting problem:

- Identify data characteristics
- Find a forecasting model that best suits the data and estimate the model
- Generate forecasts

The main problem lies in the first and second parts. What components are present in the observed series? How are they combined? What information are the estimated parameters providing? The model fits well on the training data, but why are the forecasts flat? And the questions go on.

The historical view held that optimal forecasting was more of an art than a science, as it is heavily reliant on subjective judgment. Even with the advancement of computation and technologies, identifying the best model to produce accurate forecasts is a balancing act between science and art. We should take advantage of automated processes as much as possible and allow human judgment to be strategically applied to model validation in a business context.

The `ETS()` function provides the functionality of automated model selection from the exponential smoothing family. Let's explain with an example. We will use `aus_livestock`, which has data on monthly sheep production in different Australian states and territories. When no model specification is passed into the `ETS()` function, it will find the optimal model for each series by assessing individual data characteristics, evaluating candidate models. You can apply the automated model selection as shown in the following code snippet:

```
auto_ets <- aus_livestock |>
  filter( Animal == "Sheep") |>
  model(ets = ETS(Count))
```

The code produces a mable object, as shown here:

```
> auto_ets

# A mable: 8 x 3
# Key:     Animal, State [8]
  Animal State                                ets
  <fct>  <fct>                            <model>
1 Sheep  Australian Capital Territory  <ETS(A,N,N)>
2 Sheep  New South Wales               <ETS(A,A,A)>
3 Sheep  Northern Territory           <ETS(A,Ad,N)>
4 Sheep  Queensland                   <ETS(A,Ad,A)>
5 Sheep  South Australia               <ETS(M,N,M)>
6 Sheep  Tasmania                      <ETS(M,A,M)>
7 Sheep  Victoria                     <ETS(M,Ad,M)>
8 Sheep  Western Australia             <ETS(A,N,A)>
```

As you see from the preceding output, different models have been selected for each of the series. Under the hood, a wide variety of plausible models were tested, and an optimal model has been identified that captures the existing data dynamics.

You can examine the estimated parameters from each model using the `report()` function as follows:

```
auto_ets |>
  filter(State == "Victoria") |>
  report()
```

The following output shows the full specification of the selected ETS model:

```
Series: Count
Model: ETS(M,Ad,M)
  Smoothing parameters:
    alpha = 0.766609
    beta  = 0.000163292
    gamma = 0.0004487799
    phi   = 0.9776673

  Initial states:
     l[0]      b[0]      s[0]      s[-1]     s[-2]
 931806.9 11970.24 0.6549046 0.8560137 0.9431726
    s[-3]    s[-4]    s[-5]    s[-6]    s[-7]    s[-8]
 1.130084 1.293061 1.276138 1.175191 1.254486 1.201338
   s[-9]    s[-10]   s[-11]
 0.86756 0.6886154 0.659436

  sigma^2:  0.0401
     AIC     AICc      BIC
15763.26 15764.53 15841.10
```

The `ETS()` function uses information criteria for model selection from the pool of plausible models. Information criteria is a measure that indicates how much information from the data is captured by a particular parameter set. It depends on the negative log-likelihood of a model and is penalized by the number of parameters used. Usually, three variations of information criteria are computed:

- **Akaike's Information Criterion (AIC)**: Taking $L$ as the likelihood of a model and $k$ as the total number of parameters and initial states (including the residual variance) that need to be estimated, the AIC for ETS models is defined as follows:

  $$\text{AIC} = -2log(L) + 2k$$

  The model with the minimum value of AIC often provides better forecasts.

- **Corrected Akaike's Information Criterion ($AIC_c$)**: AIC tends to select models with more parameters, which pose a problem for small data. Hence, the $AIC_c$, version is corrected for small sample bias and is defined as follows:

$$AIC_c = AIC + \frac{2k(k+1)}{n-k-1}$$

  The model with the lowest $AIC_c$ value should be selected.

- **Schwarz's Bayesian Information Criterion (BIC)**: The BIC (also known as SBIC or SC) is defined for ETS models as follows:

$$BIC = AIC + k[log(n) - 2]$$

Like the other two criteria, the minimum value of BIC is used for selecting models. The model chosen by BIC is either the same as what AIC could have selected or has fewer terms.

The steps involved in an automatic model selection routine are summarized next:

1. For each series, the `ETS()` function fits all models that are appropriate, optimizing the parameters of the models in each case.
2. The information criteria are computed for each fitted model. By default, `ETS()` uses $AIC_c$. Users can change the criterion by passing a different information criterion name in the ic argument.
3. The model with the minimum $AIC_c$ is chosen as the best model.

By default, `ETS()` uses log-likelihood for parameter estimation. Users can choose from other optimization criteria, such as mean square error (MSE), average MSE over first forecast horizons (AMSE), standard deviation of residuals (sigma), or mean absolute error (MAE).

Existing literature suggests that three models, ETS$(A, N, M)$, ETS$(A, A, M)$, and ETS$(A, A_d, M)$, suffer from numerical difficulties if input data has values close to zero. The `ETS()` function excludes these models during automated model selection. Multiplicative error models are also excluded when data contain zeros or negative values since they become numerically unstable.

Once the best exponential smoothing model is identified, you could use it to generate forecasts and calculate accuracy metrics following the coding examples presented in this chapter.

There are two limitations to ETS models:

- It cannot handle missing values. Missing values should be imputed or interpolated before fitting any ETS models.
- Exogenous regressors cannot be used in the modeling framework.

Despite these limitations, its simplicity compared to complex models, such as machine learning forecasting models, as well as its capability for producing short-term accurate forecasts, makes ETS a popular and ideal choice in many business forecasting scenarios.

# Summary

In this chapter, you have learned about prominent exponential smoothing forecasting methods and corresponding state space models. The SES method is applicable for data with no trend and seasonality. Holt's linear trend method is applied to a series with trend and no seasonal pattern. For data with both trend and seasonality, Holt-Winters' trend and seasonality methods are useful. The additive, multiplicative, and damped variations of the Holt's linear and Holt-Winters' models were also discussed. Each model was presented with mathematical equations, advantages, and disadvantages. The examples should have helped you gain thorough knowledge about how to apply the technique to your own forecasting problems. Finally, the chapter finished with an automated model selection routine of ETS, along with a brief discussion on information criteria.

In the next chapter, you will learn about ARIMA, another important statistical forecasting technique that is applicable to data with a persistent autoregressive structure.

# Further reading


- Brown, R. G. (1959). *Statistical forecasting for inventory control.* EngleWood Cliffs, NJ:Prentice-Hall.
- Gardner, E. S., & McKenzie, E. (1985). *Forecasting trends in time series. Management Science*, 31(10), 1237–1246.
- Holt, C. C. (1957). *Forecasting seasonal and trend by exponentially weighted averages.* In Reprinted in *International Journal of Forecasting* in 2004.
- Hyndman, R. J., & Athanasopoulos, G. (2021). *Forecasting: Principles and practice* (3rd edition). OTexts.
- Hyndman, R. J., Koehler, A. B., Ord, J. K., & Snyder, R. (2008). *Forecasting with exponential smoothing: The state space approach.* Springer.
- Hyndman, R. J., Koehler, A. B., Snyder, R., & Grose, S. (2002). *A state space framework for automatic forecasting using exponential smoothing methods. International Journal of Forecasting*, 18, 439–454.
- Pegels, C. (1969). *Exponential forecasting: Some new variations. Management Science*, 15(5), 311–315.
- Winters, P. R. (1960). *Forecasting sales by exponentially weighted moving averages. Management Science*, 6(3).

## Get this book's PDF version and more

Scan the QR code (or go to `packtpub.com/unlock`). Search for this book by name, confirm the edition, and then follow the steps on the page.

UNLOCK NOW

*Note: Keep your invoice handy. Purchases made directly from Packt don't require an invoice.*

# 12

# ARIMA Forecasting Models

The univariate **Autoregressive Integrated Moving Average** (**ARIMA**) models are classical statistical models that leverage the autocorrelation structure of a series. Introduced by Box et al. (1970), the ARIMA models were a revolutionary technique in time series analysis and forecasting during that era. The ARIMA model class includes a wide variety of individual models that have great applicability to data originating in many fields, such as finance, economics, retail, and supply chain management. In many business scenarios, ARIMA models are a good contender for generating accurate forecasts if the observed series is long enough and has sufficient autocorrelation structure that can be leveraged. The inferences provided by the ARIMA models and the insights from forecasts generated from them are crucial to making strategic business decisions.

As the benefits of the ARIMA models became evident in many areas, they were still deemed to be used by only experts, largely due to the subjectivity involved in the model selection process. ARIMA model selection required deep statistical knowledge, expertise in the corresponding subject matter, and a significant amount of subjective decision-making and intuition. There was a famous quote saying that ARIMA modeling was more art than science.

Over the decades, researchers developed various extensions and refinements to handle non-stationarity and seasonality, including statistical tests to bring objectivity to various parameter selections. The need for automation and simplification led to advancements in computational algorithms.

The automatic ARIMA algorithm (Hyndman et al., 2008) revolutionized the field by providing a systematic, automated approach to model selection and fitting. This algorithm has democratized the use of ARIMA models for practitioners with varying levels of expertise.

The automatic algorithm was first implemented as `auto.arima()` function in the forecast package. The fable package implements a variation of the Hyndman-Khandakar algorithm via the `ARIMA()` function. In this chapter, we will use the latter implementation in the discussion.

We will explore the following topics in this chapter:

- Section 12.1: Basics of ARIMA models
- Section 12.2: Autoregressive model (the AR in ARIMA)
- Section 12.3: Integration (the I in ARIMA)
- Section 12.4: Moving Average model (the MA in ARIMA)
- Section 12.5: General structure of an ARIMA model
- Section 12.6: ARIMA with exogenous variables (ARIMAX)
- Section 12.7: Steps in fitting an ARIMA model
- Section 12.8: Automatic ARIMA algorithm
- Section 12.9: ARIMA and forecast explainability



# Technical requirements

Let's load the packages that are required in this chapter.

```
#install.packages("pacman")
library(pacman)
p_load("conflicted"
       ,"datasets"
       ,"fable"
       ,"fabletools"
       ,"feasts"
       ,"patchwork"
       ,"tidyverse"
       ,"tsibble"
       ,"tsibbledata" )
```

```
conflicts_prefer(
  fabletools::accuracy,
  fabletools::forecast,
  dplyr::filter
)
```

The code file is available in the GitHub repo `https://github.com/PacktPublishing/Modern-Time-Series-Analysis-with-R.git` for you to download and practice.

# 12.1 Basics of ARIMA models

At its core, ARIMA is a regression model. In cross-sectional data, a regression model estimates the average relationship of the dependent variable $y$ on the set of independent variables, $x_1, x_2, \dots, x_k$. The relationship can be written mathematically as follows:

$$y_i = \beta_0 + \beta_1 x_{1,i} + \beta_2 x_{2,i} + \cdots + \beta_k x_{k,i} + \epsilon_i \qquad (12.1)$$

The estimated parameters from (12.1), $\hat{\beta}_1, \dots, \hat{\beta}_k$ measure the changes in $y$ for changes in the corresponding $x$'s. The constant term $\hat{\beta}_0$ estimates the average level of $y$ in the absence of any independent variables. If the dynamics in data cannot be fully captured by the variables and equations, then that "information" is captured by the term $\epsilon$.

To properly estimate Equation (12.1), two assumptions are critical. The first one is that the underlying distribution of each measured variable is consistent across time. That means the distributions are "stationary". The second one is that the remaining term $\epsilon$ should be a white noise series, $\epsilon \sim N(0,1)$.

The steps involved in estimating a regression model may involve (but are not limited to): transforming data to fit distributional properties, adding/deleting independent variables, or using a different formulation (linear, quadratic term), and so on. Do this in a repetitive fashion until $\hat{\epsilon} \sim N(0,1)$ is achieved.

A univariate time series has a different structure compared to cross-sectional data. A time series has temporal autocorrelation, which does not exist in cross-sectional data. On the other hand, whatever techniques are used for modeling time series should be statistically sound too.

In the previous chapter, exponential smoothing models were discussed. The exponential smoothing models use descriptions of trend and seasonality in time series, whereas ARIMA models the inter-dependency by capturing autocorrelation structure in the data.

ARIMA models are best suited for univariate time series data that is collected at regular intervals. They can capture data dynamics better when a long historical pattern is available and well suited for generating long-term forecasts. The ARIMA model has the capabilities to detect and capture seasonality quite well. It can also incorporate exogenous variables along with a forecast variable, a characteristic uncommon in other time series techniques.

ARIMA is a framework that includes a range of models, even though the term is used to denote an individual model or the full framework interchangeably. It consists of three core components. Each component is discussed in the following sections before putting them into a general structure.

## 12.2 Autoregressive model (the AR in ARIMA)

In practice, the observations $y_t$ are measurable. The autoregressive (AR) component determines how the current observation is related to past observations. This is analogous to *Equation (12.1)*, except the independent variables are replaced by the lagged values of the dependent variable. This particular relationship takes the following mathematical form:

$$y_t = c + \phi_1 y_{t-1} + \phi_2 y_{t-2} + \cdots + \phi_p y_{t-p} + \epsilon_t, \qquad (12.2)$$

where $c$ is the constant term and $\epsilon_t$ is white noise.

*Equation (12.2)* is known as Autoregressive model and is denoted as $AR(p)$.

The AR model is a powerful tool for time series analysis and forecasting due to their ability to capture temporal dependencies where past values influence future values. Finance, economics, econometrics, signal processing and control systems, and environmental data, such as pollution and temperature are a few scenarios where AR models are widely used to capture temporal dependencies in data.

The steps involved in AR model selection are as follows:

- Generate a time plot, autocorrelation function (ACF) and partial autocorrelation (PACF) plot of the series.
- Select the order $p$ by studying the patterns in PACF and ACF. For example, the number of spikes outside of the threshold in the PACF plot indicates the possible order of the AR terms. Generally, in PACF, you will notice a sharp cutoff after lag $p$ in an $AR(p)$ process. The significant spikes in ACF provide supporting evidence for determining the AR order. In the ACF plot, the oscillation of the spikes around $0$ indicates the signs of AR terms.
- Fit the AR model.

Let's explain the steps with a simulated series. The `arima.sim()` function can be used for generating a simulated series with known parameter values. First, generate an $AR(1)$ series with the AR parameter, $\phi = 0.8$. The random noise is sampled from a white noise series with $0$ mean and $1$ standard deviation using the following code:

```
set.seed(345876)
ar1_sim <- tibble(
  time_index = 1:120,
  values = c(arima.sim(model = list(ar=0.8),
                       n = 120,
                       mean = 0,
                       sd = 1))) |>
  as_tsibble(index = time_index)
```

Let's assume that you do not know how the simulated series is generated. Analyze it as you would do with real life data, oblivious of the underlying data generating process. Generate a time plot, ACF, and PACF plot to "guess" the order of an AR model using the `gg_tsdisplay()` function as shown in the following code:

```
ar1_sim |>
  gg_tsdisplay(plot_type="partial")
```

The preceding code generates *Figure 12.1*:

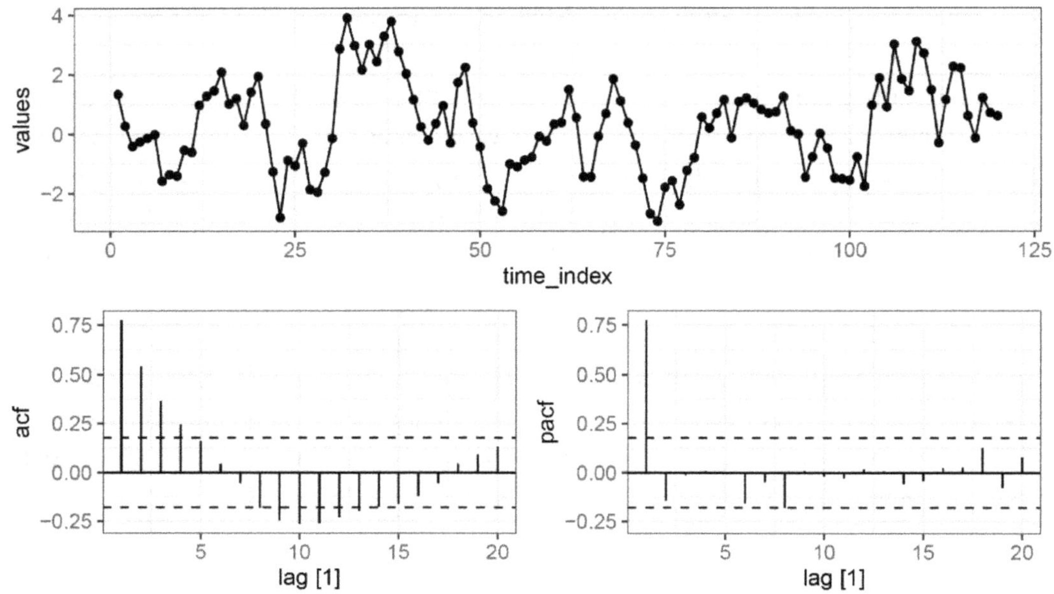


*Figure 12.1: Line plot, ACF, and PACF of a simulated series generated from an AR(1) process*

There is a sharp decline in the spikes after lag 1 in the PACF plot, suggesting an $AR(1)$ model. The ACFs are decaying slowly with four significant spikes outside of the upper limit threshold. This could be due to the lagging effect of strong inter-dependence on consecutive past values.

Next, we estimate the AR(1) model for simulated data using the `ARIMA()` function from the `fable` package.

To manually specify the order of an AR model for a non-seasonal series, pass the value of $p = 1$ inside the `ARIMA()` function, and set the other two parameters $d$ and $q$ to 0 as shown:

```
ar_manual <- ar1_sim |>
  model(manual = ARIMA(values ~ pdq(1,0,0)))

report(ar_manual)
```

Extract the model fit information using the `report()` function:

```
Series: values
Model: ARIMA(1,0,0)

Coefficients:
         ar1
       0.780
s.e.   0.056

sigma^2 estimated as 0.9266:  log likelihood=-165.66
AIC=335.33   AICc=335.43   BIC=340.9
```

The `tidy()` function extracts the model's coefficient in a tidy format:

```
> tidy(ar_manual)

# A tibble: 1 × 6
  .model term  estimate std.error statistic  p.value
  <chr>  <chr>    <dbl>     <dbl>     <dbl>    <dbl>
1 manual ar1      0.780    0.0560      13.9 7.51e-27
```

An $AR(1)$ model is fitted on the series as instructed, providing estimates of the AR parameter, $\hat{\phi}_1 = 0.78$. Even though the overall model fit is good ($AIC_c = 335.43$), the estimated AR parameter is close to what was used in the simulated series.

There are certain constraints imposed on the parameters of AR models to ensure stationarity. For an $AR(1)$ model, the parameter is bounded between $-1$ and $1$, that is, $-1 < \phi < 1$. For an $AR(2)$ model, the two parameters need to satisfy three conditions:

- $-1 < \phi < 1$
- $\phi_1 + \phi_2 < 1$
- $\phi_2 - \phi_1 < 1$

You can imagine that for higher-degree AR models, the parameter restrictions would be more complex. However, the automated AR implementation takes care of all the restrictions imposed on parameter space.

These conditions on parameters restrict AR models to being used only on stationary data. This requirement may seem counterintuitive because most real-world time series data is not stationary; it often integrates components such as trends and seasonality. Therefore, we need a mechanism to transform an "integrated" time series into a stationary series before applying an autoregressive model. This mechanism is the second component of the ARIMA modeling framework and is discussed in the following section.

# 12.3 Integration (the I in ARIMA)

Let's revisit the assumption of the multivariate regression model in *Equation (12.1)*. We know that one critical assumption for estimating parameters using a statistical technique is the distribution of each variable $(y, x_1, x_2, \ldots x_k)$, should remain the same from sample to sample. In other words, their statistical properties are "stationary" (not changing) even though actual observations may change.

This is in stark contrast to time series data. For instance, if you sample a monthly time series with a positive trend, the average in May would be lower than the average in June, which would be lower than the average in July. The statistical properties (i.e., mean) change over time. Alternatively, we can say the statistical properties depend on the time period. This phenomenon is known as non-stationarity.

Time series data is, in general, non-stationary as the underlying distribution changes due to the trend and seasonality over time. A change in variance also makes a series non-stationary. Some purely cyclical time series can be stationary because their mean and variance remain constant over time, even if the length of their cycles varies.

So, what is the problem modeling a non-stationary series? As the mean, variance, and autocorrelation structure changes, the parameters of a model would be unstable and inconsistent throughout the data. You may model spurious correlations simply because the surface pattern could hide the actual inter-dependency between observations. The model identification would be complex, hard to interpret, and above all, the models would have limited predictive power, which could lead to poor forecasting performance.

Fundamentally, you cannot confidently estimate parameters from non-stationary data. A non-stationary time series must be transformed into a stationary series before any parameter estimation takes place.

The solution to the problem is quite simple. This method is known as differencing (the opposite of integration).

- **Differencing**$(d)$: In a non-seasonal series, differencing means to subtract subsequent observations from the previous one.
- **Seasonal differencing**$(D)$: When subtraction takes place between an observation and the observation from the same season in the previous years.

The parameters, $d$ or $D$, denote the degree of differencing required to make a non-stationary series a stationary one, and take different values based on the structure of the data. Usually, $d \in \{0,1,2\}$ and $D \in \{0,1\}$.

For a stationary series $d = 0$ and $D = 0$.

First-order differencing takes place when the immediate previous observation is subtracted from the current one. This could substantially reduce (or eliminate) trends in data.

$$\begin{aligned} \text{First-order differencing for non-seasonal series } (d = 1)\text{: } y_t{}' &= y_t - y_{t-1} \\ \text{First-order differencing for seasonal series } (D = 1)\text{: } y_t{}' &= y_t - y_{t-m} \end{aligned}$$

where $m$ is the seasonal period.

A second-order difference is performed when consecutive observations are subtracted in the first-order difference series. Second-order differencing is required when the first-order differenced series has a substantial trend component in it.

Second-order differencing for non-seasonal series $(d = 2)$ is as follows:

$$\begin{aligned} y_t{}'' &= y_t{}' - y_{t-1}{}' \\ &= (y_t - y_{t-1}) - (y_{t-1} - y_{t-2}) \\ &= y_t - 2y_{t-1} + y_{t-2} \end{aligned}$$

Business data rarely takes more than second-order differencing to transform it into a stationary series. Too many differences introduce unexpected dynamics in data, which then becomes harder to model. ARIMA modeling takes only the integer values of the differencing parameters. Some advanced time series models do take a fractional differencing order. Multiple functions across many R packages are available to perform differencing. We'll use the `difference()` function from the `tsibble` package. The lag argument should always be $1$ for non-seasonal data. For seasonal differencing, the lag is equal to the seasonal periods.

In the following example, the annual USA GDP is used to show how to apply differencing on non-seasonal data. Load the `global_economy` dataset in the `tsibbledata` package, organize and compute first- and second-order differenced series, as shown in the following code:

```
data("global_economy")
usa_gdp <- global_economy |>
  filter(Code == "USA") |>
  select(Year, GDP) |>
  rename( year = Year, gdp = GDP) |>
  mutate(gdp = gdp/10^10,
         gdp_diff_1 = difference(gdp, lag = 1, differences = 1),
         gdp_diff_2 = difference(gdp, lag = 1, differences = 2))

usa_gdp
```

The result is shown in the following output:

```
# A tsibble: 58 x 4 [1Y]
    year   gdp gdp_diff_1 gdp_diff_2
   <dbl> <dbl>      <dbl>      <dbl>
 1  1960  54.3      NA         NA
 2  1961  56.3       2         NA
 3  1962  60.5       4.18       2.18
 4  1963  63.9       3.35      -0.830
 5  1964  68.6       4.72       1.37
 6  1965  74.4       5.79       1.07
 7  1966  81.5       7.13       1.34
 8  1967  86.2       4.67      -2.46
 9  1968  94.2       8.08       3.41
10  1969 102.        7.74      -0.340
# i 48 more rows
```

For illustration, you can plot the original and the differenced series for visual comparison. *Figure 12.2* shows the raw data, first- and second-order differenced series of USA GDP from 1960 to 2017:

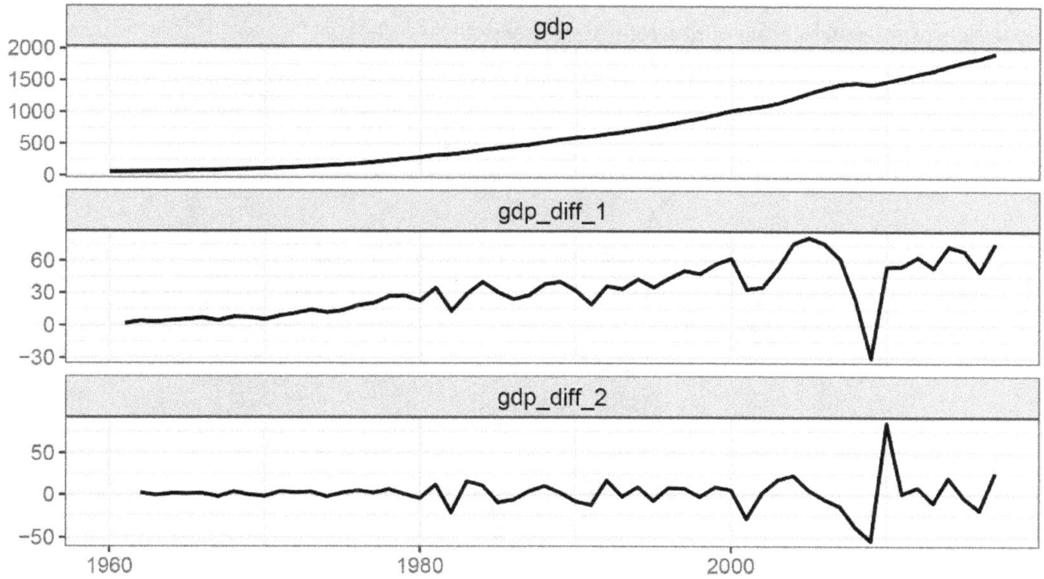


*Figure 12.2: Impact of non-seasonal differencing on USA GDP (1960-2017)*

In *Figure 12.2*, the top panel shows the non-stationarity in the actual data, as it had an upward trend, the that is, the mean/level of the series changed over time. The first difference series in the middle panel has reduced the trend but could not eliminate it completely. Also notice that the outlier in 2019, which was not apparent in the raw series before. In the bottom panel, the second-order differenced series showing the mean of the series has stabilized. The mean would be roughly `0`, irrespective of where you sample the data.

The variance of the USA GDP series has not been stabilized even after taking the second-order difference. In such scenarios, taking either log or Box-Cox transformation before differencing, could help.

Now the question is how do you objectively know whether differencing is even required? Or, if needed, which order would be the optimal one?

We've also discussed several statistical tests to identify unit roots in a time series in this book. Automated ARIMA uses the Kwiatkowski-Phillips-Schmidt-Shin (KPSS) test (Kwiatkowski et al., 1992) to first check whether non-seasonal data is stationary or not. If data turns out non-stationary, then it determines the order of differencing by repetitive application of the KPSS test on differenced data.

For a seasonal series, the order of seasonal differencing is determined by evaluating the seasonal strength of an STL decomposition. The `unitroot_ndiffs()` and `unitroot_nsdiffs()`

functions are used to identify the order of non-seasonal and seasonal differencing by applying KPSS tests or measuring seasonal strength, respectively.

Let's use the `unitroot_ndiffs()` function on the USA GDP and $log(GDP)$ to determine the order of differencing in an automated way:

```
usa_gdp <- usa_gdp |>
  mutate(gdp_log = log(gdp))

usa_gdp_diff <- usa_gdp |>
  select(year, gdp, gdp_log) %>%
  pivot_longer(cols = c(gdp, gdp_log),
               names_to = "series",
               values_to = "values") |>
  group_by(series) |>
  features(values, unitroot_ndiffs)

usa_gdp_diff
```

The following output will be produced:

```
# A tibble: 2 × 2
  series  ndiffs
  <chr>    <int>
1 gdp          2
2 gdp_log      2
```

The `unitroot_ndiffs()` function reports $d = 2$ both the original and the $log(GDP)$ series. You might think that a log transformation is redundant here. We will further explain the necessity of both log transformation and double-differencing to make the series stationary.

Generate a time plot, ACF, and PACF of the second-order differenced USA GDP series using the following code:

```
usa_gdp |>
select (year, gdp_diff_2) |>
gg_tsdisplay(plot_type = "parital") |>
labs(x ="" , y = "")
```

The code output is shown in *Figure 12.3*:

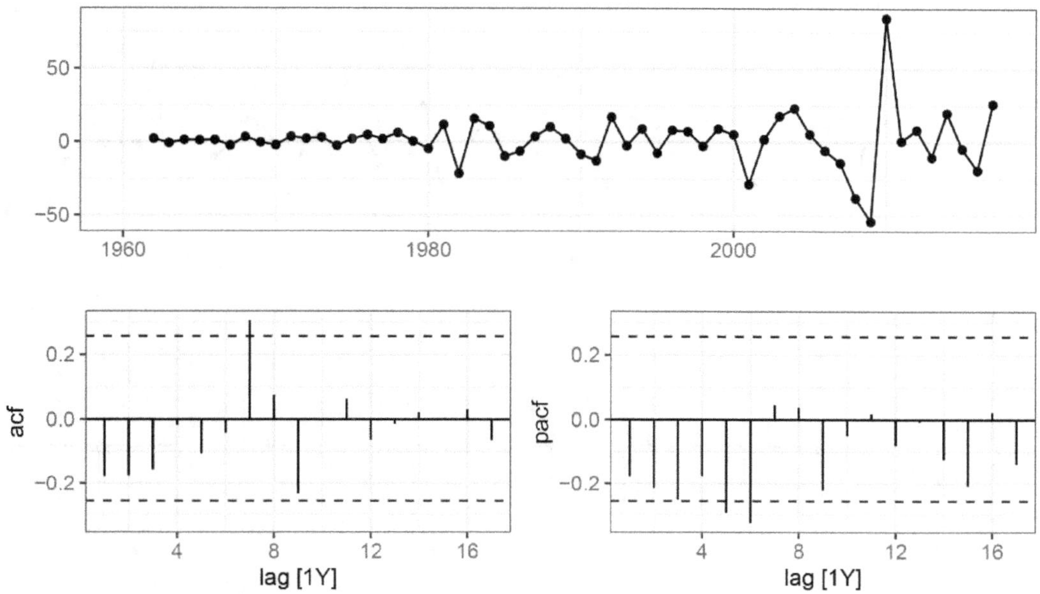


*Figure 12.3: Visualization of second-order differenced USA GDP series*

The second-order differencing makes the observations oscillate around 0. This indicates that the mean of the series has been stabilized. However, the variance of the series is not stabilized, yet evident by the large fluctuations at the end of the series. In the PACF plot, the spikes slowly increased up to lag 6. The ACF has a spike at lag 6. These suggest that the current observation depends strongly on the distant observations compared to the closest ones. This is an indication that something is still masking the actual pattern in the data.

Now, compute the second-order difference of the $log(GDP)$ and plot the ACF, and PACF of it using the following code:

```
usa_gdp <- usa_gdp |>
  mutate(gdp_log_diff_2 = difference(gdp_log, lag = 1, differences = 2))

usa_gdp |>
  select(year, gdp_log_diff_2)|>
  gg_tsdisplay(plot_type = "partial") +
  labs(x = "", y="")
```

The code will produce the plots presented in *Figure 12.4*:

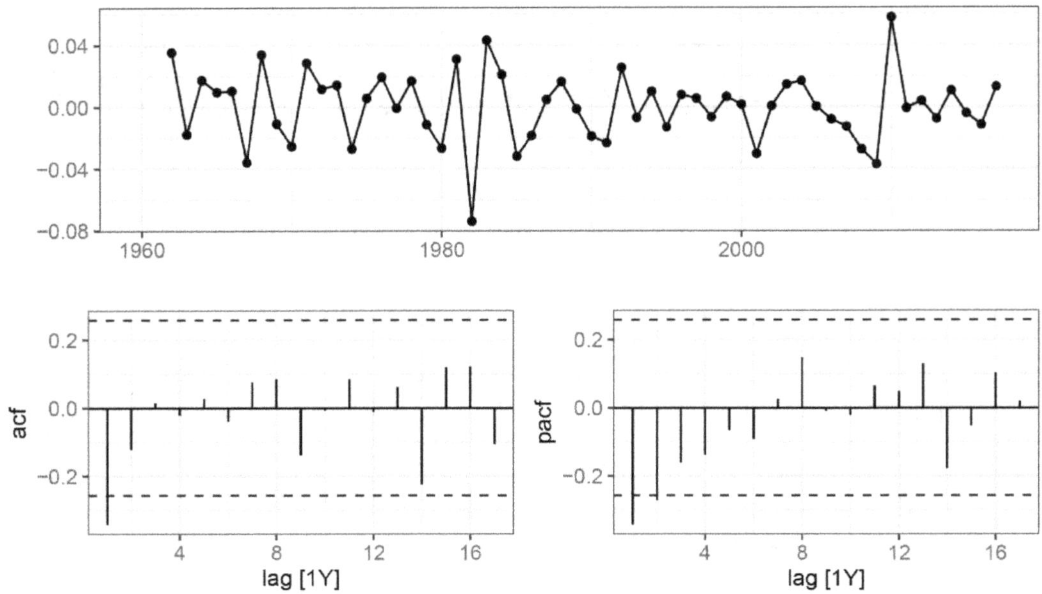


*Figure 12.4: ACF and PACF of second-order differencing on log(GDP)*

The time plot in *Figure 12.4* shows that after double-differencing, the $log(GDP)$ series has become mean and variance stationary. The PACF has two significant negative spikes at lag 1 and 2, and declined sharply after lag 2. The ACF has one negative significant spike at lag 1. Both plots suggest that the second differenced $log(GDP)$ can be modeled by two AR terms.

The `unitroot_ndiffs()` function performs differencing order selection in an automated fashion. It is still on the user's side to identify whether any transformation is needed to stabilize the variance or not.

Now you know how to handle the "integration" part in an observed time series by using proper differentiation. In the next section, we will discuss the third component, Moving Average (MA), of an ARIMA model.

# 12.4 Moving Average model (the MA in ARIMA)

In the time series context, we assume that the current observation is a mix of actual measurement and some "noise" denoted by $\epsilon$. In *Chapter 9*, we discussed **Exponential Weighted Moving Average** (**EWMA**) smoothing and filters based on MA.

Techniques like EWMA and MA filters reduce noise from data. On the other hand, an MA model captures and estimates the dependency pattern of a variable $y$ through any information available in its noise. The sources of noise in a time series could be measurement errors, system imperfections, or errors made during data processing.

If the observation $y_t$ depends on the past $q$ terms of its noise, then the model is called an $MA(q)$ model and written as:

$$y_t = c + \epsilon_t + \theta_1 \epsilon_{t-1} + \theta_2 \epsilon_{t-2} + \cdots + \theta_q \epsilon_{t-q}, \qquad (12.3)$$

where $\epsilon_t$ is a white noise process, that is, it follows a Gaussian distribution with $0$ mean and $1$ standard deviation.

It is perhaps counterintuitive that stochastic noise, though unmeasurable, can be accurately modeled and estimated. In hindsight, it is the smartness of the ARIMA model. In real life, it is not possible to measure any data-generating process without error. It is also not feasible to build a perfect model to capture all the inter-dependencies in data. There will be some "information" left over in the error series.

When the AR terms are sufficient to capture the majority of the data dynamics, the leftover information flowing to the errors is minimal, and no MA term is required. When the autocorrelation in a variable $y_t$ is weak (AR terms are not sufficient to capture inter-dependency), then more information ends up in the error series, which is then captured by the MA terms. In such situations, inter-dependency in the errors provides better information about the future path than the dependency structure in the series itself.

Let's explain how to estimate MA terms in an ARIMA model with a non-seasonal simulated series with weak autocorrelation ($\phi_1 = 0.12$) and strong dependency on the two neighboring error terms ($\theta_1 = 0.8$ and $\theta_2 = 0.45$). Use the following code to generate the simulated series and produce model order identification plots:

```
n <- 120
set.seed(98765)
ma_sim <- tibble(
  time = 1:n,
  y = c(arima.sim(
    n = n,
    model=list(ar = 0.12, ma = c(0.8,0.45)),
    mean = 0 , sd = 1 )
  )) |>
  as_tsibble(index = time)

ma_sim |>
  gg_tsdisplay(plot_type = "partial")
```

Like previous examples, the code produces the following *Figure 12.5*:

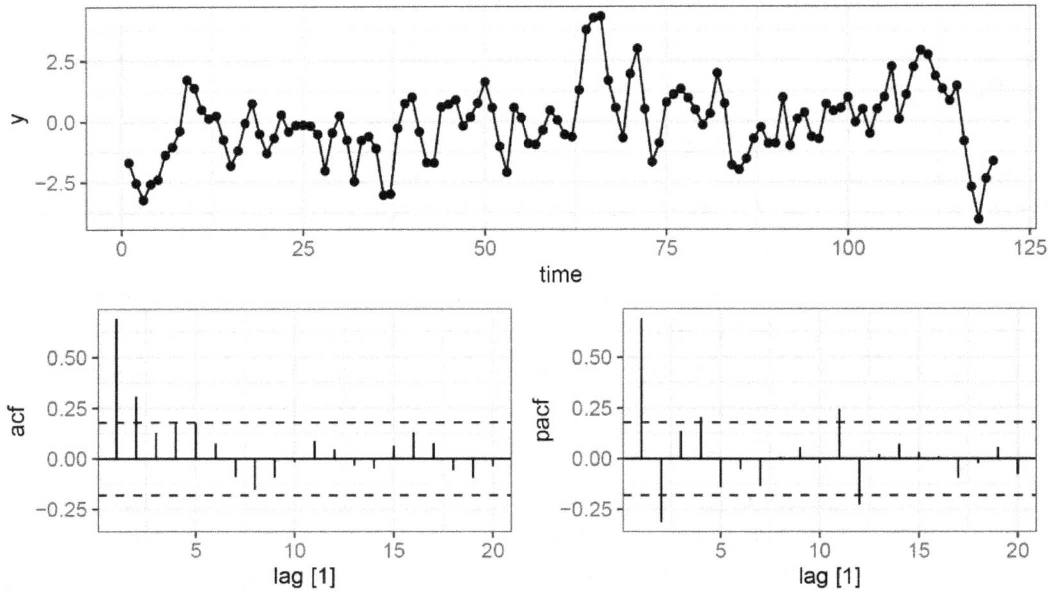


*Figure 12.5: ACF and PACF of a simulated series with weak autocorrelation and a strong dependency structure in the error*

The ACF has two significant spikes at lag 1 and 2, with a sharp cutoff after lag 1. In PACF, there are significant spikes at lag 1, 2, 4, 11 and 12, and they are oscillating around zero, suggesting this data could be better modeled with MA terms.

Before fitting any model, check whether the series needs any differencing to be stationary using the following code:

```
ma_sim |>
  features(y, unitroot_ndiffs)
```

The following output showing the `unitroot_ndiffs()` function identified that the series is stationary and not differencing is required ($d = 0$):

```
# A tibble: 1 × 1
  ndiffs
   <int>
1      0
```

Now, we will fit the following ARIMA models on the simulated series:

- *ARIMA*(1,0,2): an ARIMA model with one AR term and two MA terms
- *ARIMA*(0,0,2): an ARIMA model with two MA terms only. This could be written as a pure moving average model, *MA*(2)

You can fit multiple models on a single series as shown in the following code:

```
ma_model <- ma_sim |>
  model(arma_mod = ARIMA(y ~ pdq(1,0,2)),
        ma_mod = ARIMA(y ~ pdq(0,0,2)))

glance(ma_model)
```

The `report()` function does not work on a `mable` (model table) object with multiple models. Rather, you can `glance()` the model object to see model fit criteria, as shown in the following output:

```
# A tibble: 2 × 8
  .model   sigma2 log_lik   AIC  AICc   BIC ar_roots   ma_roots
  <chr>     <dbl>   <dbl> <dbl> <dbl> <dbl> <list>     <list>
1 arma_mod   1.04   -171.  351.  351.  362. <cpl [1]> <cpl [2]>
2 ma_mod     1.03   -171.  349.  349.  357. <cpl [0]> <cpl [2]>
```

You can also `tidy()` the model output to compare the estimated parameters from both the fitted models, as shown in the following output:

```
>tidy(ma_model)

# A tibble: 5 × 6
  .model    term  estimate std.error statistic  p.value
  <chr>     <chr>    <dbl>     <dbl>     <dbl>    <dbl>
1 arma_mod ar1       0.111    0.242      0.456 6.49e- 1
2 arma_mod ma1       0.854    0.236      3.62  4.26e- 4
3 arma_mod ma2       0.407    0.145      2.80  6.02e- 3
4 ma_mod    ma1      0.956    0.0889    10.8   2.55e-19
5 ma_mod    ma2      0.466    0.0711     6.54  1.55e- 9
```

The $AIC_c$ is smaller in the $MA(2)$ model suggesting a better overall fit. The $ARIMA(1,0,2)$ estimated the parameters closer to the actual values; however, the AR term is statistically insignificant at 5% level. In the $MA(2)$ model both the estimates are statistically significant. The coefficient of the first MA term ($\widehat{\phi_1} = 0.956$) is estimated as higher compared to the actual value. It may have captured the weak autocorrelation structure present in the series instead of estimating an AR term. The coefficient of the second MA term ($\widehat{\phi_2} = 0.466$) is closer to the actual parameter value.

It is possible to write any stationary $AR(p)$ model as an infinite $MA(\infty)$ process. On the other hand, with some constraints on the parameter space, any $MA(q)$ model can be written as an $AR(\infty)$ process. This characteristic is known as the invertible condition of MA models. As with AR terms, certain restrictions are imposed on the parameters of MA terms during estimation, which are automatically handled by the ARIMA function.

So far, we have discussed the three crucial parts of an ARIMA model. In the next section, you will find discussions on the ARIMA model following a general structure, and examples of using the fitted models to forecast.

# 12.5 General structure of an ARIMA model

Combining three different components; AR (autoregressive), I (integration), and MA (moving average), we obtain a general structure of an ARIMA model. There are two versions, depending on the frequency in the data, non-seasonal and seasonal.

### 12.5.1 Non-seasonal ARIMA model

A non-seasonal $ARIMA(p, d, q)$ model incorporates three parts:

- **Autoregressive (AR)**: Represents the dependency of an observation on a number of lagged observations. The parameter $p$ determines how many lagged observations are needed to explain the inter-dependency. $p$ is known as the order of the AR part.
- **Differencing**: Determines how many times a series needs to be differenced before it becomes stationary. The $d$ parameter determines the order of differencing.
- **Moving Average (MA)**: Represents the dependency of an observation on lagged residual errors. The $q$ parameter is the order of the MA model, which indicates how many residual error terms are used in the model.

The general equation of an $ARIMA(p, d, q)$ is written as

$$y_t{}' = c + \phi_1 y_{t-1}{}' + \cdots + \phi_p y_{t-p}{}' + \epsilon_t + \theta_1 \epsilon_{t-1} + \cdots + \theta_q \epsilon_{t-q} \qquad (12.4)$$

where, $y'_t$ is the differenced series.

*Equation (12.4)* is commonly written in short format using backshift notation (also known as the backward shift operator),

$$\left(1 - \phi_1 B - \cdots - \phi_p B^p\right)(1 - B)^d y_t = c + \left(1 + \theta_1 B + \cdots + \theta_q B^q\right). \qquad (12.5)$$

$B$ is known as the backshift notation. This one notation can be used to represent lagged observations and differencing simultaneously, such as

$$\text{first lag: } By_t = y_{t-1},$$

$$\text{shifting data back two periods: } B^2 y_t = y_{t-2}.$$

The same operator is used to represent differencing,

$$y'_t = y_t - y_{t-1} = y_t - By_t = (1 - B)y_t.$$

So, $(1 - B)^d y_t$ in Equation (12.5) represents the $d$-th order of differencing.

In the following examples, we will fit ARIMA models on data, use them in forecasting, and evaluate their forecast performance.

First, separate the USA GDP series into a training and test set, as shown in the following code:

```
usagdp_full <- usa_gdp |>
  select(year, gdp)

usagdp_train <-usagdp_full |>
  filter_index(1960~2007)

usagdp_test <- usagdp_full|>
  filter_index(2008~2017)
```

Now, fit some ARIMA models on the log USA GDP series, as we don't yet know which model will fit the data well and will also produce a better forecast. Previously, we identified that the $log(gdp)$ series became stationary after double differencing and there were two significant spikes in the PACF in *Figure 12.4*. Considering these things, we could try fitting $ARIMA(1,2,1)$ and $ARIMA(2,2,1)$ as a start. Along with these subjective choices, let the ARIMA() function identify a model in an automated fashion. Use the following code to fit all the models on the training data:

```
usagdp_arima <- usagdp_train |>
  model(arima_121 = ARIMA(log(gdp) ~ pdq(1,2,1)),
        arima_221 = ARIMA(log(gdp) ~ pdq(2,2,1)),
        arima_auto = ARIMA(log(gdp)))

usagdp_arima
```

The `usagdp_arima` has fitted a model object as shown in the following output:

```
# A mable: 1 x 3
       arima_121      arima_221      arima_auto
         <model>        <model>         <model>
1 <ARIMA(1,2,1)> <ARIMA(2,2,1)> <ARIMA(0,2,1)>
```

Extract the model fit criteria of the three models:

```
> glance(usagdp_arima)

# A tibble: 3 × 8
  .model       sigma2 log_lik   AIC  AICc   BIC ar_roots  ma_roots
  <chr>         <dbl>   <dbl> <dbl> <dbl> <dbl> <list>    <list>
1 arima_121  0.000372    117. -228. -228. -223. <cpl [1]> <cpl [1]>
2 arima_221  0.000374    118. -227. -226. -220. <cpl [2]> <cpl [1]>
3 arima_auto 0.000364    117. -230. -230. -227. <cpl [0]> <cpl [1]>
```

Let's check the significance of the fitted model's coefficients from the following output:

```
> tidy(usagdp_arima)

# A tibble: 6 × 6
  .model     term  estimate std.error statistic     p.value
  <chr>      <chr>    <dbl>     <dbl>     <dbl>       <dbl>
1 arima_121  ar1     0.0113     0.206    0.0547 0.957
2 arima_121  ma1    -0.650      0.143   -4.54   0.0000402
3 arima_221  ar1    -0.107      0.248   -0.429  0.670
4 arima_221  ar2    -0.171      0.189   -0.908  0.369
5 arima_221  ma1    -0.528      0.222   -2.38   0.0216
6 arima_auto ma1    -0.645      0.107   -6.02   0.000000275
```

The automated version selected an $ARIMA(0,2,1)$ model. This model is a parsimonious (fewer parameters) and stable (estimates are statistically significant) model. On the other hand, the estimated AR parameters in the manual model selection are statistically insignificant. They may have over-fitted the data too with slightly smaller $AIC_c$ values. Eventually, after multiple model fitting-evaluation steps, the subjective models will reduce to the same as identified by the automated selection. In that context, the automated selection saves time and manual effort needed in the ARIMA model selection process.

In practice, you should be using $ARIMA(0,2,1)$ for forecasting. For illustration, let's use all three models to forecast the next 10 observations, then compute accuracy in the training and test sets, as shown in the following code:

```
usagdp_fcst <- usagdp_arima |>
  forecast(h=10)

usagdp_fcst_eval <- bind_rows(
  accuracy(usagdp_arima),
  accuracy(usagdp_fcst, usagdp_full))

usagdp_fcst_eval |>
  select(.model, .type, ME, RMSE, MASE, ACF1)
```

Some selected accuracy metrics are provided in the following output:

```
# A tibble: 6 × 6
  .model     .type          ME   RMSE   MASE  ACF1
  <chr>      <chr>       <dbl>  <dbl>  <dbl> <dbl>
1 arima_121  Training    -1.15   9.30  0.205 0.200
2 arima_221  Training    -1.01   9.24  0.203 0.204
3 arima_auto Training    -1.14   9.33  0.206 0.204
4 arima_121  Test      -299.   330.   10.1   0.623
5 arima_221  Test      -311.   343.   10.5   0.624
6 arima_auto Test      -299.   330.   10.1   0.623
```

The forecast performance from the three models is similar, suggested by MASE and RMSE. For visual inspection, you can compare the forecasts in the training and test sets using the following code:

```
nsarima_p1 <- usagdp_fcst |>
  autoplot(usagdp_train, level = NULL,size = 0.9)+
  labs(title="Forecasts with training set")

nsarima_p2 <- usagdp_fcst |>
  autoplot(usagdp_test,level = NULL,size = 0.9)

nsarima_p1 + nsarima_p2+
  plot_layout(guides = "collect")+
  labs(title="Forecasts with test set")
```

The plots are presented in *Figure 12.6*:

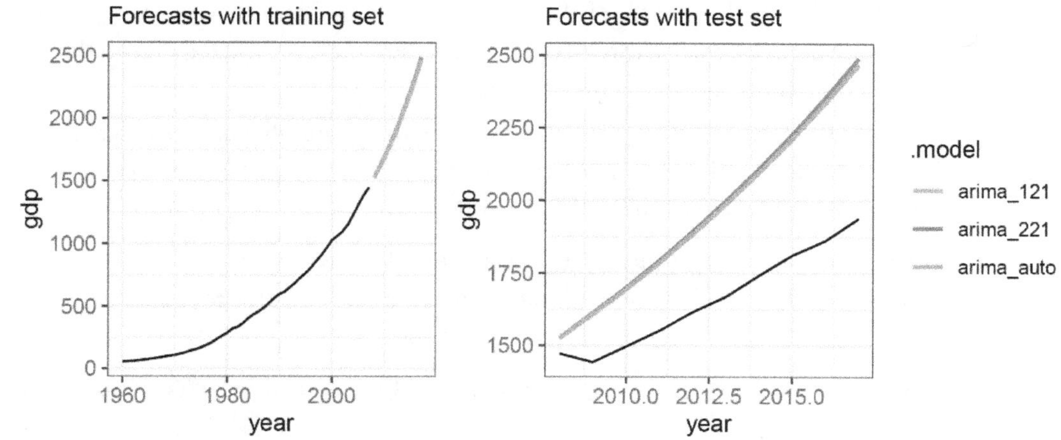


*Figure 12.6: Forecasting USA GDP with an automatically selected ARIMA(0,2,1) and manually selected ARIMA(1,2,1) and ARIMA(2,2,1) models*

It may seem that the manual ARIMA models produced good forecasts by capturing data dynamics, but they could become unreliable in the future, with unseen data. Manual ARIMA model identification involved understanding data inter-dependency and integration structure by evaluating the PACF and ACF multiple times before finalizing a model order specification. On the contrary, the automated ARIMA simplified the process completely.

So far, we have implemented a non-seasonal version of the ARIMA model. In the next section, we will look at how the ARIMA model can be extended to seasonal data.

## 12.5.2 Seasonal ARIMA models

The dependency structure in a seasonal series can be separated into the following six parts:

- Non-seasonal differencing, $d$: differencing order to make a series stationary by subtracting consecutive observations
- Non-seasonal autocorrelation structure $AR(p)$: observations at time $t$, $y_t$, depend on the immediate past values $y_{t-1}, y_{t-2}, \dots, y_{t-p}$
- Non-seasonal moving average structure, $MA(q)$: observations at time $t$, $y_t$, depend on the immediate past error terms $\epsilon_{t-1}, \epsilon_{t-1}, \dots, \epsilon_{t-q}$
- Seasonal differencing, $D$: differencing order to make the series stationary by subtracting observations from the same season in the past years

- Seasonal autocorrelation structure, $AR(P)$: observations at time $t$, $y_t$, depend on the values from the same season in past years
- Seasonal moving average structure, $MA(Q)$: observations at time $t$, $y_t$, depend on the error terms from the same season in past years

Combining all this information, $ARIMA(p,d,q)(P,D,Q)_m$, with $m$ =number of seasonal periods, gives the general notation of writing an ARIMA model that covers both non-seasonal and seasonal series. Seasonal ARIMA is also known as SARIMA.

Let's start with a simple example. Suppose we have a quarterly series (so, $m = 4$). An $ARIMA(2,1,1)(1,1,1)_4$ model for this series without a constant can be expressed using the backshift operator:

$$(1 - \phi_1 B - \phi_2 B^2)(1 - \Phi_1 B^4)(1 - B)(1 - B^4)y_t = (1 + \theta_1 B)(1 + \Theta_1 B^4)\epsilon_t. \quad (12.6)$$

Let's look at *Equation (12.6)*:

- The order of non-seasonal AR is 2. Hence, $p = 2$ and corresponding polynomials are presented by $1 - \phi_1 B - \phi_2 B^2$.
- The order of seasonal AR is 1, $P = 1$, and $(1 - \Phi_1 B^4)$ represents the seasonal AR structure.
- The order of non-seasonal differencing, $d = 1$ and represented by $(1 - B)$
- The order of seasonal differencing, $D = 1$ and represented by $(1 - B^4)$. This is a shorthand notation to represent that those past values from the same season are subtracted from the current value $(1 - B^4)y_t = y_t - y_{t-4}$.
- The order of non-seasonal MA is 1. Hence, $q = 1$ and the corresponding polynomials are presented by $1 + \theta_1 B$.
- The order of seasonal MA is 1. Hence, $Q = 1$ and the corresponding polynomials are presented by $1 + \Theta_1 B$.

Expanding *Equation (12.6)*, the general format of an $ARIMA(p,d,q)(P,D,Q)_m$ takes the form

$$\phi_p(B)\Phi_P(B^m)(1 - B)^d(1 - B^m)^D y_t = \theta_q(B)\Theta_Q(B^m)\epsilon_t. \quad (12.7)$$

We can break down the polynomials:

$$\begin{aligned}
\phi_p(B) &= 1 - \phi_1 B - \phi_2 B^2 - \cdots - \phi_p B^p \\
\Phi_P(B^m) &= 1 - \Phi_1 B^m - \Phi_2 B^{2m} - \cdots - \Phi_P B^{Pm} \\
\theta_q(B) &= 1 - \theta_1 B - \theta_2 B^2 - \cdots - \theta_q B^q \\
\Theta_Q(B^m) &= 1 - \Theta_1 B^m - \Theta_2 B^{2m} - \cdots - \Theta_Q B^{Qm}
\end{aligned}$$

Not every series needs all six orders to be determined. For example, a seasonal $AR(1)$ model for a stationary monthly series can be written as $ARIMA(0,0,0)(1,0,0)_{12}$.

The orders of seasonal differencing ($D$), $AR(P)$, and $MA(Q)$ models depend on the pattern in the seasonal lags in ACF and PACF plots. The process is described using the NSW Household Goods Retailing data in between April 1982 and December 2018:

```
retail <- aus_retail |>
  filter(`Series ID`=="A3349397X") |>
  select(Month, Turnover) |>
  rename(month = Month, turnover = Turnover)
```

First, examine the time plot, ACF, and PACF plot up to 36 lags generated using the following code:

```
retail |>
  gg_tsdisplay(plot_type = "partial", lag = 36)
```

*Figure 12.7* shows the model identification plots:

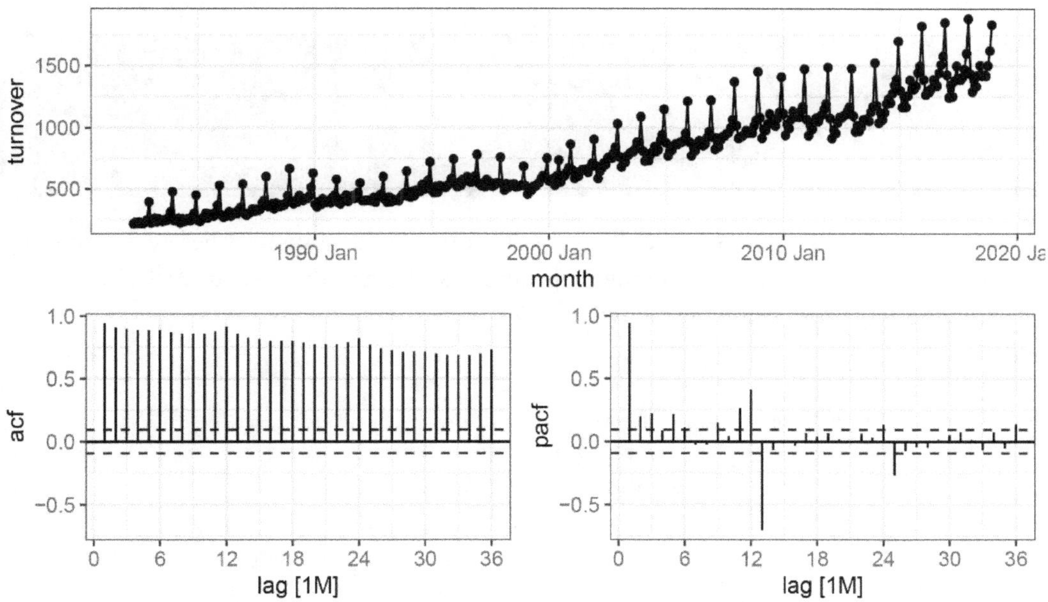


*Figure 12.7: Monthly NSW household goods retailing, 1982-2018*

The turnover series shows a strong upward trend with a strong seasonal pattern. It is non-stationary in terms of both mean and variance (variability around the level has increased as time has progressed). The ACF plot has significant spikes until 36th lag, decaying very slowly. The PACF plot has significant spikes at various lags, with prominent ones at lag 1 and 13.

A log transformation could help to stabilize the variance. A non-seasonal difference may require achieving the non-seasonal stationarity, and a seasonal difference is required to remove the inter-dependency in the seasonal periods. A seasonal difference subtracts the observations at lag 12 (in this example, frequency is $m = 12$) from the current value to stabilize the mean.

Identify the orders of non-seasonal ($d$) and seasonal differencing ($D$) of $log(turnover)$ using the `unitroot_ndiffs()` and `unitroot_nsdiffs()` functions, respectively, as shown in the following code:

```
retail |>
  features(log(turnover),
           list(unitroot_ndiffs, unitroot_nsdiffs))
```

The output is as follows:

```
# A tibble: 1 × 2
  ndiffs nsdiffs
   <int>   <int>
1      1       1
```

These unitroot tests suggest $d = 1$ and $D = 1$ to make the series stationary before determining the AR and MA orders. Let's perform these operations and generate ACF and PACF plots again for the stationary series using the following code:

```
retail <- retail |>
  mutate(turnover_log = log(turnover),
         seas_diff = difference(turnover_log, differences = 1, lag = 1),
         nonseas_diff = difference(seas_diff, lag = 12))

retail |>
  gg_tsdisplay(nonseas_diff,
               plot_type="partial", lag=36)
```

You will see the generated plots in *Figure 12.8*:

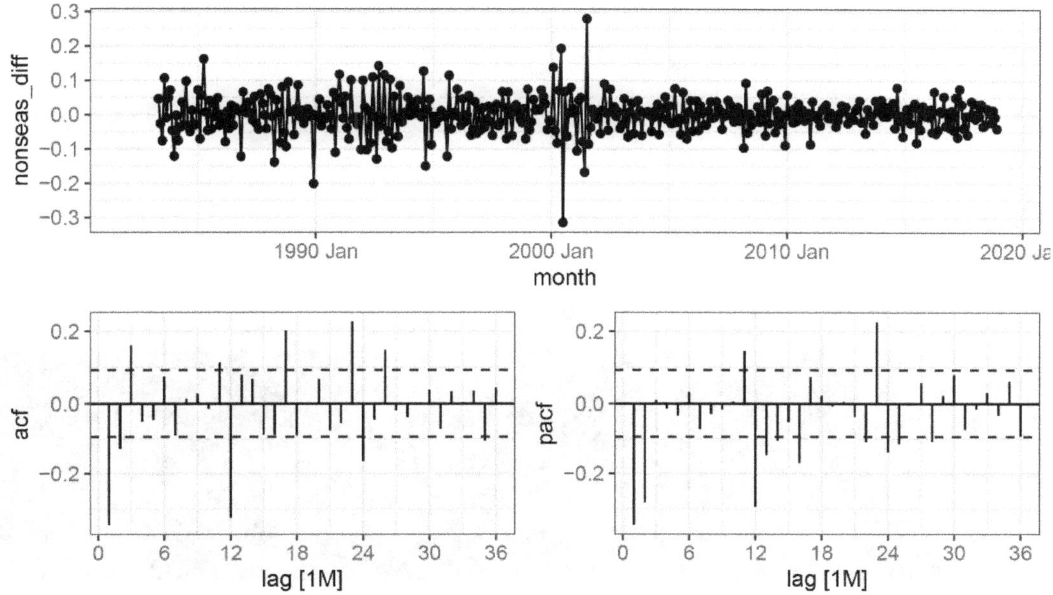


*Figure 12.8: Monthly NSW Household Goods Retailing after log transformation, non-seasonal and seasonal differences*

Now, the series has become stationary and you can determine possible AR and MA orders from the PACF and ACF in *Figure 12.8*. The patterns in the PACF and ACF plot show the following:

- Significant spikes at seasonal lags, 12 and 24, decayed sharply at lag 36. They are indicative enough to include seasonal AR or MA terms in the model.
- There are significant spikes at other lags that are oscillating around 0, indicating possible non-seasonal AR and MA terms.

We'll exclude an elaborate discussion on how manual order selection could be performed and iterations to reach the final model. Interested readers could consult other literature to learn about the techniques.

For illustration, fit two ARIMA models, one with manual order selection and the other with auto selection. Then tweak the output using the following code:

```
sarima_fit <- retail |>
  model(manual = ARIMA(log(turnover)~ pdq(2,1,1) + PDQ(1,1,2)),
        auto = ARIMA(log(turnover)))

sarima_fit |>
  pivot_longer(everything(),
               names_to="model_name",
               values_to="model_order")
```

The model fit is given in the following output:

```
# A mable: 2 x 2
# Key:     model_name [2]
  model_name                         model_order
  <chr>                                  <model>
1 manual              <ARIMA(2,1,1)(1,1,2)[12]>
2 auto       <ARIMA(1,0,1)(2,1,2)[12] w/ drift>
```

The auto model search has led to $ARIMA(1,0,1)(2,1,2)_{12}$ with a drift (constant) term. The tidied model output is shown in the following output:

```
> tidy(sarima_fit)

# A tibble: 13 × 6
   .model term     estimate std.error statistic   p.value
   <chr>  <chr>       <dbl>     <dbl>     <dbl>     <dbl>
 1 manual ar1      -0.422    0.255        -1.65   9.91e-  2
 2 manual ar2      -0.271    0.0999       -2.71   6.96e-  3
 3 manual ma1      -0.0222   0.271        -0.0820 9.35e-  1
 4 manual sar1     -0.347    0.368        -0.942  3.47e-  1
 5 manual sma1     -0.192    0.359        -0.534  5.93e-  1
 6 manual sma2     -0.303    0.210        -1.44   1.50e-  1
 7 auto   ar1       0.939    0.0203       46.4    3.41e-169
 8 auto   ma1      -0.433    0.0520       -8.34   1.05e- 15
 9 auto   sar1      0.735    0.323         2.27   2.35e-  2
10 auto   sar2     -0.251    0.0790       -3.17   1.61e-  3
11 auto   sma1     -1.28     0.331        -3.86   1.30e-  4
12 auto   sma2      0.500    0.216         2.32   2.08e-  2
13 auto   constant  0.00161  0.000250      6.42   3.60e- 10
```

In the manually selected model, some of the estimated parameters are not significant. Whereas auto ARIMA identified a numerically stable model in which all the coefficients are statistically significant.

Use the better model identified the auto selection to forecast for the next 12 months. In the following code, the last 8 years of data are used for clear visualization:

```
sarima_fcst <- sarima_fit |>
  select(auto) |>
  forecast(h = 12)

sarima_fcst |>
  autoplot(retail |>
           filter(year(month) >= "2010"),
           size = 1.2)
```

The forecasts with the 90% intervals are presented in *Figure 12.9*:

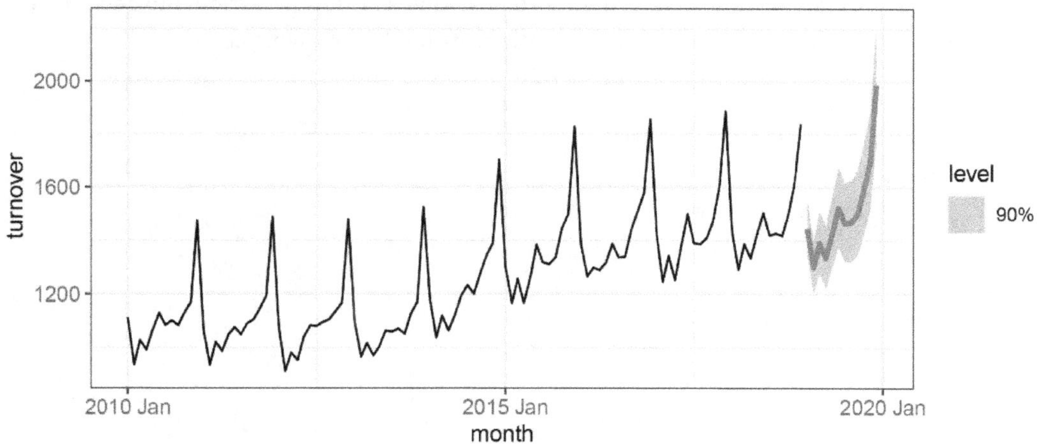


*Figure 12.9: Forecasts of monthly NSW Household Goods Retailing using an automatically selected ARIMA(1,0,1)(2,1,2)[12] model*

From the preceding discussion, you may have identified that ARIMA models have greater capabilities to capture complex data dynamics. The ARIMA class includes a wide variety of plausible models, considering all possible parameter combinations. This somehow made the model choice a process involved, time consuming task. However, the automated version takes the manual task away by suggesting an optimal model.

The ARIMA model works better with longer series that depict consistent patterns. The ARIMA model captures the underlying patterns embedded in data. Sometimes, exogenous variables that are related to the observed data, measured at the same time, could have an impact on forecasts too. ARIMA has the flexibility to add exogenous variables if they are deemed to be related to the observed series. Next, we'll discuss ARIMA model with exogenous variables, ARIMAX.

# 12.6 ARIMA with exogenous variables (ARIMAX)

One great advantage of the ARIMA model is that it can include exogenous variables $(x_1, x_2, \dots, x_k)$ along with the forecast variable $y_t$. Even though it has a different name , ARIMAX, the process is the same as fitting a standard ARIMA model. You just need to pass exogenous variables in the model function as you would do in any regression model.

As an example, we'll model the quarterly Australian electricity production data. Electricity production may depend on gas production measured in the same time periods. We'll use observations before 2008 as historical data and keep aside observations from 2009 and after as a test set. We'll examine the observed series to visually inspect the relationship between these two variables. You can perform the operations using the following code:

```
elect_gas <- aus_production |>
  select(Quarter, Electricity, Gas) |>
  rename(quarter = Quarter,
         electricity = Electricity,
         gas = Gas)

elect_gas_train <- elect_gas |>
  filter(year(quarter) <= "2008")

elect_gas_test <-elect_gas |>
  filter(year(quarter) > "2008")
```

The line plots of electricity and gas production in the training period are presented in *Figure 12.10*:

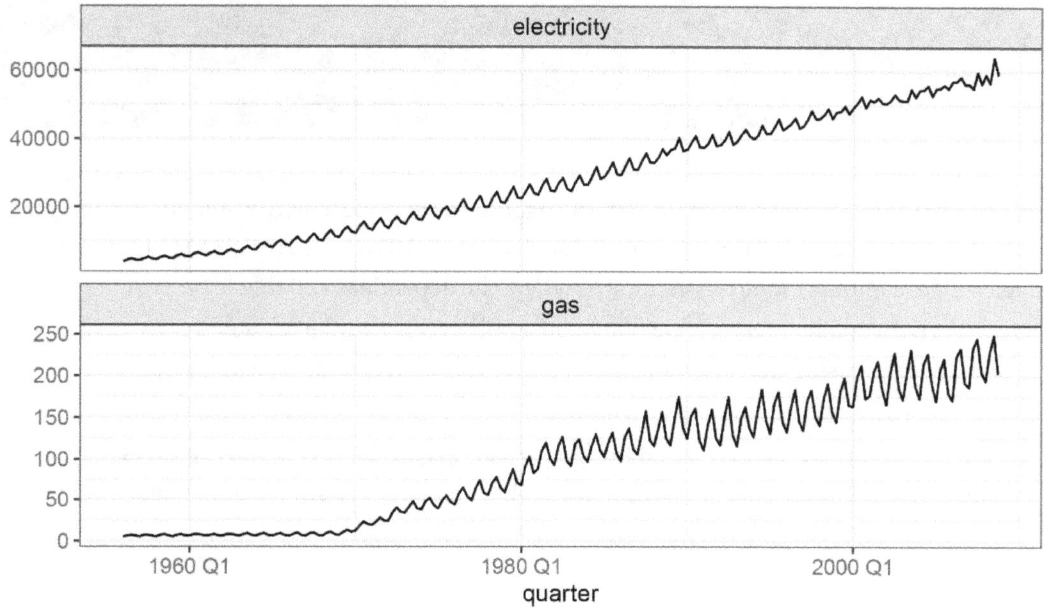


*Figure 12.10: Quarterly Australian electricity and gas production, 1956-2008*

Both series have a strong upward trend and strong seasonality. Multiple level changes are visible in the gas production series, while some disruptions in the electricity series are observed near the end of the training period.

Now, fit an ARIMA model for electricity production with automated order selection and gas as an exogenous variable as shown in the following code:

```
elect_gas_fit <-
  elect_gas_train |>
  model(arimax = ARIMA(electricity ~ gas))

report(elect_gas_fit)
```

The model fit information is shown in the following output:

```
Series: electricity
Model: LM w/ ARIMA(2,0,0)(1,1,2)[4] errors

Coefficients:
         ar1     ar2    sar1     sma1    sma2      gas  intercept
      0.4463  0.2223  0.9157  -1.4413  0.5236  37.8759   891.3641
```

```
s.e.   0.0687  0.0836  0.0795   0.1228  0.1032   9.9100   121.5084


sigma^2 estimated as 411853:  log likelihood=-1637.14
AIC=3290.28   AICc=3291   BIC=3316.98
```

For the quarterly electricity production, $ARIMA(2,0,0)(1,1,2)_4$ is selected, with gas as exogenous variable. The intercept term estimates the average amount of electricity production when gas production is $0$. The model description is Model: `LM w/ ARIMA(2,0,0)(1,1,2)[4]` errors states that a single, combined model is estimated where the dependent variable (`electricity`) is regressed on the exogenous variable (`gas`) and an intercept, and the residuals of that regression are then modeled using the $ARIMA(2,0,0)(1,1,2)_4$ process. The entire system is estimated simultaneously.

To use a model with exogenous variables, you need to pass the future values of them in the `forecast()` function. The `new_data` option will let you do that as shown in the following code:

```
elect_gas_fcst <- elect_gas_fit |>
  forecast(h = 10,
           new_data = elect_gas_test |> select(quarter, gas))

elect_gas_fcst |>
  autoplot(elect_gas_train |>
             filter(year(quarter)>= 2000))
```

The `autoplot()` code will generate a visualization of training data from 2000 and the 3-years forecasts presented in *Figure 12.11*:

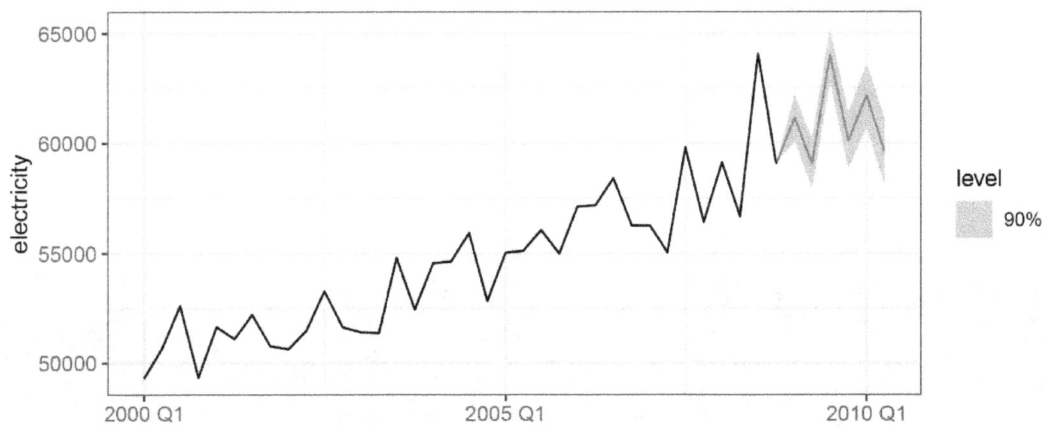


*Figure 12.11: Forecasts of quarterly Australian electricity production using ARIMA and Gas as exogenous variable*

So far, you have learned what an ARIMA model is and how to identify orders of different parameters for non-seasonal and seasonal series. You have followed how to select orders of various parameters manually and use an automatic order selection algorithm. In the next section, we'll summarize the complete process of ARIMA modeling before explaining how the automatic ARIMA algorithm works.

## 12.7 Steps in fitting an ARIMA model

The general steps involved in fitting an ARIMA model for a non-seasonal time series are summarized here.

1. Observe the time plot and identify any unusual observations.
2. If required, transform data to stabilize the variance using Box-Cox or log transformation.
3. Take differences on non-stationary data to make them stationary
4. Examine the ACF and PACF plot. A good starting point would be an $ARIMA(p, d, 0)$ or $ARIMA(0, d, q)$ model.
5. Fit your chosen model(s). If you have more than one model, choose the one with the minimum $AIC_c$ values.
6. Perform residual diagnostic checks on the errors generated from the chosen model. This involves examining the residual ACF and doing a portmanteau test. If the residuals are not white noise, go back to step 2 and try searching for a better model.
7. Once the residual is deemed white noise, use the model for forecasts.

The number of steps and iterations is amplified while dealing with a seasonal series.

The Hyndman-Khandakar algorithm (Hyndman et al., 2008) implemented automation in steps 3-5. You need to take care with other steps to obtain better forecasting performance, even when an automated algorithm is used.

The steps are summarized in *Figure 12.12* (sourced from Hyndman et al. (2021)).

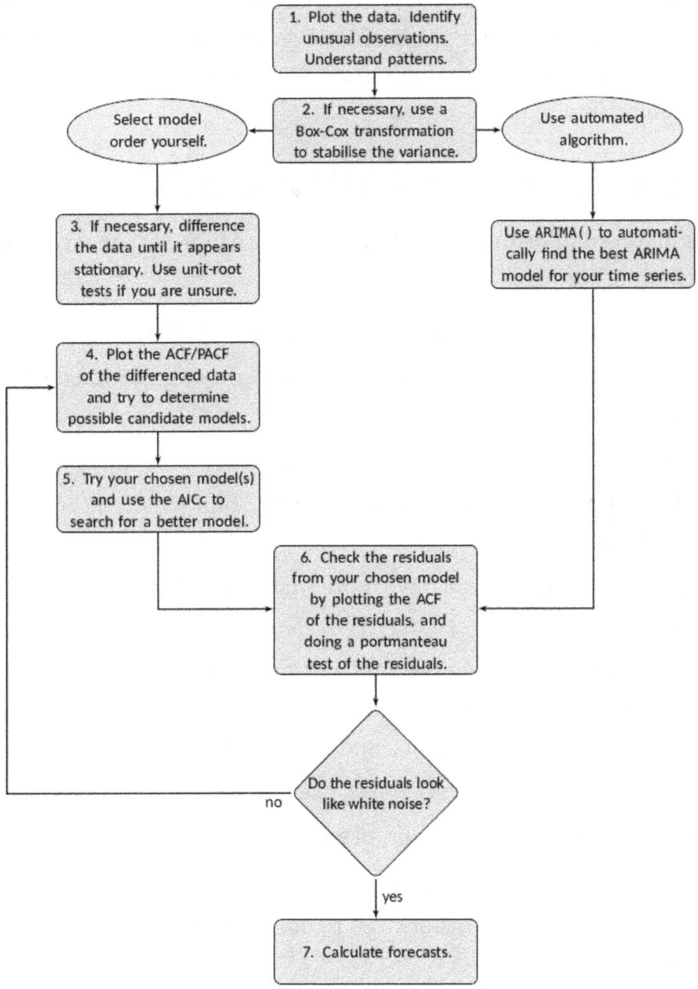


*Figure 12.12: General steps involved in using an ARIMA model for forecasting*

The Hyndman-Khandakar algorithm is implemented via the `ARIMA()` function in the `fable` package. You have seen in the examples that automated order selection is performed when no parameters are specified.

The `ARIMA()` function imposes restrictions on the orders of various components,

- $0 \leq p \leq 6$
- $0 \leq d \leq 2$
- $0 \leq q \leq 6$

- $0 \leq P \leq 2$
- $0 \leq D \leq 1$
- $0 \leq Q \leq 2$
- $p + q + P + Q \leq 6$
- $constant + d + D \leq 2$

First the order of differencing, $d$ or $D$ is determined. Many different ARIMA models are fitted for each series using MLE. The number of minimum models could be $72$ for a non-seasonal series and $278$ for a seasonal series. The model with the minimum $AIC_c$ is chosen as the best model among all the estimated models.

For a single series with moderate computational resource, the total number of models fitted by the automatic algorithm may not seem too large. However, there are scenarios in business where models for tens of thousands of series need to be updated regularly to produce forecasts. This could create a computation bottleneck in a production environment. This is where the Hyndman-Khandakar automated algorithm shines.

## 12.8 Automatic ARIMA algorithm

The stepwise search enables the automatic algorithm to find the best model by fitting a bare minimum of potential models. The steps for non-seasonal data are as follows:

1. The number of differences $0 \leq d \leq 2$ is determined using repeated KPSS tests.
2. The values of $p$ and $q$ are then chosen by minimizing the $AIC_c$ after differencing the data $d$ times. Rather than considering every possible combination of $p$ and $q$, the algorithm uses a stepwise search to traverse the model space.
    a. Four initial models are fitted:
        - $ARIMA(0, d, 0)$
        - $ARIMA(2, d, 2)$
        - $ARIMA(1, d, 0)$
        - $ARIMA(0, d, 1)$

    b. A constant is included unless $d = 2$. If $d \leq 1$, an additional model is also fitted:
        - $ARIMA(0, d, 0)$ without a constant, $c$.

    c. The best model (with the smallest $AIC_c$ value) fitted in step (a) is set to be the "current model".

d. Variations on the current model are considered:

- vary $p$ and/or $q$ from the current model by $\pm 1$;
- include/exclude $c$ from the current model.

The best model considered so far (either the current model or one of these variations) becomes the new current model.

e. Repeat Step 2(c) until no lower $AIC_c$ can be found.

A visual illustration of stepwise searching is presented in the *Figure 12.13*.

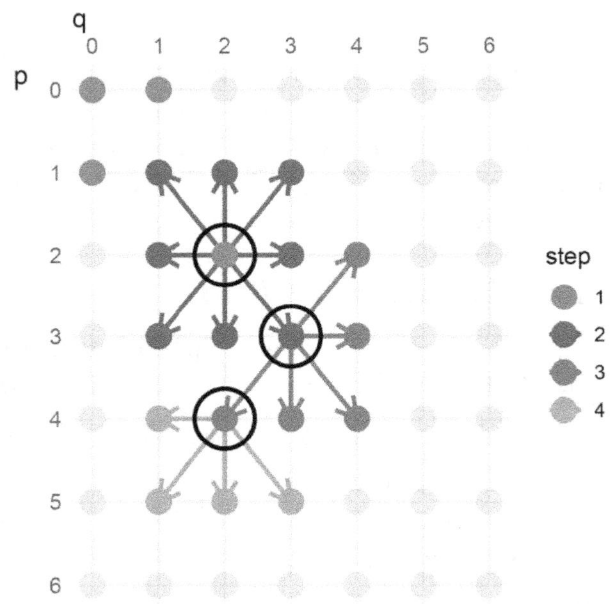


*Figure 12.13: Illustration of the stepwise automatic ARIMA algorithm (sourced from Forecasting: Principles and Practice (3rd ed))*

The benefit of the stepwise search is that it considerably reduces the total number of potential models to fit. In every step, the algorithm updates the "current model" according to the minimum $AIC_c$ value, and then searches in the nearest neighbors of the current model for a better one. By default, the `ARIMA()` function performs a stepwise search and switches to a new "current model" as soon as a better model is found, instead of going through all the neighboring models. The `greedy=FALSE` option enables a full nearest neighboring search. In addition to that, some approximations are performed to further speed up the procedure. If a minimum $AIC_c$ model is not found due to approximation, or even due to the stepwise search, users can turn off the stepwise search by setting `stepwise=FALSE`. There are options to use other information criteria, such as $AIC$ or $BIC$, for selecting a model.

The `trace=TRUE` arguments print the fitted models information on console. You could experiment with various arguments and observe the difference in potential model fits with and without stepwise search option.

Apart from R packages, the auto ARIMA algorithm is implemented in a lot of proprietary software dedicated to forecasting, such as AWS forecasting algorithms (Autoregressive Integrated Moving Average (ARIMA) Algorithm-Amazon Forecast), Google Cloud's BigQuery ML `ARIMA_PLUS` (Cheng et al., 2021) to name a few.

While complex modeling processes can be automated, it is equally important to understand the impact of generated forecasts in a business context. Next, we discuss the power of the ARIMA model on forecast explainability.

# 12.9 ARIMA and forecast explainability

Forecast explainability refers to the ability to understand and interpret the underlying reasons behind forecasts generated from a model. It involves identifying and communicating the key factors and data patterns that influence the forecasted outcomes, making the forecasting process more transparent and trustworthy.

ARIMA models support forecast explainability by providing clear and interpretable components such as autoregressive terms, differencing, and moving averages that reflect the underlying patterns in data. By analyzing these components, users can understand the influence of past values, trends, and noise on future predictions, making the forecasting process more transparent and understandable.

The ARIMA model is a useful technique to generate long-term forecasts for series with a long history. The `ARIMA()` function is highly useful to find an optimal model from a pool of a large number of potential models. There are scenarios where even the best-fitted ARIMA model produces undesirable forecasts. Those situations would be easy to comprehend if you are aware of the impact of various parameters on forecast values obtained by ARIMA models.

## 12.9.1 Impact of the constant term on forecasts

The constant $c$ and the differencing order $d$ combined have an important effect on the long-term forecasts generated by ARIMA models.

- If $c = 0$ and $d = 0$, the long-term forecasts will go to zero.
- If $c = 0$ and $d = 1$, the long-term forecasts will go to a non-zero constant.
- If $c = 0$ and $d = 2$, the long-term forecasts will follow a straight line.

- If $c \neq 0$ and $d = 0$, the long-term forecasts will go to the mean of the data.
- If $c \neq 0$ and $d = 1$, the long-term forecasts will follow a straight line.
- If $c \neq 0$ and $d = 2$, the long-term forecasts will follow a quadratic trend. This is not permitted in `fable::ARIMA()`.

## 12.9.2 Prediction intervals from ARIMA models

The prediction intervals from ARIMA models only consider the variations in errors. Other sources, such as variations in parameter estimation and model order selection, are not accounted for. This is a reason why ARIMA-based prediction intervals tend to be narrow.

By default, the computation of prediction intervals assumes residuals are normally distributed and uncorrelated. The prediction interval may be incorrect if these assumptions are not satisfied. It is crucial to always perform residual diagnostics using residual's ACF and histogram to validate the core assumptions and use the prediction intervals.

The prediction interval computed from ARIMA models also assumes that the past pattern in the data will continue into the forecast period. If there is a significant change at the end of the series compared to the full historical data, it is suggested to use prediction intervals with caution.

## 12.9.3 Impact of differencing orders on prediction intervals

Generally, prediction intervals from ARIMA models widen as the forecast horizon increases. The value of the differencing order $d$ has an impact on prediction intervals. The higher the value of $d$, the more rapidly the prediction intervals increase in size.

When $d = 0$, the standard deviation of the long-term forecast will be converged to the standard deviation of the historical data, so the prediction intervals will essentially stay the same.

## 12.9.4 Impact of AR order on cyclic forecasts

If data has cyclical behavior, then it is necessary to set $p \geq 2$ to obtain cyclic forecasts. This also requires some additional conditions to impose on parameters. The cyclical behavior for an $AR(2)$ model occurs if $\phi^2 + 4\phi_2 < 0$. If historical data shows a cyclical pattern, but it is not reflected in forecasts generated from the best model, which has $p = 1$, then you know what to fix to obtain better forecasts.

# Summary

In this chapter, you learned about ARIMA forecast models. The various components of the ARIMA framework were discussed, along with worked-out coding examples. Three variations of ARIMA models were discussed; non-seasonal, seasonal, and ARIMA with exogenous variables. You learned how to manually identify orders of AR, differencing, and MA parts by studying time plots, ACF and PACF plots. The automated ARIMA algorithm was discussed in detail so that you understand what happens behind the scenes.

Generating forecasting from a good model is one thing. Properly using the forecasted value for business decision making is another thing. The latter requires understanding how and when a model's parameters could impact forecasts. This was discussed in the ARIMA model context. Now you should be confident to use the ARIMA model in your future forecasting project.

In this part of the book, some basic to advanced statistical forecasting models were covered. In the next part, we will discuss some machine learning models that can be used for forecasting.

# Further reading


- *Autoregressive integrated moving average (ARIMA) algorithm-Amazon Forecast.* `https://doi.org/https://docs.aws.amazon.com/forecast/latest/dg/aws-forecast-recipe-arima.html`
- Box, G. E. P., & Jenkins, G. M. (1970). *Time series analysis: Forecasting and control.* Holden-Day.
- Cheng, X., Shen, W., Chen, H., & Shen, C. (2021). *ARIMA_PLUS: Large-scale, accurate, automatic and interpretable in-database time series forecasting and anomaly detection in google BigQuery*. Arxiv.org. `https://arxiv.org/pdf/2510.24452.pdf`.
- Hyndman, R. J., & Athanasopoulos, G. (2021). *Forecasting: Principles and practice* (3rd edition). OTexts.
- Hyndman, R. J., & Khandakar, Y. (2008). *Automatic time series forecasting: The forecast package for R. Journal of Statistical Software*, 27(3), 1–22. `https://doi.org/10.18637/jss.v027.i03`
- Kwiatkowski, D., Phillips, P. C. B., Schmidt, P., & Shin, Y. (1992). *Testing the null hypothesis of stationarity against the alternative of a unit root: How sure are we that economic time series have a unit root? Journal of Econometrics*, 54(1-3), 159–178.

# Join our community on Discord

Join our community's Discord space for discussions with the author and other readers:

`https://discord.gg/G4YmB6DQvf`

# 13

# Advanced Computational Methods for Forecasting

In this chapter, we will explore several computational predictive methods for univariate time series forecasting. This field has gained substantial attention in recent literature and in applied forecasting due to the ever-rising popularity of various machine learning (ML) and Deep Learning (DL) based prediction algorithms. ML, which includes DL, is collectively referred to as machine learning, is an influential field within artificial intelligence that focuses on creating algorithms that allow computers to learn from data and make informed, data-driven decisions. Rather than being explicitly programmed for specific tasks, ML algorithms detect patterns within data and leverage these patterns for predictions. These algorithms have proven highly effective across various domains, from less sensitive applications, such as targeted advertising for tech giants to highly sensitive tasks such as predicting the preoperative risk of death for cardiac surgery patients.

Professor Efron (2020), in his commemorative paper following the International Prize in Statistics (often regarded as the statistics equivalent of the Nobel Prize, Fields Medal, or Turing Award) noted: "*if media attention is the appropriate metric, then pure prediction algorithms are our era's statistical stars.*" If you've eagerly anticipated this chapter or felt compelled to start reading the book from here, that excitement is understandable.

Over the past 40 years, forecasting competitions have made a compelling case for the effectiveness of ML in forecasting tasks. Since the first formal competition in 1984, early contests, such as M1 and M2, the first and second Makridakis forecasting competitions (Makridakis et al. (1982) and Makridakis et al. (1993)), which benchmarked forecasting methods across diverse real-world time series, showed that simpler methods, such as exponential smoothing, often outperformed more complex statistical models, such as ARIMA. Ensembles of multiple forecasting methods consistently improved accuracy and reduced uncertainty. This trend continued in the M3 competition, where simple methods and forecast ensembles excelled. However, the M4 competition in 2018 marked a pivotal shift: sophisticated models that blended statistical and ML approaches outperformed all others, reversing previous findings. By the time of the M5 competition, the top 50 forecasting methods were based on ML, highlighting the alignment of ML with certain problem spaces in forecasting.

This chapter focuses on applying ML techniques to time series forecasting, offering key insights, and demonstrating efficient implementations using various R libraries. While an in-depth understanding of ML isn't required, a familiarity with fundamental concepts in algorithm development and validation techniques is assumed. For those new to ML, we recommend consulting introductory textbooks, such as Gareth et al. (2013) and Hastie et al. (2009). These resources will help you build a solid foundation in supervised learning techniques, including essential optimization methods, such as gradient descent and back-propagation. Additionally, understanding model tuning through cross-validation is crucial, especially in time series forecasting, where maintaining the sequence of data during validation is vital, as discussed in earlier chapters.

The chapter is organized as follows:

- Section 13.1: Understanding the connection between forecasting and prediction
- Section 13.2: DL forecasting: Neural networks
- Section 13.3: ML forecasting
- Section 13.4.1: Forecasting as a curve fitting: The Prophet method
- Section 13.5: Forecast ensemble

## Technical requirements

Let's load the packages that are required in this chapter.

```
#install.packages("pacman")
library(pacman)
p_load("conflicted"
       ,"datasets"
       ,"fable"
       ,"fabletools"
       ,"feasts"
       ,"Mcomp"
       ,"rsample"
       ,"patchwork"
       ,"tidyverse"
       ,"tsibble"
       ,"tsibbledata"
       ,"forecastML"
       ,"glmnet"
       ,"hqreg"
       ,"fable.prophet" )

conflicts_prefer(
  fabletools::accuracy,
  fabletools::forecast,
  fabletools::components,
  dplyr::filter,
  tsibble::difference,
  tsibble::index,
  dplyr::last,
  dplyr::select
)
```

The code file is available in the GitHub repo `https://github.com/PacktPublishing/Modern-Time-Series-Analysis-with-R.git` for you to download and practice.

## 13.1 Understanding the connection between forecasting and prediction

Before you dive into prediction-based algorithms for forecasting, it is important to understand the difference between the terms *prediction* and *forecasting*. ML techniques primarily aim to enhance the accuracy of predictions for a variable of interest. These advanced methods leverage algorithms to make informed predictions about unknown outcomes based on known input data. The terms prediction and forecasting are often mistakenly used interchangeably. While they may seem synonymous in everyday language, they carry distinct meanings in academic literature. In time series analysis, a statistical forecast refers specifically to predicting the future behavior of an observed random variable based on its past states. This is rooted in the theory that an observed sequence is the outcome of a random process, where randomness results from the interaction of many independent factors in a linear fashion.

Dynamical systems theory, however, suggests that what may appear as random behavior could arise from deterministic systems with a few factors interacting non-linearly. This opens the possibility of reframing forecasting as a supervised learning problem, where ML techniques can be applied to capture complex non-linear relationships and enhance prediction accuracy.

Supervised learning involves building a model from a finite set of observations to establish the relationship between input and output variables. Once developed, the model can be used for one-step or multi-step forecasting. In the context of forecasting, when past values of a time series are available, the forecasting task can be framed as a standard prediction regression problem using lagged values of the time series and the predictions from the regression model can be considered as forecasts (Bontempi et al., 2013). Several approaches have been explored, and this chapter will cover two important techniques useful for univariate forecasting that have gained substantial popularity in recent applications, the neural network prediction and ML-based prediction for forecasting.

In its simplest form, **Time Series Linear Models** (**TSLMs**), as discussed in *Chapter 10*, explicitly measure dependencies over time by estimating deterministic trends and seasonal components. However, in supervised learning, more advanced methods, particularly those based on ML and DL, have proven extremely effective for handling tabular data, and these will be the focus of this chapter.

Forecasting through prediction involves designing an input/output process, where the output is a scalar and the input is a vector. This relies on the availability of a collection of observed pairs, typically referred to as a training set. In forecasting, the training set is derived from historical time-series data, where the input matrix, $X_{[(T-p-h)\times p]}$, represents the lag dependencies, and the output vector, $y_{[(T-p-h)\times 1]}$, represents the forecasted values. Here, $T$ is the number of observations in the training set, $p$ is the lag dependency, and $h$ is the forecast horizon. This setup reflects the notation used earlier in the book with the supervised learning algorithms that rely on input-output pairs in the form $\{(\mathbf{x}_i, \mathbf{y}_i)\}_{i=1,\ldots,n}$ where $n = T - p - h$.

It is important to note that while the input matrix, $X$, explicitly accounts for lag dependence in the above notation, additional covariates could also be incorporated into the supervised learning process. However, to maintain simplicity and consistency with prior chapters, we omit this part of the notation. Although the choice of lag dependency can be left arbitrary and optimized by modern ML methods, it would be beneficial to use the knowledge gained from earlier chapters to identify relevant lags. This approach helps design a prediction problem aligned with domain knowledge.

Additionally, supervised learning can be used to perform multi-step-ahead forecasts either by setting the forecast horizon, $h > 1$, or by using a rolling one-step-ahead forecasting method. In the latter case, as forecasting progresses through successive steps, more forecast values are used to derive further predictions.

In summary, the process explained in this section gives the conceptual and technical basis for applying predictive modeling tools for forecasting applications. In subsequent sections, you will learn how to apply predictive modeling-based methods using a mostly automated workflow for forecasting.

## 13.2 DL forecasting: Neural network

DL is a specialized area of ML that leverages artificial neural networks to model and solve complex problems. Inspired by the structure and functioning of the human brain, it enables computers to learn from large volumes of data. Neural networks are structured as layers of interconnected nodes (often referred to as neurons), and their purpose is to predict outcomes based on input data. These networks excel in recognizing intricate patterns within data, making them particularly effective for tasks, such as forecasting. In recent years, DL algorithms have gained significant popularity due to advancements in computational techniques that allow for faster training and convergence at scale (Bengio et al., 2017). One of the key advantages of DL over traditional ML algorithms is its ability to automatically extract complex features from vast datasets. Moreover, these complex features can be shared across different parts of the network, enhancing its ability to perform sophisticated prediction tasks.

A neural network consists of three primary layers:

- **Input layer**: This is the first layer, where data features are fed into the network. Each node in this layer represents a unique feature from the input data.
- **Hidden layers**: In this middle layer computations are performed, allowing the network to learn complex feature representations. Each node in a hidden layer applies a

weighted sum to its inputs and processes the result through a non-linear activation function, such as the Sigmoid function. The more hidden layers a network has, the better it can capture complex relationships within the data.

- **Output layer**: The final layer generates the network's predictions. In regression tasks, this layer typically consists of a single neuron with a linear activation function, providing a continuous output value.

*Figure 13.1* illustrates a simple neural network, comprising an input layer with four nodes, a hidden layer with two nodes, and a single output node. This type of structure is commonly referred to as a multi-layer feed-forward network:

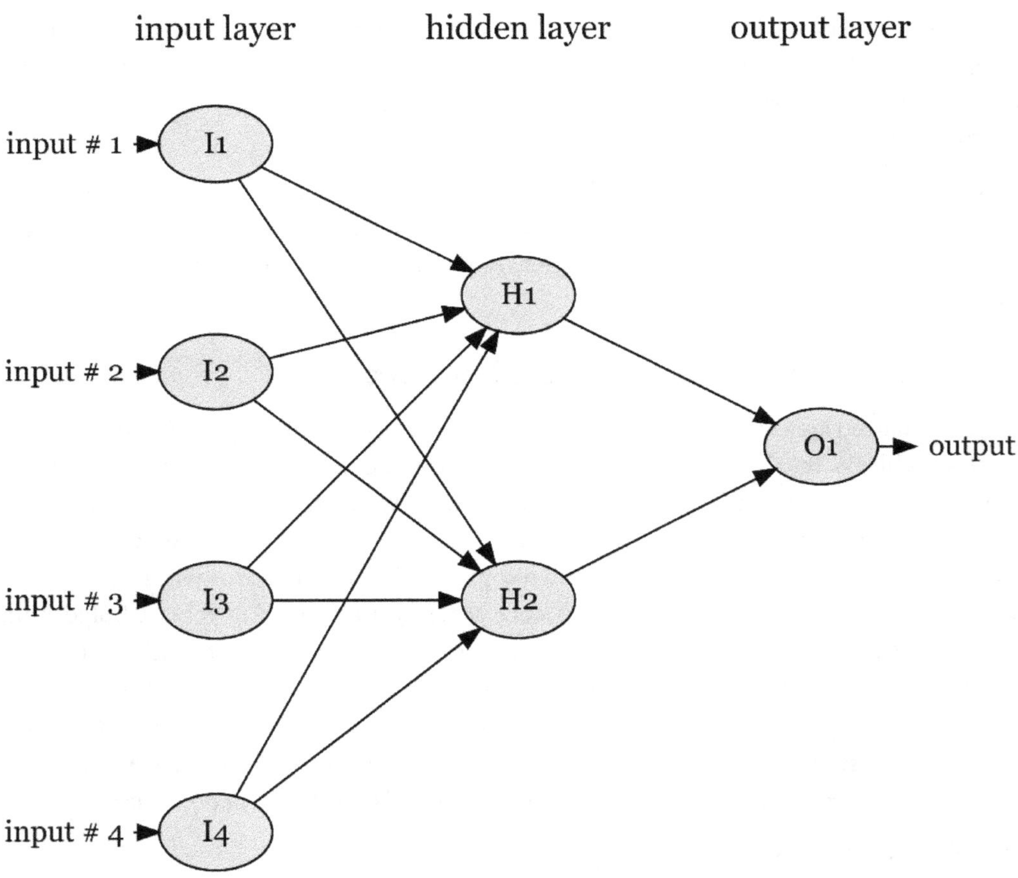


*Figure 13.1: A single hidden layer neural network.*

The simplest form of a neural network consists solely of an input layer and an output layer, devoid of any hidden layers. This structure is analogous to a linear regression model without interaction terms, where the relationships between variables are purely linear.

The core components required for a neural network are the weights and biases, along with the activation function. Weights are parameters that control the strength of the connections between neurons across different layers of the network. When an input is introduced to a neuron, each feature of the input is multiplied by a corresponding weight, which determines the significance of that feature in shaping the final decision. Bias, another important parameter within each neuron, allows the model to adjust the activation function curve. This bias is added to the weighted sum of the inputs before the result is passed through the activation function. The inclusion of bias enables the network to model data that doesn't necessarily pass through the origin, essentially handling situations where the input is zero, but the output is not. Once the weighted sum of the inputs and bias is calculated, the result is passed through an activation function, such as ReLU, Sigmoid, or tanh. The activation function is crucial in introducing non-linearity to the model, allowing it to capture and represent intricate patterns within the data that would otherwise be missed by a purely linear model.

The learning process of a neural network, known as forward and backward propagation, hinges on the continual adjustment of weights and biases. In forward propagation, input data traverses through the network's layers, where each neuron computes a weighted sum, applies an activation function, and forwards the output to the next layer. This sequence continues until the network produces a prediction. The accuracy of this prediction is evaluated using a loss function, which measures the deviation between predicted and actual values. In regression tasks, a common loss function is the **Mean Squared Error** (**MSE**), which penalizes larger errors more heavily.

The process of adjusting the network's weights and biases involves minimizing the loss function. This is achieved by computing the gradient of the loss with respect to each weight and bias, and then updating these parameters using an optimization algorithm, such as gradient descent. Through this iterative process, the model gradually improves, reducing the error and enhancing the accuracy of its predictions.

Training a multi-layer feed-forward neural network for regression involves finding the set of weights, $w_{ij}$ (which are equivalent to the coefficients attached to these predictors), and biases, $b_j$, in the network that minimizes the difference between the network's predicted outputs and the actual target values. This process is iterative and involves several key steps, including forward propagation, loss calculation, and backpropagation with optimization.

The training process is initialized by assigning some weights and biases, often with small random values. In the feed forward step for each neuron in the hidden layers, compute the weighted sum of inputs, add a bias term, and apply the activation function to produce the output for that neuron. For neuron $j$ in layer $l$, the process is as follows:

$$z_j^{(l)} = \sum_i w_{ij}^{(l)} a_i^{(l-1)} + b_j^{(l)}$$

$$a_j^{(l)} = \sigma(z_j^{(l)})$$

where,

- $w_{ij}^{(l)}$ are the weights from neuron $i$ in layer $l-1$ to neuron $j$ in layer $l$
- $a_i^{(l-1)}$ is the activation of neuron $i$ in the previous layer
- $b_j^{(l)}$ is the bias term for neuron $j$ in layer $l$
- $\sigma(\cdot)$ is the activation function

The output from the last hidden layer is passed to the output layer. The output layer typically uses a linear activation function in regression tasks, producing the final predicted value. For the network presented, we could avoid the index for layer $l$, but that would lose the generalization of the neural network.

The predicted output from the network is compared to the actual target value using an MSE loss function:

$$\text{Loss}_{MSE} = \frac{1}{n}\sum_{i=1}^{n}(y_i - \hat{y}_i)^2$$

where,

- $y_i$ is the actual target value
- $\hat{y}_i$ is the predicted value from the network
- $n$ is the number of examples

The most fascinating part of neural network learning is the backpropagation of the loss to update the weights and biases in an iterative way. The gradients of the loss function with respect to each weight and bias in the network are calculated using the chain rule of calculus. This is the backward step where errors are propagated from the output layer back through the network. For each weight, $w_{ij}^{(l)}$, the gradient is defined as follows:

$$\frac{\partial \text{Loss}_{MSE}}{\partial w_{ij}^{(l)}} = \frac{\partial \text{Loss}_{MSE}}{\partial a_j^{(l)}} \cdot \frac{\partial a_j^{(l)}}{\partial z_j^{(l)}} \cdot \frac{\partial z_j^{(l)}}{\partial w_{ij}^{(l)}}$$

where $\frac{\partial \text{Loss}_{MSE}}{\partial a_j^{(l)}}$ is the error term for neuron $j$ in layer $l$.

The weights and biases are updated in the direction that minimizes the loss, typically using an optimization algorithm, such as **Stochastic Gradient Descent** (**SGD**). The update rule for each weight, $w_{ij}^{(l)}$, can be expressed as follows:

$$w_{ij}^{(l)} = w_{ij}^{(l)} - \eta \cdot \frac{\partial \text{Loss}_{MSE}}{\partial w_{ij}^{(l)}}$$

where $\eta$ is the learning rate, controlling the size of the update step.

The forward propagation, loss calculation, and backpropagation are repeated for a specified number of iterations (epochs) or until the loss converges to a minimum. During training, the network gradually adjusts its weights and biases to minimize the loss function, improving its predictions on the training data. The choice of random initial weights can impact the convergence and accuracy measure of the forecasts. To average out the fluctuations of random initialization the final forecasts are derived by averaging forecasts over multiple initializations of a feed-forward neural network.

An advantage of employing DL methodologies for time series forecasting lies in their capacity to automatically engineer complex features within their hidden layers that can improve forecasting accuracy. This approach is particularly effective when the underlying data exhibits sharable or transferable features. In univariate forecasting, sharable or transferable features, refer to fundamental, abstract patterns that a DL model automatically learns within a single time series. These are not specific data points but general mathematical properties, such as recognizing robust autoregressive relationships (the dependence on recent past values), identifying the general shape of a seasonality and cycle (e.g., lynx population boom every 11 years), or detecting long-term trends. By identifying these universal building blocks of time series data in its hidden layers, the model can apply this learned knowledge efficiently across different time periods, improving forecast accuracy without needing explicit human instruction for feature engineering.

## 13.2.1 Forecasting using neural networks

The R programming language offers several built-in time-series datasets that are useful for exploring the application of neural networks in forecasting. One notable example is the lynx dataset, available in the datasets package, which was previously used to illustrate seasonality in an earlier chapter. However, this dataset reveals that it does not exhibit traditional seasonal patterns. Instead, the lynx data reflects cycles that repeat approximately every 10 years, driven by the natural boom-and-bust dynamics of the Canadian lynx population. These population fluctuations are closely tied to the availability of its food supply, which, in turn, is influenced by long-term weather cycles.

Let's visualize the autocorrelation structure for the `lynx` series using the following code:

```
lynx_tsb <- lynx |>
  as_tsibble() |>
  rename(year = index,
         lynx_count = value)

lynx_tsb |>
  gg_tsdisplay(plot_type = "partial", lag = 36)
```

The ACF, the PACF, and a time plot of the original series are presented in *Figure 13.2*:

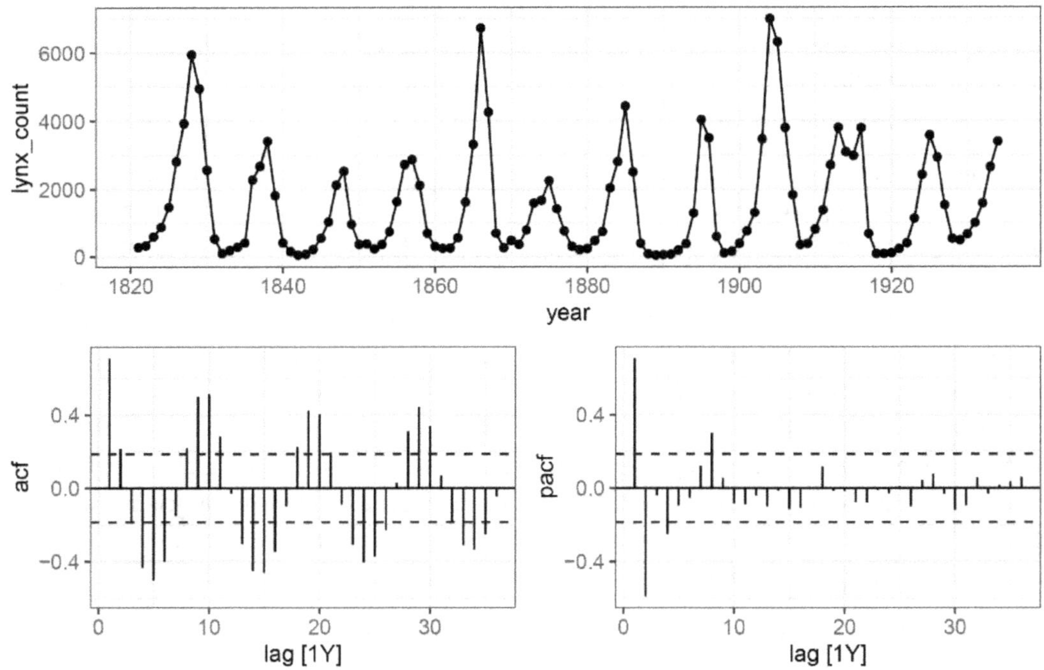


*Figure 13.2: ACF and PACF of the lynx data.*

The PACFs and ACFs in *Figure 13.2* show the cyclical behavior clearly. They show strong lag 1 dependency and a cycle repeating every 10 years.

This dataset lacks a significant trend and seasonality, which is evident from the time plot. This immediately eliminates at least one of the statistical forecasting models we discussed earlier from consideration. Can you guess which one? If not, do not worry—we will identify it shortly.

Let us build both statistical forecasting models (using ETS and ARIMA) and neural network-based forecasting for this data. Note the following:

- We will use data up to 1920 for training/model development, reserving the remaining data for validation. Split the training (from 1821 to 1919) and test set (from 1920 to 1934), as shown in the following code:

```
lynx_train <- lynx_tsb |> filter(year < 1920)
lynx_test <- lynx_tsb |> filter(year >= 1920)
```

- It is important to note that the lynx population count is constrained by a lower bound of 0. To preserve this non-negative characteristic in forecasting, we will apply a square root transformation (sqrt) to the series. For the **Neural Network Autoregressive (NNETAR)** forecasting, we will use an input layer with 8 inputs, a hidden layer with 4 neurons, and an output layer. The `NNETAR()` function from the `fable` package automates this process. It selects the number of nodes in the input layer based on the estimated order of the linear AR model, which is chosen by minimizing the **Akaike Information Criterion (AIC)**. By default, the size of the hidden layer is half the size of the input layer, rounded to the nearest integer. For seasonal time series, unless specified otherwise by the user, the default choice of seasonal lag is 1.
- A primary benefit of utilizing automated DL forecasting functions, such as NNETAR is the inherent ability to bypass the manual selection of model parameters. The function can automatically initialize these settings. Nonetheless, advanced users can leverage the algorithm's flexibility by manually adjusting the neural network's structure through parameter input. This design accommodates both novice users seeking a straightforward, naïve application and expert users requiring granular control over the model architecture.

  Fit the ETS, ARIMA, and NNETAR models on the `sqrt(lynx_train)` using the following code:

```
set.seed(12345)
lynx_mods <- lynx_train |>
  model(ets_model = ETS(sqrt(lynx_count)),
        arima_model = ARIMA(sqrt(lynx_count)),
        nnetar = NNETAR(sqrt(lynx_count)))

lynx_mods
```

The following output shows which models are selected:

```
# A mable: 1 x 3
     ets_model             arima_model       nnetar
       <model>                 <model>      <model>
1 <ETS(M,A,N)> <ARIMA(2,0,2) w/ mean> <NNAR(8,4)>
```

Unlike statistical models, ML and DL models usually do not provide estimated coefficients that require significant tests. You can use the selected model directly for forecasting.

- As a DL algorithm, NNETAR forecasting also allows for the inclusion of additional explanatory variables during training and forecasting, provided that future values of these variables are known.
- To ensure the reproducibility of the results, seed values are used during training and forecasting, as NNETAR initializes weights randomly. This is essential for achieving consistent results, particularly when simulations are involved. Now, compute forecasts for the next 30 years from all three fitted models using the following code:

```
set.seed(67890); lynx_fcst <- lynx_mods |>
  forecast(h = nrow(lynx_test)*2)
```

The forecasts are generated longer than the test period to evaluate which model could capture the long-term cyclical behavior.

In the NNETAR forecasting process, the network is applied iteratively. The one-step-ahead forecasts use the available historical data, while the two-step-ahead forecasts use the previous one-step-ahead forecast as an input, along with the historical data. This process continues until all required forecasts are generated.

Once forecasts are computed from the models, you can visually evaluate their performance. The following code plots the training data and mean forecasts generated from each model. We add vertical lines to highlight the start and end of a test period in the plots presented in *Figure 13.3*:

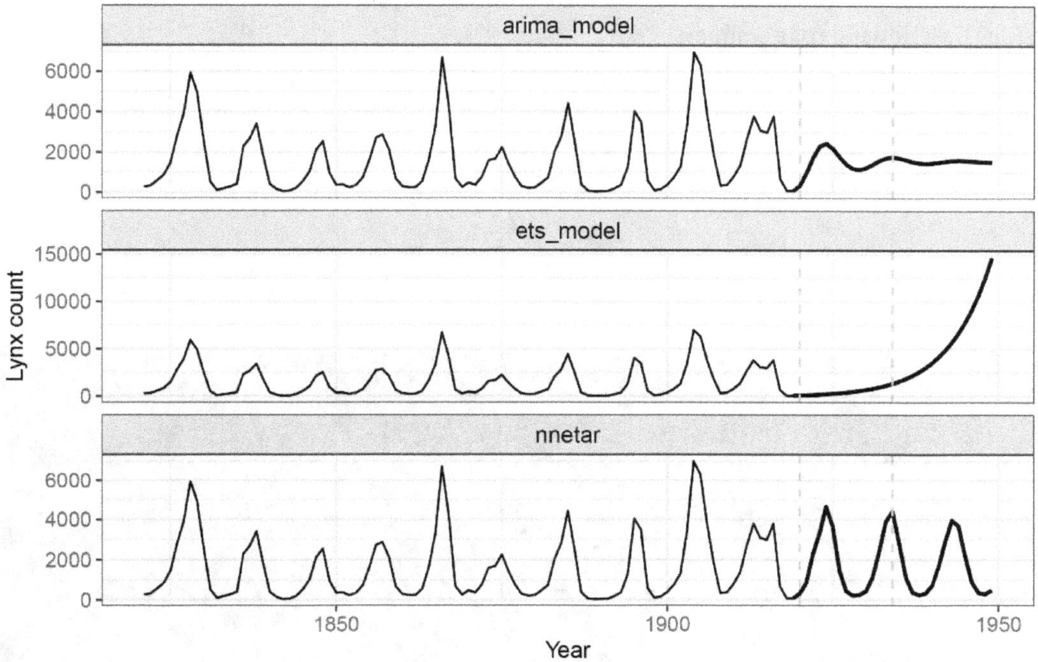


*Figure 13.3: Forecasts of lynx count by the ETS, ARIMA, and NNETAR models*

The mean forecasts by different approaches for the next 30 years are shown in *Figure 13.3*. As expected, the ETS model's forecast appears unrealistic. The observed series lacks both trend and seasonality, the missing key elements that an ETS model relies on for accurate predictions. Clearly, an exponential smoothing model is not suitable for this type of time series. On the other hand, the ARIMA model captures the short-term cyclic patterns well but flattens in the long term. NNETAR effectively captures asymmetry in the cycles, where the declining phase of the cycle is steeper than the increasing phase. This highlights a key difference between the NNETAR and linear ARIMA models: while both can capture cyclicity, the linear ARIMA model produced symmetric cycles, whereas NNETAR captured the asymmetric cycles.

NNETAR appears to capture the cycles and gives visually convincing forecasts over the forecasting horizon. This visually compelling performance is also mirrored in the forecast accuracy measures.

Compute the forecast accuracy metrics in both the training and test sets for all the models using the following code snippet:

```
bind_rows(
  accuracy(lynx_mods),
  accuracy(lynx_fcst |> filter(year <= 1934),
           lynx_test)) |>
  select(.model, .type, ME, RMSE, MAPE, ACF1)
```

The accuracy metrics are presented in the following output:

```
# A tibble: 6 × 6
  .model      .type         ME  RMSE  MAPE    ACF1
  <chr>       <chr>      <dbl> <dbl> <dbl>   <dbl>
1 ets_model   Training -242.   1318. 119.    0.374
2 arima_model Training   83.4   856.  61.7 -0.0395
3 nnetar      Training   31.0   294.  24.1 -0.106
4 arima_model Test      -43.4   893. 103.    0.578
5 ets_model   Test      984.   1449.  49.6  0.689
6 nnetar      Test     -472.   1166. 108.    0.686
```

All accuracy measures support the superior performance of NNETAR in the training. However, the superior performance of NNETAR did not hold in the test dataset according to any forecast accuracy measure.

To understand this inconsistency in forecast performance of NNETAR, plot the test period observations, forecasts, and 95% prediction intervals from ARIMA and NNETAR using the following code:

```
lynx_fcst |>
  filter(.model  %in% c("arima_model", "nnetar")) |>
  autoplot(lynx_tsb, level = 95, size = 0.9) +
  facet_wrap(~.model, ncol = 1, scales = "free_y") +
    labs(y = "Lynx count", x = "") +
    theme(legend.position = "none")
```

*Figure 13.4* presents a visual comparison of forecasts from the fitted ARIMA and NNETAR models:

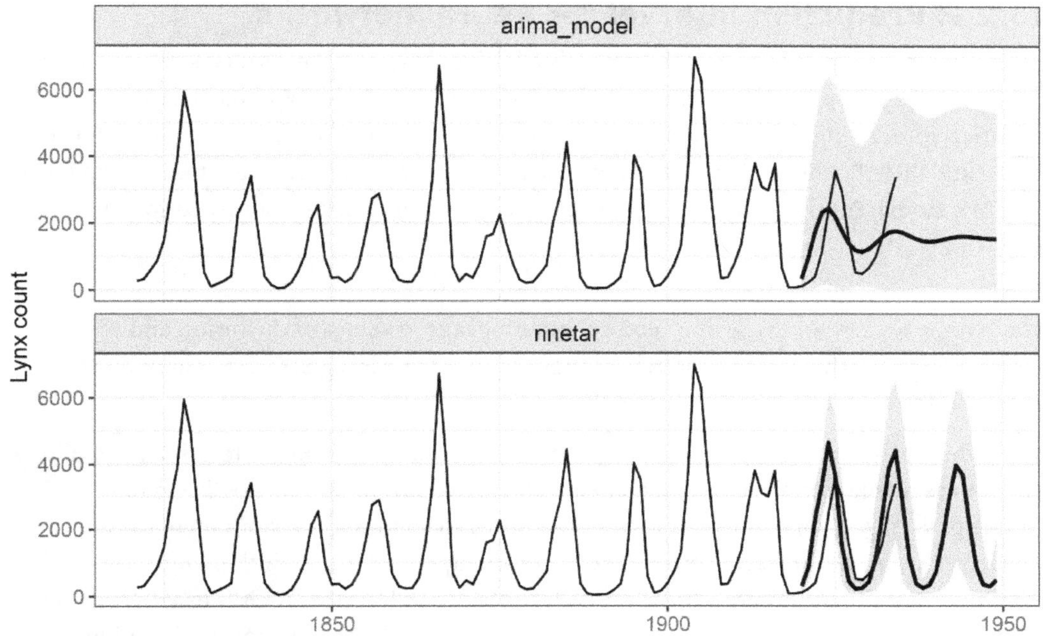


*Figure 13.4: Forecasts with a 95% prediction interval from the ARIMA and NNETAR models on lynx count*

A close inspection of *Figure 13.4* reveals that the forecasts from the ARIMA model are contained within the interval over both short and long term. However, the NNETAR forecasts fall outside of the interval on multiple occasions.

It is important to note that this comparison does not prove or disprove the superiority of one method over another. This is to give you a greater understanding of what to consider in selecting a method to address specific business problems. This comparison reaffirms known facts that, for shorter time series, a statistical method can have very strong performance, and it is also computationally very fast.

The performance of a model could vary at different lengths of the forecast horizon. If you reuse the provided code to run an example using `sunspots.year` data from the `datasets` package, keeping most recent 30 data points as a test sample, you will observe a competitive performance of NNETAR over ARIMA. In that scenario, NNETAR benefits from learning complex behavior from more data points. We leave that as an exercise for you.

### 13.2.2 Prediction interval for neural networks

The prediction interval for NNETAR demands additional explanation as it is derived using a simulation-based approach compared to the closed-form analytical approach used in statistical forecasting models. This is due to the non-existence of an underlying probabilistic model in NNETAR. The prediction interval is derived under an assumed functional form for the NNETAR model. For the lynx example, the underlying model can be assumed as follows:

$$y_t = f(\mathbf{y}_{t-1}) + \epsilon_t$$

where $\mathbf{y}_{t-1} = (y_{t-1}, y_{t-2}, \ldots, y_{t-8})'$ is a vector containing lagged values of the series, and $f$ is a neural network with 4 hidden nodes in a single layer. The error, $\epsilon_t$, series is assumed to be homoscedastic (and possibly also normally distributed).

Under this formulation, it is possible to generate future sample paths by iteratively sampling $\epsilon_t$ from its assumed normal distribution or empirical distribution. By repeatedly simulating the sample path, say, 10,000 times, we will be able to get an empirical distribution of the forecast for each value of the forecast horizon. The lower and upper quantiles of these forecast distributions will serve as prediction intervals for NNETAR forecasts. This is exactly what was used in our example and that is the reason behind using a seed in our code while deriving forecasts using the `NNETAR()` function.

However, DL methods typically operate as black boxes in forecasting applications. The presence of non-zero weights connecting the hidden layers to the output layer may present a substantial challenge in interpreting the specific influence of these auto-curated features in many business contexts. This can be a critical drawback in situations requiring actionable insights into why a particular forecast was generated. To facilitate a clearer, more interpretable understanding of the relationship between input features and the resulting forecast, one might consider exploring alternative ML approaches, which will be detailed in the following section.

# 13.3 ML forecasting

Supervised ML is a blend of art and science that involves creative feature engineering, careful tuning of the algorithmic parameters and intelligent validation to derive the best prediction. In addition to this, top performing ML algorithms benefit from fine-tuning their configuration, often referred to as tuning hyperparameters. The list of algorithms used in ML is already huge, and that list is ever-growing. No wonder there is no R package that works like a breeze to perform forecasting using all ML algorithms. The `forecastML` package (Redell, 2022), inspired by Bergmeir et al. (2018), can help to curate and manage features in an efficient way. This also gives a convenient wrapper for virtually any ML algorithm available in R to be ported into forecasting. The package also provides some fine control over combining forecasts over different forecast horizons and visualization.

It is important to realize that the purpose of this section is not to serve as a user manual for a specific R package. This section also does not intend to cover the details of various ML methods and compare them. It is already well known to practitioners of ML algorithms that tabular data methods based on ensemble learning algorithms, for example, LightGBM and Random Forest, show superior predictive performance compared to a single linear model. So, forecasting based on properly tuned and validated ensemble learning algorithms is expected to perform better. However, anyone with some level of knowledge of these algorithms will realize that writing code for implementing alternative methods will be lengthy even for simple use cases. This is due to the difference in the way these algorithms use input data, and the difference in the list of hyperparameters that require tuning these algorithms. To avoid those complexities and to keep things concise, we will illustrate the core idea using a regularized regression algorithm. This will offer both an intuitive understanding and a foundation for developing forecasts using other ML algorithms with suitable customization.

The `forecastML` package offers direct forecasting that comprises three steps:

1. Developing a series of horizon-specific forecast models using the chosen ML algorithm
2. Assessing model generalization performance across a variety of holdout datasets
3. Selecting those models that consistently performed the best at each forecast horizon and combining them to produce a single ensemble forecast

You will find a detailed example of an ML model used for forecasting next.

## 13.3.1 Application of ML in forecasting

All ML models work better when a feature set related to the target variable needs prediction and is curated in the context of the prediction problem. Using ML models for forecasting is no exception. When using an ML algorithm for forecasting univariate time series, a feature set could be curated based on the lagged values of the series. The complexity arises when the resulting dataset needs to be prepared in a way that it is suitable for feeding into prediction algorithms to serve the purpose of forecasting as described in *Section 13.1*.

The `forecastML` package provides a flexible framework for building such ML pipelines tailored for time-series forecasting. The package also allows you to use any ML model for time-series forecasting, such as `randomForest` and `xgboost`. You can also train different models for different forecast horizons using custom model wrappers. In this application, we will use a single model, quantile regularized regression (via the `glmnet()` function) for all forecast horizons.

For illustration, we will use `AirPassengers` data from the `datasets` package. Let's create a plot of the observed series and examine what time-series characteristics are present in it. A time plot of the monthly airline passenger count is presented in *Figure 13.5*:

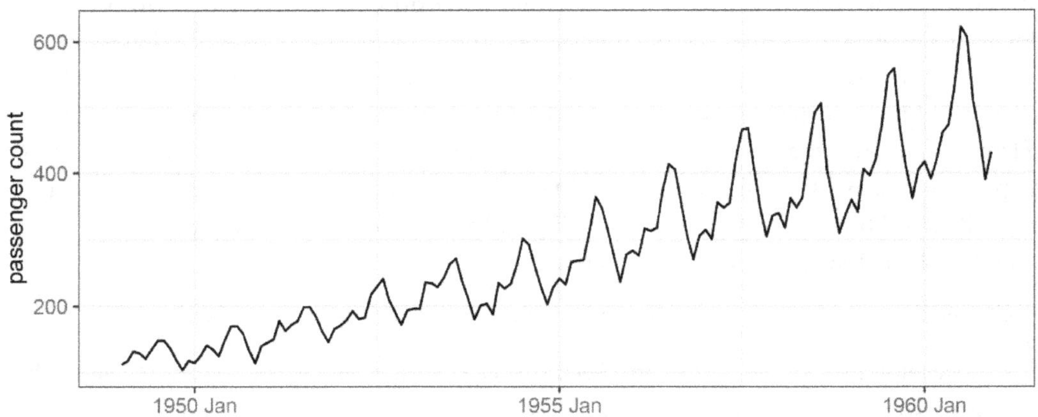


*Figure 13.5: Monthly airline passenger numbers, 1949 - 1960 (AirPassengers)*

The observed series in *Figure 13.5* exhibits a strong upward trend and strong monthly seasonality. The seasonality is combined multiplicatively as the variance of the seasonality increased as time progressed. Any ML model will first create some features to capture trend and seasonality. Let's initiate the process by creating year and month features from the AirPassengers data using the following code:

```
apdf <- AirPassengers |>
  as_tsibble() |>
  rename(passengers = value) |>
  mutate(year = lubridate::year(index),
         month = lubridate::month(index)) |>
  tibble() |>
  select( - index)
```

The year feature helps us to capture the trend in the data, whereas the month feature will capture the seasonality.

Before proceeding to fitting an ML model for forecasts, it is crucial to understand the steps involved in the whole process and what information is required at various steps. The `forecastML` package provides some useful functions that we can leverage, such as for data preparation, how to train the model in sliding windows, and so on. Next, we highlight some important decision variables required for the proper use of an ML model in the context of forecasting. The code is presented in small chunks for illustration and to help explain the logic. The code chunks are not supposed to be executed independently but should be combined in a single run. The end-to-end code to run a regularized regression on the AirPassengers data is provided at the end of the section.

- **Multi-horizon forecasting**: Like statistical models, an ML model can produce multi-horizon forecasts. However, the process is not as intuitive as the statistical models. You need to create proper lagged features from the observed data that align with the multiple forecast horizons. The forecastML package allows you to forecast multiple time steps ahead simultaneously, such as 1- step ahead, 3- steps ahead, and so on. In our example, there are 144 observations in the AirPassenger data. We will keep the first 132 for training, and the remaining 12 observation for the test set. To show you how forecasts from an ML model perform on a longer horizon, we will use up-to 36-steps-ahead forecasting (that is, $h = 36$):

```
trn_idx <- 132
horizons <- 36
```

- **Automatic lag creation**: Once the training, test and forecast horizons are set, the next step is to create lagged features from the observed time series, which are critical for many ML algorithms. The look back window in the lagged features needs to be at least as big as the largest forecast horizon. We use 39 in our example. The `create_lagged_df()` function from the `forecastML` package allows you to curate a set of lagged features for each forecast horizon separately. You can control the number of look-back periods (lags) using the lookback option. We have three columns in our main training data, passengers, year, and month. Each training set will have the target column (passengers) and lagged features computed from the three variables. The following code will produce 36 training sets, one for each forecast horizon:

```
trn_idx <- 132
data_train <- create_lagged_df(
  apdf |> slice(1:trn_idx),
  type = "train",
  method = "direct",
  outcome_col = 1,
  lookback = 1:(max(horizons) + 3),
  horizons = 1:horizons)
```

Each training set has 93 rows ($132 - 39$). The number of columns (total number of lagged features) reduces as the forecast horizon increases. Keeping one column aside for the target variable, the training set for horizon 1 has 117 features ($39 \times 3$), whereas the training set for the 36th horizon has only 12 lagged features, as shown in the following output:

```
dim(data_train[[1]])
# [1]  93 118

dim(data_train[[36]])
# [1] 93 13
```

- **Dynamic features**: You can add time-varying features, such as the world's population, and economic indicators, to improve your forecasts. These features change with time and provide additional context to the model. We don't have such features in this exercise.
- **Static features**: These are features that do not change over time (for example, geographic location, or product category). They are treated differently and can be used to improve the model's performance across different segments. We don't have such features in this exercise.
- **Choice of ML algorithm**: As mentioned earlier, `forecastML` provides the flexibility to apply a single ML algorithm across all horizons or multiple ML models on individual

horizons. You have a wide variety of models to choose from, depending on the context of the forecast problem. In this example, we will use the regularized regression algorithm for all forecast horizons. A brief discussion on this algorithm is presented in *Section 13.3.2*. You may need to write a custom function to apply the model of your choice in the training sets as we have done in the following code snippet:

```
model_fn_init <- function(data){
  x <- as.matrix(data[, -1, drop = FALSE])
  y <- as.matrix(data[, 1, drop = FALSE])
  model <- glmnet(x, y, type.measure = "mse")
}
```

- **Custom models and pipelines**: You can define your own custom model wrapper functions, allowing for full control over the ML model and the tuning process. This enables advanced users to implement customized, state-of-the-art models.
  For example, we are interested in cross-validation for finding the optimal model for regularized regression. We also want to find prediction intervals for the fitted regularized regression. To achieve the latter, we use the estimated 5% and 95% regularized regression quantiles. This requires us to customize the model function accordingly. We use the `glmnet` package (Friedman et al., 2024) for the mean prediction model and the `hqreg` package (Yi, 2022) for quantile prediction models. The custom wrapper function is presented in the following code snippet:

```
model_fn_cust <- function(data, outcome_col = 1) {
  x <- as.matrix(data[, -outcome_col, drop = FALSE])
  y <- as.matrix(data[, outcome_col, drop = FALSE])
  if(tau_vec[[k]] == 0.5){
    model <- cv.glmnet(
      x,
      y,
      type.measure = "mse",
      nfolds = cv_nfolds
    )
  }else{
    model <- cv.hqreg(
      x,
      y,
      type.measure = "mse",
      tau = tau_vec[k],
      method = "quantile",
```

```
        nfolds = cv_nfolds
      )
    }
  }
```

- **Sliding windows for training**: The `forecastML` package also allows you to create training windows from the lagged data, enabling you to take advantage of both short- and long-term historical data. This feature is useful for expanding window strategies in time series forecasting.
  We will use all available data points to create time-contiguous validation datasets for model evaluation. We will use functions from the `forecastML` package, such as `create_windows()` and `train_model()`, as well as the custom function we presented earlier (`model_fn_cust()`) for sliding windows training. The `create_windows()` function with `window_length=0` makes sure to train the model on the entire dataset. The `train_model()` function trains the user-defined model across forecast horizons and validation datasets. When the custom `model_fn_cust()` function is supplied inside `train_model()`, the complete process involves model training for each forecast horizon with 10-fold cross-validation. A seed is specified to ensure the reproducibility of the results. The following code performs these operations:

```
windows <- create_windows(data_train,
                          window_length = 0)
set.seed(54321);model_results <-
  train_model(data_train,
              windows,
              model_name = "glmnetQuant",
              model_function = model_fn_cust)
```

- **Future data handling**: The `forecastML` package allows you to make predictions on future data and seamlessly generate forecasts for multiple time periods in the future. All the features used for model training should be available in the forecast periods so that forecast values can be computed. The `create_lagged_df()` function with `type="forecast"` curates lagged features for the forecast periods. Then we need to apply the `predict()` function on the trained model object and pass the future features dataset to generate forecasts. Lastly, the `combine_forecasts()` function from the `forecastML` package aggregates (median) across the horizon specific models to provide one forecast value. The following code snippet shows you how to perform these operations:

```
feature_data <- create_lagged_df(
  apdf |> slice(1:trn_idx),
  type = "forecast",
  method = "direct",
  outcome_col = 1,
  lookback = 1:(max(horizons) + 3),
  horizons = 1:horizons)

# Prediction function for ML model
predict_fn <- function(model, data) {
  data_pred <- as.data.frame(
    predict(model, as.matrix(data)))
}

data_forecasts <- predict(
  model_results,
  prediction_function = list(predict_fn),
  data = feature_data )

# combine forecasts
data_forecasts <- combine_forecasts(data_forecasts)
```

The complete code for forecasting using regularized regression with 10-fold cross-validation for model tuning and using quantile regression to derive the 95% prediction interval is presented for convenience.

*A note on execution*: The execution time for the following code is dependent on available computing power. It is highly recommended to conduct a preliminary small-scale test run to estimate performance before a full execution. This can be achieved by utilizing reduced parameters, such as specifying a single quantile value (e.g., `tau_vec <- 0.5`) and a shorter forecast horizon (e.g., `horizons <- 12`), and configuring the `hqreg::cv.hpreg()` function with a smaller number of cross-validation folds (e.g., `nfolds=4`).

```
forecasts_store <- list(NA)
tau_vec <- c(0.05, 0.5, 0.95)
horizons <- 36
trn_idx <- 132
cv_nfolds <- 10
```

```
for (k in 1:length(tau_vec)) {
  # Step 1: Create data and build model
  data_train <- create_lagged_df(
    apdf |> slice(1:trn_idx),
    type = "train",
    method = "direct",
    outcome_col = 1,
    lookback = 1:(max(horizons) + 3),
    horizons = 1:horizons )

  # ML model: wrapper for mean and quantile forecast
  model_fn_cust <- function(data) {
    x <- as.matrix(data[, -1, drop = FALSE])
    y <- as.matrix(data[, 1, drop = FALSE])
    if (tau_vec[[k]] == 0.5) {
      model <- cv.glmnet(
        x, y,
        type.measure = "mse",
        nfolds = cv_nfolds)
    }else{
      model <- cv.hqreg(
        x, y,
        type.measure = "mse",
        tau = tau_vec[k],
        method = "quantile",
        nfolds = cv_nfolds)
    }
  }
  windows <- create_windows(
    data_train,
    window_length = 0)

  set.seed(54321);model_results <-
    train_model(
      data_train,
      windows,
      model_name = "glmnetQuant",
      model_function = model_fn_cust)

  # Step 2: forecast
  feature_data <- create_lagged_df(
```

```
    apdf |> slice(1:trn_idx),
    type = "forecast",
    method = "direct",
    outcome_col = 1,
    lookback = 1:(max(horizons) + 3),
    horizons = 1:horizons )

  # Prediction function for ML model
  predict_fn <- function(model, data) {
    data_pred <- as.data.frame(
      predict(model, as.matrix(data))
    )
  }

  data_forecasts <- predict(
    model_results,
    prediction_function = list(predict_fn),
    data = feature_data )

  # Step 3: combine
  forecasts_store[[k]] <- combine_forecasts(data_forecasts)
}
```

After the code runs, organize the results (detailed code is available in the GitHub repo), which will produce a plot of training (from 1954), test, and forecast values. The resulting forecasts are plotted in *Figure 13.6*:

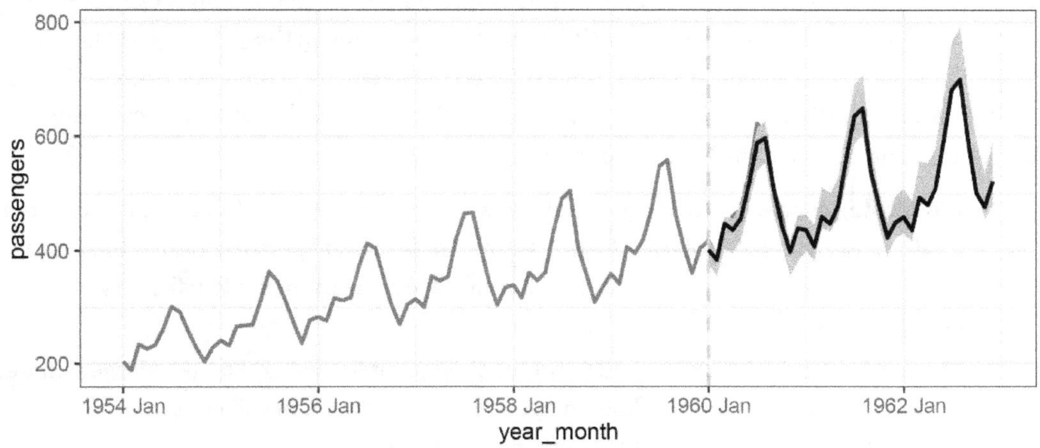


*Figure 13.6: ML-based forecast and 95% prediction intervals for AirPassengers data*

*Figure 13.6* shows the forecast performance of the regularized regression in the 12 months in the test set. We can also evaluate the performance in the additional 24 months period. The bold solid line is the mean forecast, whereas the shaded area shows the region bounded by 2.5% and 97.5% quantile forecasts, which is the ML equivalent of prediction intervals. The forecasts seem to extend the trend and seasonality well. However, the variance in the seasonality increases as time progresses, which needs further evaluation.

`forecastML` stands out for its flexibility, allowing users to apply ML methods to time series forecasting with ease. Its ability to handle multiple forecast horizons, dynamic or static features, and custom model integration makes it a powerful tool for complex forecasting problems, while its time-aware cross-validation ensures that models are properly evaluated.

We intentionally avoid comparisons of ML-based forecast with statistical alternatives as one can always think of new, innovative feature engineering in an attempt to improve forecasting performance. It is worth mentioning that these methods can be used as a black box as long as you know how to write code for tuning a ML model and deriving predictions without being a master of time series. The best performing forecasting algorithm for the M5 competition (Makridakis et al., 2022) was developed using ML methods. We intentionally avoid further discussion of M5's winning method as the primary focus of the competition was forecasting large data with multiple time series, which is outside the scope of this book. You can use an ML-based forecasting model for univariate time series if you have identified that a linear model might not fit your data well.

## 13.3.2 Quantile regularized regression: A brief summary

Quantile regularized regression combines quantile regression with Elastic Net regularization. It aims to estimate conditional quantiles of a response variable while controlling for model complexity using elastic net regularization. The core of regularized regression is explained in Zou et al. (2005) and Friedman et al. (2010), and implementation in the `glmnet` package (Friedman et al., 2024). Quantile prediction using regularized regression is implemented in the `hqreg` package (Yi, 2022). This kind of regression model is particularly useful as it can handle two common problems in the business domain:

- **Multicollinearity**: Features are correlated among themselves and as a result estimated model parameters are volatile. At its extreme (in the case of perfect multicollinearity), standard regression model estimation will not work due to the rank deficiency in the design matrix.
- **The $n << p$ problem**: This refers to the small $n$ (sample size) and large $p$ (large number of parameters) problem, also known as the $n << p$ problem. In such a scenario, standard regression fails due to fewer data points compared to the number of parameters needed to be estimated for modeling.

In our example, the lag features were created from the observed series, year, and month columns, which have 93 rows (instances) with 118 columns. Correlations between these features often exceeding $0.8$, exhibiting multicollinearity, hence, the choice of quantile regularized regression as an ML model for forecasting, rather than other, more popular models, such as LightGBM.

Let's see how regularized regression works. For simplicity, we will use the notion, $Y$ for the outcome variable for the prediction and $X$ for all features. We will also avoid time subscript to simplify our exposition and use $i$ as subscript to denote instance/observation as used in standard ML literature. In general, a quantile regression aims to estimate the $\tau$-th quantile. The objective function for quantile regression is as follows:

$$Q(\beta) = \sum_{i=1}^{n} \rho_\tau (Y_i - X_i\beta)$$

where $Y_i$ is the response for the $i$-th observation, $X_i$ is the predictor vector for the $i$-th observation, and $\rho_\tau(u)$ is the check function, defined as follows:

$$\rho_\tau(u) = \begin{cases} \tau u & \text{if } u \geq 0 \\ (\tau - 1)u & \text{if } u < 0 \end{cases}$$

The elastic net regularization is a mix of $L1$ (Lasso) and $L2$ (Ridge) regularization. The elastic net penalty term is as follows:

$$\Omega(\beta) = \alpha \sum_{j=1}^{p} |\beta_j| + \frac{1-\alpha}{2} \sum_{j=1}^{p} \beta_j^2$$

where:

- $\alpha \in [0,1]$ controls the balance between $L1$ and $L2$ penalties
- $\sum_{j=1}^{p} |\beta_j|$ is the $L1$ (Lasso) penalty
- $\frac{1-\alpha}{2} \sum_{j=1}^{p} \beta_j^2$ is the $L2$ (Ridge) penalty as used in any ML algorithm.

Combining quantile regression with elastic net regularization, the objective function becomes the following:

$$L(\beta) = \sum_{i=1}^{n} \rho_\tau (Y_i - X_i\beta) + \lambda \left( \alpha \sum_{j=1}^{p} |\beta_j| + \frac{1-\alpha}{2} \sum_{j=1}^{p} \beta_j^2 \right)$$

where $\lambda \geq 0$ is the regularization parameter controlling the overall strength of the regularization, and $\alpha \in [0,1]$ balances between $L1$ and $L2$ regularization.

The term $\sum_{i=1}^{n} \rho_\tau (Y_i - X_i\beta)$ is the sum of the quantile-specific losses. It captures the deviation of the predicted values from the observed values, weighted according to the quantile $\tau$.

The term $\lambda\left(\alpha \sum_{j=1}^{p}|\beta_j| + \frac{1-\alpha}{2}\sum_{j=1}^{p}\beta_j^2\right)$ penalizes large coefficients and controls model complexity. The regularization parameter $\lambda$ scales the strength of the penalty.

Quantile elastic net regularized regression extends quantile regression by incorporating elastic net regularization. The combined objective function balances the quantile-specific loss with a regularization term that mixes $L1$ and $L2$ penalties. This approach is useful for high-dimensional data and can handle cases where predictors are highly correlated or when a balance between sparsity and shrinkage of coefficients is desired.

It is important to emphasize that a strong foundation is crucial for effectively applying these techniques to real-world problems. Relying solely on ML approaches without a thorough understanding of the underlying algorithm may not yield the best return on investment.

Additionally, many research and business problems extend beyond making predictions. In practice, the application of predictive algorithms often focuses on delivering high-quality forecasts. However, this can be a limiting factor, especially when the ability to quickly compute prediction intervals becomes critical, which is often why traditional statistical models remain a preferred choice for their computational efficiency in this area.

This concludes our discussion of the core ML method applied to forecasting within this chapter. We will now transition to an alternative methodology that has gained significant traction, predominantly among data science practitioners, ML users, and applied researchers.

## 13.4 Forecasting as a curve fitting: The Prophet method

Prophet is an open-source forecasting tool developed by the Meta research team (Taylor et al., 2018). It is designed for time-series forecasting, particularly for data with strong seasonal patterns and historical data spanning several seasons. This model was introduced originally for forecasting daily data incorporating weekly and yearly seasonality, plus holiday effects and was later extended to cover more types of seasonal data. It works best with time series that have strong seasonality and several seasons of historical data. Some key features of the Prophet method are as follows:

- **Additive model**: Prophet uses an additive model where non-linear trends are fit with yearly, weekly, and daily seasonality, plus holiday effects. You can get good forecasts with minimal effort.
- **Ease of use**: It is designed to be user-friendly and fully automated, making it accessible even for those who are not experts in time series forecasting. It automatically detects

and models yearly, weekly, and daily seasonality which is great for data with strong seasonal patterns.

- **Robustness**: The model is robust to missing data and outliers, which means it can still provide accurate forecasts even if your data is not perfect. Prophet can model non-linear trends and can handle changes in trends, which is useful for data with complex patterns.
- **Customization**: While it works well out of the box, it also allows for manual tuning by data professionals. You can add your own seasonalities, holidays, and other special events to improve the forecast.

The Prophet model is based on a decomposable time-series model, which can be expressed as follows:

$$y(t) = g(t) + s(t) + h(t) + \epsilon_t$$

where:

- $g(t)$ represents the trend component, which can be either linear or logistic.
- $s(t)$ captures the seasonal effects, which can include yearly, weekly, and daily seasonality.
- $h(t)$ accounts for holiday effects, which are modeled as dummy variables.
- $\epsilon_t$ is the error term, representing any unexplained variability (white noise).

The trend, $g(t)$, is modeled as a piece-wise linear function with automatic changepoint detection. This allows the model to adapt to changes in the trend over time. Alternatively, a logistic growth model can be used to set an upper bound on the trend, which is useful for modeling saturating growth. Unlike a linear trend, the logistic growth curve accounts for growth that increases quickly at first, then slows down as it approaches an upper limit (carrying capacity). The general form of the logistic growth model is as follows:

$$g(t) = \frac{C}{1 + \exp\big(-k(t - m)\big)}$$

where:

- $C$ is the carrying capacity (the maximum value that $g(t)$ can reach)
- $k$ is the growth rate
- $m$ is the midpoint or the time at which the growth rate is highest
- $t$ is the time

In the context of Prophet, the logistic growth model is used to forecast time-series data with an upper limit that growth approaches asymptotically.

Seasonal effects, $s(t)$, are modeled using the Fourier series. The Fourier series helps to model periodic fluctuations (such as daily, weekly, or yearly seasonality) by using a sum of sine and cosine terms. By default, Prophet uses order $10$ for yearly seasonality and order 3 for weekly seasonality. For a seasonal period $P$ (e.g., $365.25$ days for yearly seasonality), the seasonal component is modeled as follows:

$$s(t) = \sum_{n=1}^{N} \left( a_n \cdot \sin\left(\frac{2\pi nt}{P}\right) + b_n \cdot \cos\left(\frac{2\pi nt}{P}\right) \right)$$

where:

- $N$ is the number of Fourier terms to use
- $a_n$ and $b_n$ are the coefficients to be estimated (parameters of the model)
- $P$ is the period of the seasonality (e.g., $365.25$ for yearly, $7$ for weekly)

You can set the number of Fourier terms, $N$, which determines how many sine and cosine pairs are used to model the seasonality. A higher value of $N$ captures more detail, while a lower value smooths the seasonal component.

Holidays, $h(t)$, are included as dummy variables, allowing the model to account for specific events that can affect the time series.

The model is estimated using a Bayesian approach to allow for automatic selection of the change points and other model characteristics. In Bayesian estimation, priors can be specified for all parameters, which helps incorporate prior knowledge into the model. The default priors are often sufficient for many applications, but they can be adjusted based on specific needs. It is worth realizing that Prophet is a curve-fitting exercise, which is inherently different from time-series models that explicitly account for the temporal dependence structure in the data.

## 13.4.1 Prophet forecasting

Forecasting by the Prophet method can be performed using the `fable.prophet` package in R (O'Hara-Wild et al., 2024). For this example, we use monthly food retail turnover in Australia from the `tsibbledata` package. We use data from January 2000 to December 2015 for deriving forecasts until December 2018. To evaluate forecasting performance against the Prophet model, we also fit ETS and ARIMA models, as detailed in the following steps:

1. At first, split the training and test sets for Prophet tuning and forecast evaluation using the following code:

```
aus_food_retail <- tsibbledata::aus_retail |>
  filter(Industry == "Food retailing") |>
  index_by(Month) |>
  summarise(National_turnover = sum(Turnover, na.rm=T)) |>
  filter(Month > yearmonth("1999 Dec"))

aus_food_retail_train <- aus_food_retail |>
  filter(Month < yearmonth("2016 Jan"))

aus_food_retail_test <- aus_food_retail |>
  filter(Month >= yearmonth("2016 Jan"))
```

2. The `prophet()` function from the `fable.prophet` package allows you to fit a Prophet model with some users' input, such as the shape of growth (linear versus logistic), how seasonality is incorporated (additive versus multiplicative), and the number of terms in the partial Fourier sum when estimating seasonality, among other arguments. For our example series, we use a linear growth term and multiplicative seasonality with four Fourier series. The automated ARIMA and ETS models will be used on the same data. After fitting, use the estimated models to compute three-year forecasts. You can perform model fitting and compute forecasts using the following code:

```
foodretail_fit <- aus_food_retail_train %>%
  model(
    arima = ARIMA(National_turnover),
    ets = ETS(National_turnover),
    prophet = prophet(
      National_turnover ~ growth("linear") +
      season(period = 12, order = 4, type = "multiplicative" )
      )
  )

foodretail_fc <- foodretail_fit %>%
  forecast(h = "3 years")
```

3. The components extracted from a Prophet fitted model can be plotted for visual inspection using the following code:

```
foodretail_fit |>
  select(prophet) |>
  fabletools::components() |>
  autoplot()
```

*Figure 13.7* presents a plot of components extracted from the fitted Prophet model:

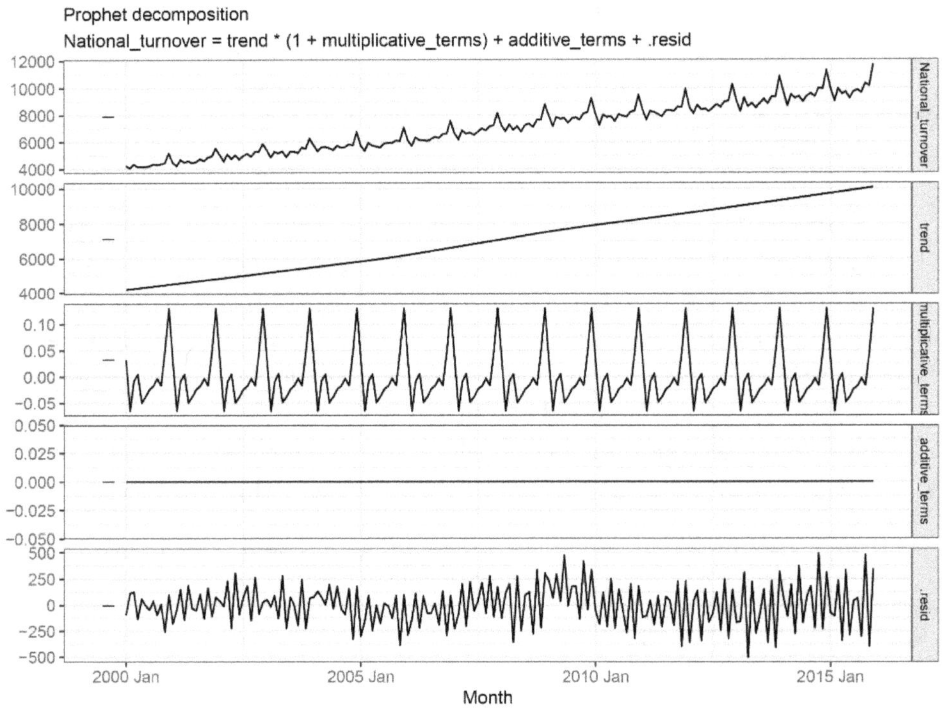


*Figure 13.7: Prophet's decomposition of national food retail turnover*

The components in *Figure 13.7* show that the estimated Prophet model captured the trend and seasonal components well.

4. Let's compute forecast accuracy metrics in the training and test sets:

```
bind_rows(
  accuracy(foodretail_fit),
  accuracy(foodretail_fc,aus_food_retail)
```

```
) |>
  select(.model, .type, RMSE,MPE, MASE, ACF1)
```

The results are shown in the following output:

```
# A tibble: 6 × 6
  .model  .type     RMSE      MPE  MASE     ACF1
  <chr>   <chr>    <dbl>    <dbl> <dbl>    <dbl>
1 arima   Training 111.  -0.0394  0.230  0.0102
2 ets     Training  95.8  0.00259 0.205  0.00735
3 prophet Training 204.  -0.0672  0.462 -0.616
4 arima   Test     194.  -1.23    0.466 -0.0837
5 ets     Test     194.  -1.40    0.469 -0.00199
6 prophet Test     327.  -1.31    0.776 -0.727
```

In this example, both the ARIMA and ETS models show superior performance in the training and test sets compared to the Prophet model. You could try improving the Prophet model by choosing different parameter combinations.

5. Let's examine the residual diagnostic plots from the Prophet model:

```
foodretail_fit |>
  select(prophet) |>
  gg_tsresiduals()
```

6. The residual plots are presented in *Figure 13.8*:

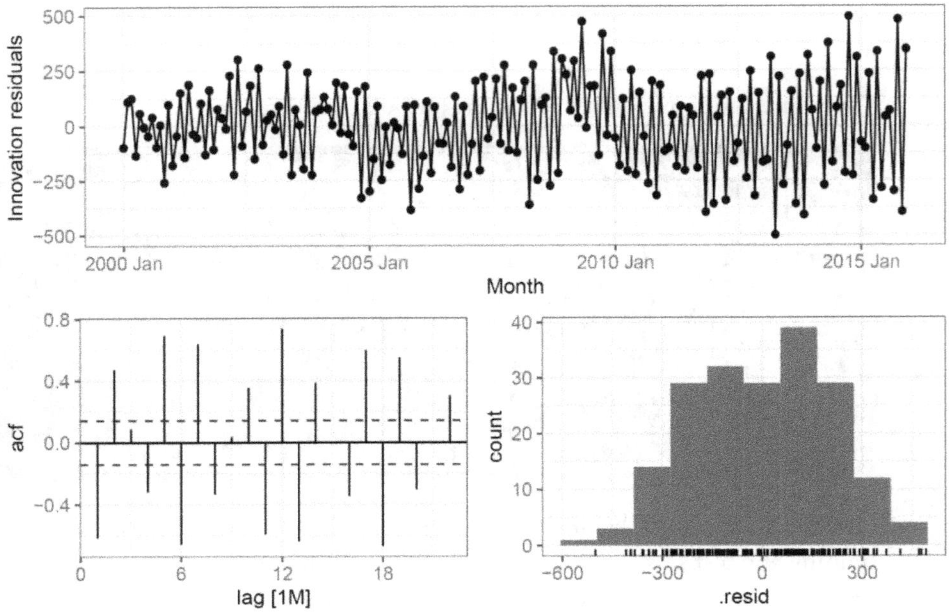


*Figure 13.8: Residual diagnostics plots from the fitted Prophet method on food retail turnover data*

The significant spikes in the ACFs of the Prophet model residuals in *Figure 13.8* suggest that the Prophet method left a considerable amount of serial correlation in the residuals. This is a common shortcoming of the Prophet method. It might work when the characteristics of observed data can be properly captured with combinations of parameters; however, Prophet shows inferior forecasting performance compared to alternative models in business forecasting scenarios in practice.

Throughout this book, we have explored the application of various forecasting models and methods. A consistent observation is that no single method is uniformly superior across all contexts. This realization leads to the powerful concept of employing multiple forecasting tools in conjunction. By combining the results from diverse individual methods, you can often derive a composite forecast that outperforms many of its constituent parts. The next section will dive into the ensemble forecasting approach.

# 13.5 Forecast ensemble

So far, we've explored various approaches to forecasting. We started with simple naïve methods for data that has no apparent patterns. We then studied statistical models, such as ETS and ARIMA, for data with consistent discernible patterns. In this chapter, we expanded into more advanced computation-heavy techniques, such as DL methods, ML algorithms, and the Prophet method. Applying these methods across diverse datasets and comparing alternative approaches on the same data underscored an important insight: there is no single "best" forecasting model that universally applies to all situations.

For example, as shown in *Figure 13.3*, the NNETAR forecast may appear visually suitable for capturing cyclical patterns in the lynx dataset; however, it significantly deviates from the true values in short-term forecasting, both in point and interval estimates. Similarly, *Figure 13.6* illustrates that ML-based forecasts on the `AirPassengers` series, which exhibits strong trends and seasonality, also face challenges in producing accurate point and interval forecasts.

Differences in forecasts produced by various methods are well documented in the literature, with established explanations for why certain approaches may outperform others. Generally, these differences relate to how effectively each method leverages the available information within the data, as well as the richness of the historical data itself. Factors, such as the length of historical records and the strength of systematic components (e.g., trends or seasonality) can impact each model's performance. As you gain expertise, you will be better equipped to choose methods that are conceptually well suited for specific forecasting scenarios and to fine-tune models for improved accuracy.

Although no single method consistently delivers the "best" forecast, you can enhance forecast accuracy by applying a simple yet powerful technique: averaging forecasts from multiple models. This may sound surprising, but combining forecasts from various approaches applied to the same time series often yields more accurate results. This ensemble technique is widely endorsed in applied forecasting. The averaging process can be implemented in various ways; here, we demonstrate it using a straightforward arithmetic mean, though more sophisticated weighted averages may also be employed to further refine forecasts.

We illustrate how to perform a forecast ensemble from multiple models on the `AirPassengers` series in the following steps:

1. **Split data into training and test sets**: The `AirPassengers` series is organized into a `tsibble` object. For model training, we will use observations from January 1949 to December 1956. The remaining four years are reserved for testing:

```
ap_tsb <- AirPassengers |>
  as_tsibble() |>
  rename(month = index, passenger_count = value)
ap_train <- ap_tsb |>
  filter(month < yearmonth("1957 Jan"))
ap_test <- ap_tsb |>
  filter(month >= yearmonth("1957 Jan"))
```

2. **Fit multiple forecasting models**: Five different forecasting models are fitted: seasonal naïve, TSLM, ETS, ARIMA, and Prophet. You can choose any number of forecasting models depending on the context of the forecasting problem at hand. As a general idea, the selection of the model should leverage the strengths and support the weaknesses of each other. Fit the models on the training data, and at the same time, you can make an ensemble of the fitted models by taking average of them, as shown in the following code:

```
ap_models <- ap_train |>
  model(
    snaive = SNAIVE(passenger_count),
    tslm = TSLM(log(passenger_count) ~ trend() + season()),
    ets = ETS(log(passenger_count) ~ season("A")),
    arima = ARIMA(log(passenger_count)),
    prophet = prophet(
      passenger_count ~ growth("linear") +
      season(period = 12, order = 4, type = "multiplicative"))
    ) |>
  mutate(ensemble = (snaive + tslm + ets + arima + prophet) / 5 )

ap_models
```

This will produce the following output:

```
# A mable: 1 x 6
    snaive    tslm          ets                                    arima
   <model> <model>      <model>                                  <model>
1 <SNAIVE>  <TSLM> <ETS(A,A,A)> <ARIMA(2,0,0)(0,1,1)[12] w/ drift>
# 2 more variables: prophet <model>, ensemble <model>
```

Following previous examples shown throughout the book, you can evaluate each model's goodness of fit, the significance of estimated parameters and residual diagnostic checks, and update model specifications if required.

3. **Generate forecasts**: Once you have identified the candidate models and combined them into an ensembled object, next, generate forecasts using them. The following code will generate monthly forecasts for the next four years:

```
ap_fcst <- ap_models |>
  forecast(h = "4 years") |>
  mutate(.model = str_c(.model, "_forecast"))

ap_fcst |>
  filter(row_number() <= 6 | row_number() > (n()-6))
```

This will produce a `fable` object with 6 keys (models) and 288 rows ($6 \times 48$). Let's see the first 6 rows and the last 6 rows from the `fable` object as shown in the following output:

```
# A fable: 12 x 4 [1M]
# Key:     .model [2]
   .model                month passenger_count .mean
   <chr>                 <mth>          <dist> <dbl>
 1 snaive_forecast   1957 Jan    N(284, 1071)   284
 2 snaive_forecast   1957 Feb    N(277, 1071)   277
 3 snaive_forecast   1957 Mar    N(317, 1071)   317
 4 snaive_forecast   1957 Apr    N(313, 1071)   313
 5 snaive_forecast   1957 May    N(318, 1071)   318
 6 snaive_forecast   1957 Jun    N(374, 1071)   374
 7 ensemble_forecast 1960 Jul         608.542  609.
 8 ensemble_forecast 1960 Aug        599.8902  600.
 9 ensemble_forecast 1960 Sep        535.9316  536.
10 ensemble_forecast 1960 Oct        465.2943  465.
```

```
11 ensemble_forecast 1960 Nov        411.8913  412.
12 ensemble_forecast 1960 Dec         468.112  468.s
```

For the fitted models, the `.mean` column provides the point forecast values, and `passenger_count` column provides the distribution around that mean. For ensemble forecast, both of the columns provide the point forecasts.

4. **Forecast evaluation**: Evaluate the performance of each of the candidate models. First, visually inspect forecasts from each of five models alongside the ensembled approach. For better visualization, the last two years of the training data (solid grey line), test set (solid grey line), and forecasts from each model (solid black line) are plotted in *Figure 13.9* which we can inspect visually:

```
ap_fcst |>
  mutate(.model = factor(.model,
          levels = unique(ap_fcst$.model))) |>
  autoplot(level = NULL, color = "black", linewidth = 0.9 ) +
  scale_color_manual(values = c("black"= "black")) +
  autolayer(ap_tsb |> filter(month >= yearmonth("1955 Jan")),
            color = "grey", linewidth = 0.7) +
   geom_vline(xintercept = as.Date(yearmonth("1957 Jan")),
             color = "grey", linetype = "dashed", linewidth = 0.8) +
  facet_wrap(.model~.) +
  labs(x  = "",
       y = "Air passengers count (in 000's)") +
  theme(legend.position = "none")
```

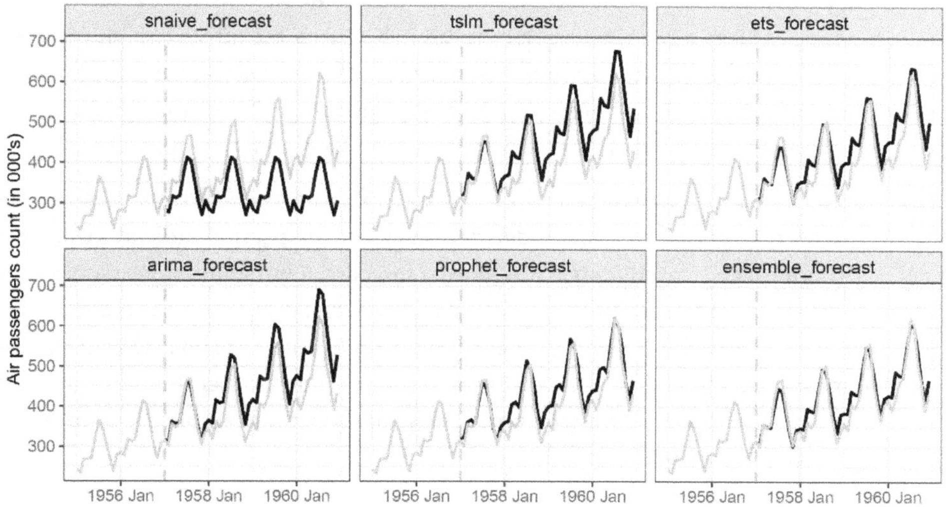


*Figure 13.9: Visual comparison of five candidate models with ensemble forecasts on AirPassenger series*

*Figure 13.9* reveals variations in the performance of individual methods on the same observed data, some showing instances of underestimation or overestimation. However, the ensemble forecast, calculated as the average of the five approaches, demonstrates notably strong performance, effectively smoothing out these discrepancies.

Let's quantify the forecast performance by computing accuracy metrics in the test set:

```
ap_fcst |>
  accuracy(ap_tsb) |>
  select(.model, .type, RMSE, MAPE, MASE) |>
  arrange(MASE)
```

The result is shown in the following output:

```
# A tibble: 6 × 5
  .model            .type  RMSE  MAPE  MASE
  <chr>             <chr> <dbl> <dbl> <dbl>
1 ensemble_forecast Test   23.4  4.73 0.643
2 prophet_forecast  Test   32.2  6.70 0.896
3 ets_forecast      Test   39.4  8.08 1.12
4 arima_forecast    Test   55.1 11.2  1.60
5 tslm_forecast     Test   59.6 12.4  1.75
6 snaive_forecast   Test   97.8 19.6  2.92
```

We assess the forecast performance of the candidate models with a particular emphasis on the MASE as a key evaluation measure. The ensemble forecast achieves the highest accuracy across all evaluation metrics, demonstrating a significant improvement over the second best forecast generated by the Prophet model. The magnitude of this improvement underscores the effectiveness of the ensemble approach.

It may be tempting to remove the poorest-performing forecast and rerun the ensemble. However, this is not recommended here, as visual inspection indicates that snaïve is the only approach to producing a low-biased forecast. Removing snaïve would result in all components of the ensemble being positively correlated, which would reduce the effectiveness of the ensemble. The benefit of ensemble methods arises from averaging components with lower correlations. For a more robust ensemble, consider excluding one of the positively correlated forecasting method, for example the Prophet method, shown in the following code:

```
# fit model
ap_mod_revised <- ap_train |>
  model(
    snaive = SNAIVE(passenger_count),
    tslm = TSLM(log(passenger_count) ~ trend() + season()),
    ets = ETS(log(passenger_count) ~ season("A")),
    arima = ARIMA(log(passenger_count))    ) |>
  mutate(ensemble = (snaive + tslm + ets + arima ) / 4)

# forecast
ap_fcst_revised <- ap_mod_revised |>
  forecast(h = "4 years") |>
  mutate(.model = str_c(.model, "_forecast"))

# accuracy
ap_accuracy_revised <- ap_fcst_revised |>
  accuracy(ap_tsb) |>
  select(.model, .type, RMSE, MAPE, MASE) |>
  arrange(MASE)

ap_accuracy_revised
```

The revised accuracy metrics as provided in the following output:

```
# A tibble: 5 × 5
  .model              .type  RMSE  MAPE  MASE
```

```
  <chr>              <chr> <dbl> <dbl> <dbl>
1 ensemble_forecast Test   21.9   4.39 0.601
2 ets_forecast      Test   39.4   8.08 1.12
3 arima_forecast    Test   55.1  11.2  1.60
4 tslm_forecast     Test   59.6  12.4  1.75
5 snaive_forecast   Test   97.8  19.6  2.92
```

5. **Generate prediction intervals**: The prediction intervals for forecast ensembles can be derived using simulation. Simulation works on model-based forecasts; this is another reason to exclude the Prophet method from our candidate models earlier while revising ensemble forecasts. We will generate 1,000 future sample paths for all forecasting which can be used to calculate distributions of the future sample path. The `generate()` function from the `fabletools` package performs the simulation while the `set.seed()` function ensures reproducibility. The `dist_sample()` function from the `distributional` package help summarizing the result from the simulation, which is finally stored as a `tidy forecast` object (`fable`), as shown in the following code:

```
set.seed(98765); ap_sims <- ap_mod_revised |>
  fabletools::generate(h = "4 years", times = 1000) |>
  mutate(.model = str_c(.model, "_prediction_interval")) |>
  tibble() |>
  group_by(month, .model) |>
  summarise(
    dist = distributional::dist_sample(list(.sim))) |>
  ungroup() |>
  as_fable(index = month, key = .model,
           distribution = dist,
           response = "passenger_count")
```

Let's visualize the prediction intervals for the ensemble method. The plot in *Figure 13.10* shows the observed series from 1954, forecasts from ensemble and the prediction intervals around it:

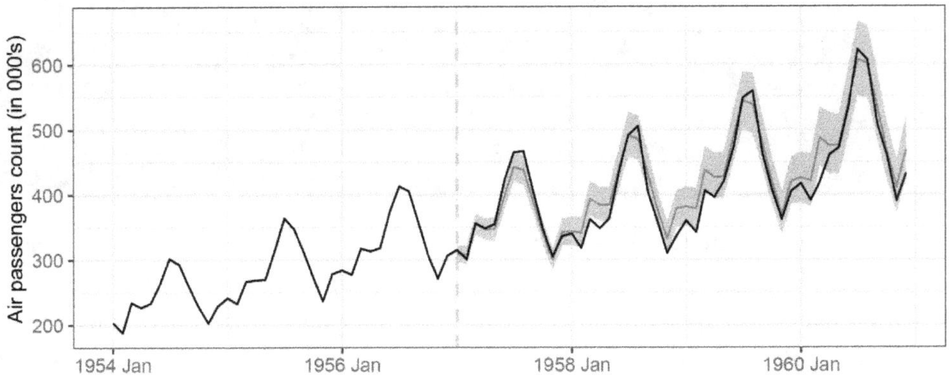


*Figure 13.10: Prediction intervals for forecast ensemble on AirPassenger data.*

Visual inspection suggests that the 95% intervals contain true observations for both short- and long-term forecasting.

It is possible to compare the accuracy of the prediction intervals from candidate models using a quantitative measure. One such measure is known as the Winkler score. It measures how well an interval forecast captures the actual observed value.

For a given prediction interval, $[L, U]$, bounded lower at $L$ and bounded upper at $U$, the Winkler score $W$ is calculated at time point $h$ as follows:

$$W = \begin{cases} U - L & \text{if } L \leq y_h \leq U \\ (U - L) + \frac{2}{\alpha}(L - y_h) & \text{if } y_h < L \\ (U - L) + \frac{2}{\alpha}(y_h - U) & \text{if } y_h > U \end{cases}$$

where:

- $y_h$ is the actual observed value
- $\alpha$ is the desired coverage of the interval (say 95%)

The score rewards narrower intervals (lower scores) that contain the actual observation but heavily penalizes intervals that miss the observed value by a significant margin. For multistep forecasts, the average of the Winkler score is used. This score can be calculated using the accuracy function by specifying the interval accuracy measure shown, in the following code snippet:

```
interval_accuracy <- ap_sims |>
  accuracy(ap_tsb,
           measures = interval_accuracy_measures,
```

```
            level = 95) |>
  select(-pinball, -scaled_pinball) |>
  arrange(winkler)

interval_accuracy
```

The result is presented in the following output:

```
# A tibble: 5 × 3
  .model                        .type winkler
  <chr>                         <chr>   <dbl>
1 ensemble_prediction_interval Test     127.
2 ets_prediction_interval      Test     190.
3 arima_prediction_interval    Test     356.
4 tslm_prediction_interval     Test     777.
5 snaive_prediction_interval   Test    1653.
```

The forecasts ensemble produced the lowest Winkler score for prediction intervals compared to other candidate models. This further supports the idea that the forecasts ensemble provides superior performance over the individual candidate models.

# Summary

In this chapter, we explored advanced time-series forecasting techniques, including DL, ML, curve fitting (Prophet), and ensemble methods for forecasting. The examples used highlighted both the potential and limitations of these approaches.

DL is effective for various types of time-series data, provided there is sufficient historical data. A key advantage of DL is that it eliminates the need for manual feature engineering by the user.

ML approaches offer flexibility in choosing algorithms, but they require feature engineering, making them generally more time-intensive in terms of model training. Additionally, deriving prediction intervals can be more technically demanding.

DL, particularly through feed-forward neural networks, and curve fitting using Prophet are both well integrated within the tidy forecasting ecosystem. However, deriving prediction intervals using DL can be time-consuming due to its dependency on simulation.

When using ML for forecasting, the flexibility of algorithm selection must be balanced against the effort required for model tuning and code development.

It is important to note that no single method consistently outperforms others, including traditional statistical forecasting techniques. Forecasting ensemble can improve the accuracy of forecasting, and this can be achieved by simply averaging forecasts from different approaches on the same data.

This concludes our discussion on approaches useful for univariate time series forecasting. The concepts covered so far provide a solid foundation for analyzing multiple time series. The field of analyzing and forecasting multiple time series is becoming increasingly important in various business and economic applications. In the next chapter, we will explore some modern methods applied in these contexts.

# Further reading


- Bengio, Y., Goodfellow, I., & Courville, A. (2017). *Deep learning* (Vol. 1). MIT Press Cambridge, MA, USA.
- Bergmeir, C., Hyndman, R. J., & Koo, B. (2018). *A note on the validity of cross-validation for evaluating autoregressive time series prediction. Computational Statistics & Data Analysis*, 120, 70–83.
- Bontempi, G., Ben Taieb, S., & Le Borgne, Y.-A. (2013). *Machine learning strategies for time series forecasting. Business Intelligence: Second European Summer School, eBISS 2012, Brussels, Belgium, July 15-21, 2012, Tutorial Lectures 2, 62–77.*
- Efron, B. (2020). *Prediction, estimation, and attribution. International Statistical Review*, 88, S28–S59.
- Friedman, J., Hastie, T., & Tibshirani, R. (2010). *Regularization paths for generalized linear models via coordinate descent. Journal of Statistical Software*, 33(1), 1.
- Friedman, J., Hastie, T., Tibshirani, R., & al., et. (2024). *glmnet: Lasso and elastic-net regularized generalized linear models.* `https://cran.r-project.org/web/packages/glmnet/index.html`
- Gareth, J., Witten, D., Hastie, T., & Tibshirani, R. (2013). *An introduction to statistical learning: With applications in R. Spinger.*
- Hastie, T., Tibshirani, R., & Friedman, J. H. (2009). *The elements of statistical learning: Data mining, inference, and prediction (Vol. 2). Springer.*
- Makridakis, S., Andersen, A., Carbone, R., Fildes, R., Hibon, M., Lewandowski, R., Newton, J., Parzen, E., & Winkler, R. (1982). *The accuracy of extrapolation (time series) methods: Results of a forecasting competition. Journal of Forecasting*, 1(2), 111–153.
- Makridakis, S., Chatfield, C., Hibon, M., Lawrence, M., Mills, T., Ord, K., & Simmons, L. F. (1993). *The M2-competition: A real-time judgmentally based forecasting study. International Journal of Forecasting*, 9(1), 5–22.

- Makridakis, S., Spiliotis, E., & Assimakopoulos, V. (2022). *M5 accuracy competition: Results, findings, and conclusions. International Journal of Forecasting*, 38(4), 1346–1364.
- O'Hara-Wild, M., Taylor, S., & Letham, B. (2024). *fable.prophet: Prophet modelling interface for 'fable'*. `https://cran.r-project.org/web/packages/fable.prophet/index.html`
- Redell, N. (2022). *forecastML: Time series forecasting with machine learning methods.* `https://cran.r-project.org/web/packages/forecastML/index.html`
- Taylor, S. J., & Letham, B. (2018). *Forecasting at scale. The American Statistician*, 72(1), 37–45.
- Yi, C. (2022). *hqreg: Regularization paths for lasso or elastic-net penalized huber loss regression and quantile regression.* `https://cran.r-project.org/web/packages/hqreg/index.html`
- Zou, H., & Hastie, T. (2005). *Regularization and variable selection via the elastic net. Journal of the Royal Statistical Society Series B: Statistical Methodology*, 67(2), 301–320.

# 14

# Forecasting Models for Multiple Time Series

In the earlier chapters of this book, we primarily explored theories, methods, and tools focused on analyzing a single time series. While this covers most business applications, there is an increasing demand for analyzing multiple time series simultaneously. This type of analysis examines and models data involving more than one time-dependent variable, proving particularly valuable when variables are *interconnected and influence one another over time*. This approach enables a more comprehensive understanding of complex systems in which interrelated variables evolve together.

In this chapter, we'll focus on two important use cases:

- Modeling and forecasting bidirectional relationships in multiple time series, where feedback loops exist
- Forecasting multiple time series that are interdependent, but follow group or hierarchical relationships rather than a strict bidirectional influence

The discussion will be presented with detailed, worked-out examples and code.

The chapter is organized as follows:

- Section 14.1: The dynamics of multiple interconnected time series
- Section 14.2: Vector Autoregression model
- Section 14.3: Hierarchical and grouped time series
- Section 14.4: Machine learning for multiple time series forecasting

Note

**Your purchase includes a free PDF copy + exclusive extras**

Your purchase includes a DRM-free PDF copy of this book, 7-day trial to the Packt+ library (no credit card required), and additional exclusive extras. See the *Free benefits with your book* section in the *Preface* to unlock them instantly and maximize your learning.

# Technical requirements

Let's load the packages that are required in this chapter.

```
# install.packages("pacman")
library(pacman)
p_load("conflicted"
       ,"datasets"
       ,"fable"
       ,"fabletools"
       ,"feasts"
       ,"here"
       ,"patchwork"
       ,"tidyverse"
       ,"tsibble"
       ,"tsibbledata"
       ,"vars"
       ,"WDI"
       ,"vars" )

conflicts_prefer(
  fabletools::accuracy
  ,fabletools::forecast
  ,dplyr::filter
  ,dplyr::select
  ,dplyr::lag
  ,fable::VAR
  ,feasts::unitroot_kpss
)
```

The complete code file is available in the GitHub repo `https://github.com/PacktPublishing/Modern-Time-Series-Analysis-with-R.git` for you to download and practice.

For fitting a **Vector Autoregression** (**VAR**) model, we use functions from two packages, `fable` and `vars`, some of which carry the same name. We prefer functions from `fable` package (as shown for conflict management) and use them without namespace calling. We use namespace calling for any function used from the `vars` package.

# 14.1 The dynamics of multiple interconnected time series

When analyzing multiple time series, it is crucial to understand the nature of the relationships between them. These relationships can be classified as follows:

- Direction: either unidirectional or bidirectional
- Structure: either hierarchical or grouped time series

A unidirectional relationship occurs when future values of one time series depend on its own past values, and potentially on the past values of other, independent (exogenous) variables. Crucially, the exogenous variables are not influenced by the time series of primary interest. For example, consider forecasting a supermarket's total sales. These sales may depend on the company's marketing expenditure for various product categories. The marketing expenditure acts as an exogenous regressor because the sales volume does not typically influence the planned marketing budget in the same immediate time frame. Models such as ARIMA with exogenous regressors, as discussed in *Section 12.6*, can effectively capture and model this type of one-way dependency.

In contrast, bidirectional relationships occur when time series mutually influence one another, creating a feedback loop. This introduces a "lag-lead effect" where changes in one series are followed by changes in the other, and vice versa. In economic scenarios, complementary or substitute goods within a sector rarely have a one-way relationship. For example, in many major economies, natural gas demand and electricity demand are bidirectionally linked due to the complementary effect. An increase in the demand for electricity necessitates greater use of natural gas for generation, which in turn increases the demand for natural gas as a commodity, influencing its price and supply dynamics. Both variables influence each other.

While less common in typical business forecasting, bidirectional modeling is a crucial part of government policy analysis for understanding complex macroeconomic and financial systems with multiple time series with bidirectional relationships. For example, agencies such as the Treasury and Reserve Bank of Australia studied interconnected time series such as **Gross**

**Domestic Product** (**GDP**), interest rates, inflation, and unemployment. These exhibit both short- and long-term lag-lead relationships. Raising interest rates is a policy decision designed to curb inflation, but it also impacts business growth, employment levels, and GDP. Analyzing these variables in isolation offers limited insights, underscoring the need for bidirectional approaches to assess comprehensive policy impacts.

Putting uni/bidirectional modeling aside, a distinct class of use cases has garnered significant attention in modern business applications, driven by advancements in large-scale data collection capabilities. For instance, large supermarkets are increasingly interested in understanding demand across hundreds or thousands of product categories. These categories often exhibit short-term, point-in-time relationships (not lag-lead) due to factors, such as the substitution effect, complementing effect, and price elasticity.

Take, for example, the consumption of berries in a geographically diverse country, such as Australia. Seasonal demand can vary significantly between states, such as Queensland and Victoria, due to differences in climate. Consumers' taste varies in tropical versus subtropical regions, influencing short-term demand for berries. While these product categories may not exhibit long-term equilibrium relationships, their short-term dynamics are of critical importance for inventory management and planning. This area of forecasting has seen substantial academic development over the past two decades.

For modeling such relationships, specialized models are required. We will discuss the Vector Autoregression model in the following section.

## 14.2 Vector Autoregression model

Vector Autoregression (VAR) is a statistical model used to capture the relationship between multiple time series variables. Unlike univariate models, VAR treats all variables as endogenous or dependent, allowing each variable to depend not only on its own past values but also on the past values of other variables. This makes it ideal for analyzing systems where variables are interrelated, such as economic and financial indicators. VAR models are widely used in econometrics to assess dynamic impacts and forecast multiple variables simultaneously. This class of model was advocated in the macroeconomic literature through seminal works by Sims (1980) and Sims (1987).

### 14.2.1 Mathematical background of the VAR(1) model

The simplest VAR model will have at least two time series with at least one lag order bidirectional relationship. A two-dimensional VAR of order 1, written as VAR$(1)$, consists of two time series, $y_1$ and $y_2$, where the current values of both $y_1$ and $y_2$ are modeled as a linear function of their own past values and the past values of each other. Let us denote two time series at time $t$ as $y_{1,t}$ and $y_{2,t}$. A VAR$(1)$ model can be written as:

$$
\begin{aligned}
y_{1,t} &= c_1 + \phi_{11} y_{1,t-1} + \phi_{12} y_{2,t-1} + \epsilon_{1,t} \\
y_{2,t} &= c_2 + \phi_{21} y_{1,t-1} + \phi_{22} y_{2,t-1} + \epsilon_{2,t}
\end{aligned}
$$

where:

- $y_{1,t}$ and $y_{2,t}$ are the values of these time series at time $t$
- $c_1$ and $c_2$ are constant terms (intercepts)
- $\phi_{11}, \phi_{12}, \phi_{21}, \phi_{22}$ are the coefficients that measure the influence of the lagged values of $y_1$ and $y_2$ on each other. Specifically:
    - $\phi_{11}$ represents the impact of $y_{1,t-1}$ on $y_{1,t}$
    - $\phi_{12}$ represents the impact of $y_{2,t-1}$ on $y_{1,t}$
    - $\phi_{21}$ represents the impact of $y_{1,t-1}$ on $y_{2,t}$
    - $\phi_{22}$ represents the impact of $y_{2,t-1}$ on $y_{2,t}$
    - $\epsilon_{1,t}$ and $\epsilon_{2,t}$ are white noise error terms with mean zero and constant variance.

The estimation of the VAR(1) model involves estimating the parameters $c_1, c_2, \phi_{11}, \phi_{12}, \phi_{21}, \phi_{22}$. This is usually done using Ordinary Least Squares (OLS) applied to each equation separately, as each equation is linear. This is achieved by minimizing the sum of squared residuals:

$$\hat{\epsilon}_{1,t} = y_{1,t} - \left(\hat{c}_1 + \hat{\phi}_{11} y_{1,t-1} + \hat{\phi}_{12} y_{2,t-1}\right)$$

The estimation involves organizing your time series data into two vectors $y_1$ and $y_2$, and creating the lagged versions of the variables $y_{1,t-1}$ and $y_{2,t-1}$. For the first equation:

$$y_{1,t} = c_1 + \phi_{11} y_{1,t-1} + \phi_{12} y_{2,t-1} + \epsilon_{1,t}$$

Regress $y_{1,t}$ on $y_{1,t-1}$ and $y_{2,t-1}$ to estimate $c_1, \phi_{11}$, and $\phi_{12}$.

For the second equation:

$$y_{2,t} = c_2 + \phi_{21} y_{1,t-1} + \phi_{22} y_{2,t-1} + \epsilon_{2,t}$$

Regress $y_{2,t}$ on $y_{1,t-1}$ and $y_{2,t-1}$ to estimate $c_2, \phi_{21}$, and $\phi_{22}$.

In a VAR(1) model with two variables, the total number of coefficients to be estimated is six. When the model is extended to VAR(2), the lagged values of order 2 for all variables are introduced as additional predictors on the right-hand side of the equations. Specifically, for a system with three time series, each equation will have six predictors: three lagged values from order 1 and three from order 2.

More generally, for a VAR model of order $p$ with $K$ time series, the number of parameters to be estimated across the system is $K + pK^2$. This illustrates how rapidly the number of parameters can increase in a VAR system as the number of time series or the lag order grows, posing potential challenges in model estimation and overfitting.

The expression for higher-order VAR can be cumbersome for large $p$ and $K$. However it is possible to simplify the notation by using a vector and matrix representation. We can rewrite the system of equations for VAR(1) in matrix form for generalization

$$\mathbf{Y}_t = \mathbf{c} + \mathbf{\Phi}\mathbf{Y}_{t-1} + \boldsymbol{\epsilon}_t$$

where,

- $\mathbf{Y}_t = \begin{bmatrix} y_{1,t} \\ y_{2,t} \end{bmatrix}$
- $\mathbf{c} = \begin{bmatrix} c_1 \\ c_2 \end{bmatrix}$
- $\mathbf{\Phi} = \begin{bmatrix} \phi_{11} & \phi_{12} \\ \phi_{21} & \phi_{22} \end{bmatrix}$
- $\boldsymbol{\epsilon}_t = \begin{bmatrix} \epsilon_{1,t} \\ \epsilon_{2,t} \end{bmatrix}$

Such a representation is particularly useful for simplifying the notation for a VAR(p) model with $K$ time series. A VAR(p) model can be rewritten as $p$, VAR(1) models that represent the first $p$ lags of each variable in the system as regression predictors for each of the $K$ series.

For example, the VAR(2) model can be recast as the VAR(1) model for compact statement

$$\mathbf{Y}_t = \mathbf{c} + \mathbf{\Phi}_1\mathbf{Y}_{t-1} + \mathbf{\Phi}_2\mathbf{Y}_{t-2} + \boldsymbol{\epsilon}_t$$

$$\begin{bmatrix} Y_t \\ Y_{t-1} \end{bmatrix} = \begin{bmatrix} \mathbf{c} \\ 0 \end{bmatrix} + \begin{bmatrix} \Phi_1 & \Phi_2 \\ I & 0 \end{bmatrix} \begin{bmatrix} Y_{t-1} \\ Y_{t-2} \end{bmatrix} + \begin{bmatrix} \boldsymbol{\epsilon}_t \\ 0 \end{bmatrix}$$

where, $I$ is an identity matrix.

## 14.2.2 Inspect time series for stationarity

To illustrate VAR modeling, we will utilize GDP data for Australia and the United States covering the period from 1960 to 2022. This dataset has been obtained from the World Bank's World Development Indicators (WDI), and is accessible via the `WDI` package in R.

Download GDP for Australia and USA between 1960 and 2022 and make it a `tsibble` object using the following code:

```
gdp_data <- WDI(
  indicator = 'NY.GDP.PCAP.KD',
  country = c('AU', 'US'),
  start = "1960", end = "2022" ) |>
  tibble() |>
  select(country, year, GDP = 'NY.GDP.PCAP.KD')
```

To facilitate ease of analysis, the data has been pre-processed and prepared specifically for modeling. The GDP figures for 2022 have been used as the baseline, allowing for a clearer interpretation of the estimated coefficients across the time series for both countries. Each series is normalized by its value in 2022. This data can be interpreted as an index with a 2022 baseline. The operation is performed in the following code:

```
GDP_norm <- gdp_data |>
  left_join(
    gdp_data |>
      filter(year == max(year)) |>
      rename(gdp2022 = GDP) |>
      select(-year)
  ) |>
  mutate(GDP_index = (GDP/gdp2022)*100) |>
  select(-gdp2022) |>
  as_tsibble(key = country, index = year)
```

We start with a visual inspection of the normalized GDP of Australia and the USA to analyze their components and understand the possible bidirectional relationship. Let's inspect the time plots of the GDP index for Australia and the USA presented in *Figure 14.1*:

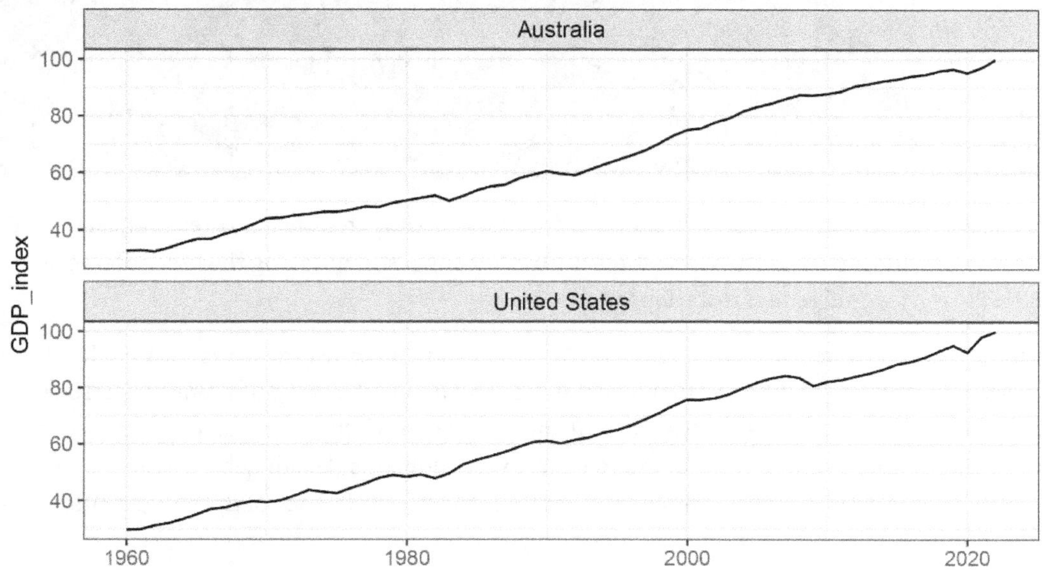


*Figure 14.1: GDP (normalized) for Australia and the United States*

The plot shows a strong upward trend in the GDP index for both countries. There is a potential bidirectional relationship, as they moved in close parallel—a large increase or decrease in one GDP series was followed by the other one. This might indicate that they are correlated. However, it is problematic to measure correlation between two time series when they are non-stationary as they could provide unreliable inference and an inaccurate interpretation.

In real life analysis, we do not rely on visual inspection only, as this could be misleading. The unit root tests, such as `unitroot_kpss()`, discussed in *Chapter 8* will help to detect whether these series are stationary or not. You also used the `unitroot_ndiffs()` function earlier to automatically identify the optimal amount of differencing required to make a non-stationary series stationary.

Let's apply both functions on the GDP series to check stationarity, and order of differencing if needed, using the following code:

```
GDP_norm |>
  features(GDP_index,
           list(unitroot_kpss, unitroot_ndiffs))
```

The result is presented in the following output:

```
# A tibble: 2 × 4
  country       kpss_stat kpss_pvalue ndiffs
  <chr>             <dbl>       <dbl>  <int>
1 Australia          1.66        0.01      1
2 United States      1.68        0.01      1
```

Given the strong trend observed in the data, it is unsurprising that the KPSS test indicates that the series are non-stationary. The test results further suggest that taking the first difference (`ndiffs = 1`) would render it stationary.

If the series were stationary, we could fit a VAR to the data directly (known as a VAR in levels). Another approach would be to difference the data and fit a VAR model, commonly referred to as VAR in differences. However, we are more interested in modeling the year-over-year growth in the GDP index, a common phenomenon in business domains. An advantage of modeling year-over-year growth is that it inherently captures the first difference of the series by its formulation.

Now, compute the growth in the GDP series for each country using the following code:

```
GDP_change <- GDP_norm |>
  group_by(country) |>
  mutate(GDP_growth = ((GDP_index/lag(GDP_index) - 1)*100)) |>
  ungroup() |>
  as_tsibble(key = country, index = year)
```

You can confirm whether the year-on-year growth in GDP became stationary by design by performing unit root tests and estimating the non-seasonal differencing order, applying the following code:

```
GDP_change |>
  features(GDP_growth,
           list(unitroot_kpss,
                unitroot_ndiffs))
```

The following output showing the GDP growth has become stationary as shown in the following output:

```
# A tibble: 2 × 4
  country       kpss_stat kpss_pvalue ndiffs
  <chr>             <dbl>       <dbl>  <int>
1 Australia         0.258         0.1      0
2 United States     0.332         0.1      0
```

*Figure 14.2* shows the year-on-year growth in the GDP index:

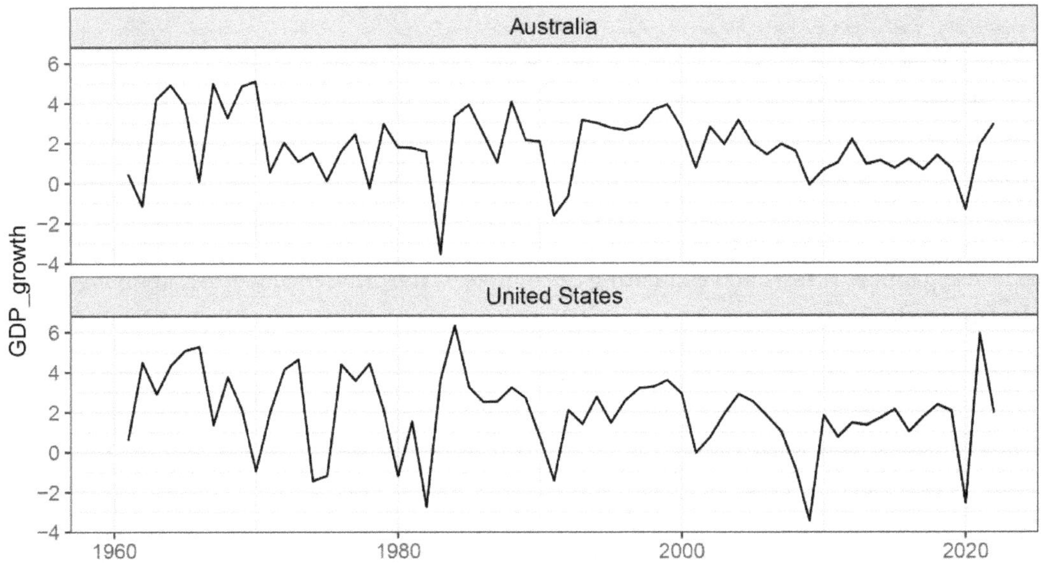


*Figure 14.2: Year-on-year growth in GDP Index for Australia and the United States*

Observing *Figure 14.2*, we could highlight that once stationarity is removed, the possible bidirectional relationship between the GDP index growth between Australia and the USA are more prominent.

As the year-over-year growth in the GDP is now stationary, we will be able to fit a VAR in level for this time series. We will discuss how to identify the order of a VAR model in the next subsection.

### 14.2.3 Identifying VAR model orders

A crucial step in fitting a VAR model is determining the appropriate order of the lag dependencies between the observed time series. One approach is to approximate the lag structure through visual inspection, which can provide initial insights into the bidirectional relationships between variables over time. A more systematic method is to utilize a **Cross-Correlation Function** (**CCF**) plot. You have used ACF plots to detect inter-dependency on an univariate series. A CCF plot provides similar information for two-time series by displaying both lag and lead correlations between them. This tool helps to identify significant temporal relationships by highlighting how one time series correlates with past or future values of another, thereby aiding in the selection of the optimal lag order for the VAR model.

A CCF plot between the GDP growth series of Australia and the USA is obtained by first computing the cross-correlation between the two series using the `CCF()` function from the `feasts` package, and plotting them. You can perform the operations using the following code:

```
ccf_plt_growth <- GDP_change |>
  select(GDP_growth) |>
  pivot_wider(names_from = country, values_from = GDP_growth) |>
  CCF() |>
  autoplot() +
  labs(y = "Correlation", x = "Lag",
       title = "Cross correlation plot") +
  theme_bw()

ccf_plt_growth
```

The CCF is presented in *Figure 14.3*:

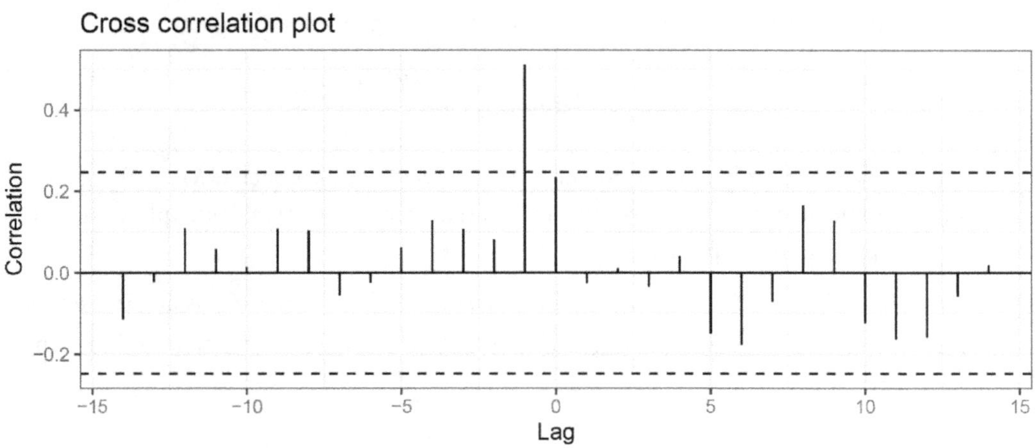


*Figure 14.3: Cross-correlation plot for growth in GDP in Australia and the USA*

Unlike ACF, the CCF computes the cross-correlation in positive and negative lags, meaning it captures the strength and direction of the leading indicator. If a significant spike is at a negative lag, the primary series leads the secondary series. The CCF plot in *Figure 14.3* shows a clear and statistically significant spike at lag $-1$, suggesting a potential lead-lag effect where the changes in one country's GDP occur one time-period before changes are detected in the other country's GDP.

To assess this relationship more rigorously, a more formal approach involves evaluating the information loss associated with different lag order selections for the VAR model. Several information criteria can guide this selection, and in this analysis, we will employ two widely used criteria: the **Akaike Information Criterion** (**AIC**) and the **Bayesian Information Criterion** (**BIC**). These criteria provide a quantitative measure of model fit, balancing model complexity with predictive accuracy. They are expressed as follows:

$$\text{AIC}(p) = \log\left(\det(\hat{\Sigma}_p)\right) + \frac{2K^2p}{T}$$

$$\text{BIC}(p) = \log\left(\det(\hat{\Sigma}_p)\right) + \frac{K^2p\log(T)}{T}$$

where

- $\hat{\Sigma}_p$ is the estimated covariance matrix of residuals.
- $K$ is the number of variables.
- $p$ is the order of the VAR model.
- $T$ is the number of observations.

The small sample corrected version, $AIC_c$, takes the following form:

$$\text{AIC}_c(p) = \log\left(\det(\hat{\Sigma}_p)\right) + \frac{2K^2p(p+1)}{T - Kp - 1}$$

The $AIC_c$ penalizes the number of parameters more heavily as $T$ approaches $Kp$, making it particularly beneficial when $T$ is not significantly larger than $Kp$. For small samples, $AIC_C$ often finds the more parsimonious model.

Next, we will organize the `GDP_change` variable to a suitable format before passing it the VAR (`vars()`) function from the `fable` package to compute $AIC_c$ and $BIC$ for order selection using the code shown:

```
# Create dataset for fitting VAR model
GDP_growth_dt <- GDP_change |>
  select(GDP_growth) |>
  pivot_wider(names_from = country, values_from = GDP_growth) |>
  as_tsibble() |> na.omit()

var_fit <- GDP_growth_dt |>
  model(
    var_aicc = VAR(vars(Australia , `United States`),
                   ic = "aicc"),
    var_bic = VAR(vars(Australia , `United States`),
```

```
                    ic = "bic")
  )
```

The following output shows that both the information criteria suggest the same model specification:

```
# A mable: 1 x 2
            var_aicc           var_bic
             <model>           <model>
1 <VAR(1) w/ mean> <VAR(1) w/ mean>
```

For our data there is no model more parsimonious than VAR(1) that can adequately capture the data dynamics as identified by $AIC_C$ and $BIC$.

```
var_fit |>
  select(var_aicc) |>
  report()
```

We can examine the estimated coefficients in the selected model in the following output:

```
Series: Australia, United States
Model: VAR(1) w/ mean

Coefficients for Australia:
      lag(Australia,1)  lag(United States,1)  constant
                0.0791                0.4085    0.8971
s.e.            0.1152                0.0951    0.3124

Coefficients for United States:
      lag(Australia,1)  lag(United States,1)  constant
               -0.0856                0.1878    1.8064
s.e.            0.1602                0.1323    0.4348

Residual covariance matrix:
              Australia United States
Australia        2.2076        0.5304
United States    0.5304        4.2746

log likelihood = -237.57
AIC = 495.14    AICc = 499.54   BIC = 516.25
```

The VAR(1) model provides two sets of coefficient estimates, one for each series in the system. The coefficients show the short-term relationship and immediate impact of a unit change in an independent variable on the dependent variable, holding other variables constant. We can interpret the coefficients as follows:

- The growth in the GDP index of Australia depends on its own past value and USA's past value. The positive coefficient, `lag(Australia,1)`=0.079, suggests that Australia's value in the previous period has a positive but small effect on its current value. The `lag(United States,1)`=0.408 coefficient is large, indicating a strong positive influence of USA's GDP index on Australia's current GDP index.
- The USA's GDP index also depends on its past value and Australia's past value; however, the influence of this dependency on the current value is smaller or negative.
- The residual covariance matrix shows the joint variability of the error terms for the two equations for the same time period.

Before this model can be used for forecasting, we need to check whether any dependencies are still left in the model's residual. The VAR model has two sources of residuals (from two different models) and their ACF plots could reveal such information. Generate ACFs from the residuals obtained from the two equations in our VAR(1) model of the GDP using the following code:

```
var_fit |>
  select(var_aicc) |>
  augment() |>
  ACF(.innov) |>
  autoplot() +
  labs(x = "")
```

The residual diagnostic plots are presented in *Figure 14.4*:

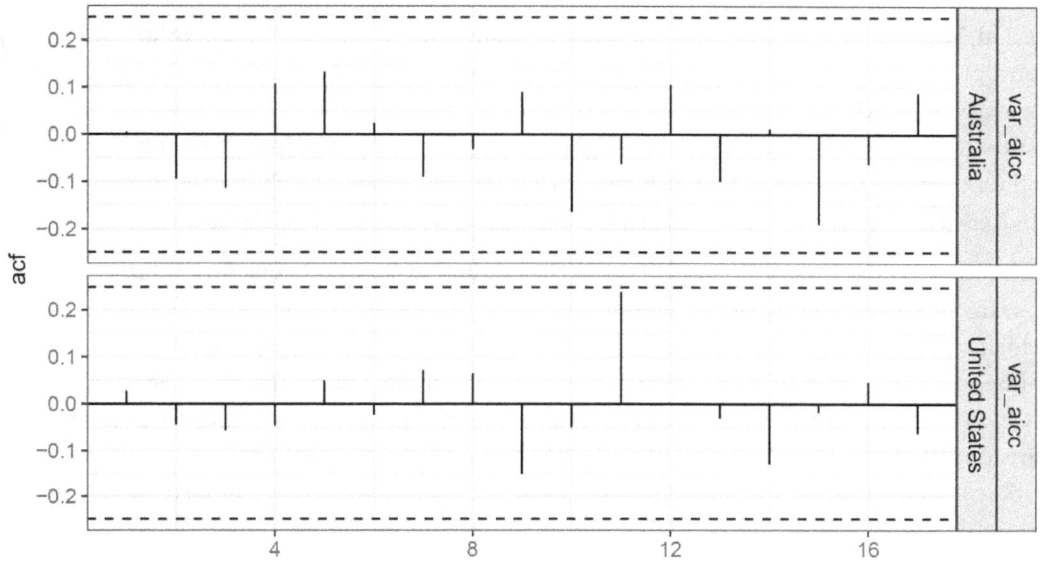


*Figure 14.4: Residual ACFs from the fitted VAR(1) model on the growth GDP index of Australia and the USA*

There are no significant spikes seen in either of the equations' residuals in *Figure 14.4*, providing enough evidence that there is no unaccounted autocorrelation in the residuals, indicating that the model sufficiently explains the temporal dependencies in the data.

## 14.2.4 Interpretation of VAR model coefficients

The estimated parameters of a VAR model provide valuable insights into the underlying bidirectional relationships within the system. By examining these parameters, particularly the slope coefficients associated with lagged variables, we can gain a deeper understanding of the temporal dependencies in the dynamic system. The direction of the slope indicates whether the effect is positive or negative. This analysis is a crucial starting point for determining whether the influence of a variable operates with a lag or leads to changes in other system components over time.

In the example of the VAR model of GDP using two countries, Australia and the United States, we observed that Australia reveals a strong positive relation between the prior year's GDP growth in the United States and the current year's GDP growth in Australia (estimated coefficient = $0.408$). This suggests that economic expansion in the USA, is likely to boost Australia's economic performance in the subsequent year. Conversely, the impact of Australia's prior year GDP growth on USA GDP growth is smaller and negative (estimated coefficient = $-0.085$), indicating a more limited influence in the opposite direction.

When analyzing this relationship using a VAR model with only two time series and a lag of one period, VAR(1), the dynamics between the two economies are relatively straightforward to conceptualize. However, when the model is extended to a VAR($p$) framework involving $k$ time series, where each series potentially affects all others in the system, the interpretation of estimated coefficients becomes increasingly complex. In such scenarios, disentangling the influence of one time series on another poses a significant challenge. This leads to a classic chicken-and-egg dilemma: which series is the true driver of the other's behavior?

This problem is further compounded when the original scales of the time series differ, as is the case in our analysis. The GDP of the United States is several times larger than that of Australia, which can complicate both the analysis and interpretation of cross-country economic influences within a multivariate model. However, this problem is mitigated in the analysis as the GDP data was normalized before fitting a model. In general, for VAR, understanding the direction of causality and the relative impacts between these economies thus requires more sophisticated methods beyond simple coefficient analysis as we will see in the following sub-section.

## 14.2.5 Granger causality test for a VAR(1) model

The Granger causality test evaluates whether the lagged values of one time series help predict another time series in a VAR model (Granger, 1969). In a two-dimensional VAR(1) model, we check if $y_2$ Granger-causes $y_1$ and vice versa.

For a two-variable VAR(1) model the coefficient $\phi_{12}$ shows the effect of $y_2$'s lag on $y_1$ and the coefficient $\phi_{21}$ measures the effect of $y_1$'s lag on $y_2$.

The hypotheses for Granger causality tests are as follows:

- Does $y_2$ Granger-cause $y_1$?
    - Null hypothesis ($H_0$): $\phi_{12} = 0$ (i.e., $y_2$'s lag does not affect $y_1$).
    - Alternative hypothesis ($H_1$): $\phi_{12} \neq 0$ (i.e., $y_2$'s lag affects $y_1$).
- Does $y_1$ Granger-cause $y_2$?
    - Null hypothesis ($H_0$): $\phi_{21} = 0$. (i.e., $y_1$'s lag does not affect $y_2$).
    - Alternative hypothesis ($H_1$): $\phi_{21} \neq 0$. (i.e., $y_1$'s lag affects $y_2$)

The Granger causality test can be performed by fitting a VAR(1) model to the time series data and estimating the coefficients through the following steps:

- Step 1: Restricted model (univariate):
    - To test if $y_2$ Granger-causes $y_1$, estimate a restricted model without $y_2$'s lag:

$$y_{1,t} = c_1 + \phi_{11} y_{1,t-1} + \epsilon_{1,t}$$

- Step 2: Unrestricted model (a full VAR(1) model):
  - Estimate the unrestricted VAR(1) model that includes both $y_{1,t-1}$ and $y_{2,t-1}$:

$$y_{1,t} = c_1 + \phi_{11}y_{1,t-1} + \phi_{12}y_{2,t-1} + \epsilon_{1,t}$$

- Step 3: F-test for model comparison:
  Use an F-test to compare the restricted and unrestricted models. The test statistic is:

$$F = \frac{\left(\frac{RSS_{restricted} - RSS_{unrestricted}}{p}\right)}{\left(\frac{RSS_{unrestricted}}{T-k}\right)}$$

  where,

  - $RSS_{restricted}$: Residual sum of squares from the restricted model.
  - $RSS_{unrestricted}$: Residual sum of squares from the unrestricted model.
  - $p$: Number of lagged terms.
  - $T$: Number of observations.
  - $k$: Number of parameters in the unrestricted model.
- Step 4: Decision:
  - If the F-test or p-value is significant, reject $H_0$, meaning $y_2$ Granger-causes $y_1$ (i.e., $\phi_{12} \neq 0$)
  - If the test is not significant, fail to reject $H_0$
- Step 5: Testing the other direction: To test if $y_1$ Granger-causes $y_2$, repeat the preceding steps for the second equation.

Unfortunately, the tidy model objects derived for VAR order selection and estimation so far are not suitable for performing the Granger causality test. We use the core function for fitting VAR from the `vars` package in the tidy framework explicitly before performing the Granger causality test. The following code shows you how to perform the test on the GDP system:

```
varmat <- GDP_growth_dt |>
  as_tibble() |>
  na.omit() |>
  select(-year) |>
  as.matrix()

varfit <- vars::VAR(varmat, p = 1)
OzCausesUS <- vars::causality(varfit, cause = "Australia")$Granger
USCausesOZ <- vars::causality(varfit, cause = "United.States")$Granger
```

```
causality_test_rs <- as_tibble(
  cbind(Hypothesis = c(OzCausesUS$method, USCausesOZ$method),
  `p-value` = round(c(OzCausesUS$p.value, USCausesOZ$p.value),2)))
```

The test result is presented in *Table 14.1:*

| **Hypothesis** | **p-value** |
|---|---|
| Granger causality H0: Australia do not Granger-cause United.States | 0.59 |
| Granger causality H0: United.States do not Granger-cause Australia | 0 |

*Table 14.1: Granger causality test on the GDP growth*

The results indicate that growth in GDP in the United States Granger-causes subsequent growth in the Australian GDP, as evidenced by a statistically significant p-value of less than 0.001. This finding suggests that economic expansion in the USA has predictive power for Australia's GDP growth in the following year. Conversely, the analysis reveals that Australian GDP growth does not have a significant influence on the USA GDP growth ($p - value > 0.05$).

When multiple time series are interrelated, joint modeling through techniques such as VAR provides valuable insights. This approach allows for a more comprehensive examination of how fluctuations in one time series may impact another, enabling the study of potential adverse or reciprocal effects between the series using causality analysis, which is the focus of the next section.

## 14.2.6 Impulse Response Analysis

**Impulse response analysis** (**IRA**) is a valuable technique employed to assess the impact of a shock to one time series on the entire system over a specified time horizon. In our analysis, we established that growth in the GDP of the United States has a positive effect on the GDP growth of Australia. A pertinent question arises: What would be the repercussions for Australia's GDP growth if there were a shock in the USA GDP growth, particularly in the context of an event such as the COVID-19 pandemic? In a business domain, an application of IRA is analyzing how a sudden marketing campaign (that is, impulse) affects web traffic, and subsequently sales and server load (the responses).

To explore the impact of impulses, we utilize the **impulse response function** (**IRF**), which allows us to quantify the dynamic responses in one source while the shocks originate from another source of information.

For illustration, we will continue to use the GDP data of Australia and the USA, and measure the response in the Australian economy to shocks originating from the USA economy. We'll use the `irf()` function from the `vars` package to perform an IRF on the GDP data. We will investigate the impact of the initial shock for 10 steps into the future, passed in the `n.ahead` argument to the function. The following code snippet computes two IRF models, one for each country:

```
# IRF: US -> Australia
irf_fit_us_2_oz <- vars::irf(
  varfit,
  impulse = "United.States",
  response = "Australia",
  n.ahead = 10,
  ortho = TRUE
)

# IRF: Australia -> US
irf_fit_oz_2_us <- vars::irf(
  varfit,
  impulse = "Australia",
  response =  "United.States",
  n.ahead = 10,
  ortho = TRUE
)
```

Among many elements returned by the `irf()` function, we are interested in the following:

- **Impulse response coefficients**: Represents the dynamic reaction of a variable to a sudden, isolated shock in another variable over time
- **Confidence intervals**: Lower and upper bounds of the coefficients that provide information on their statistical significance

The impulse response coefficients from the two fitted models are presented in *Table 14.2*:

| Year | USA -> Australia | Australia -> USA |
|---|---|---|
| 0 | 0.000 | 0.357 |
| 1 | 0.832 | -0.060 |
| 2 | 0.222 | -0.034 |

| Year | USA -> Australia | Australia -> USA |
|---|---|---|
| 3 | 0.018 | -0.006 |
| 4 | -0.006 | 0.000 |
| 5 | -0.003 | 0.000 |
| 6 | 0.000 | 0.000 |
| 7 | 0.000 | 0.000 |
| 8 | 0.000 | 0.000 |
| 9 | 0.000 | 0.000 |
| 10 | 0.000 | 0.000 |

*Table 14.2: Impulse response coefficients of Australia and the USA GDP data*

The coefficients in the first column show how Australia (the response variable) reacts over 11 time periods to a 1-unit unanticipated shock occurring in the USA. There is no immediate impact on Australia in the first year. In year 2, a 1-unit increase in the USA at period 1 causes Australia to increase by approximately 0.83 units in period 2. The positive effect persists but begins to decay in the year 3-4, turns slightly negative in the year 5-6 before converging toward zero. The coefficients of impact response on the USA GDP can be interpreted in the same manner.

IRs are most effectively illustrated through graphical representations that display the responses of endogenous time series within a VAR model over time. We organize the results to generate the impulse response plots, showcasing the effects of fluctuations in the USA GDP on the Australian GDP. The resulting plots provide a side-by-side comparison of the impulse responses for both Australia and the United States, enabling a clear visualization of the interconnectedness of their economic growth trajectories.

The impulse response plots are presented in *Figure 14.5* in which the estimated coefficients (solid black line) and the confidence intervals (shaded area) are plotted:

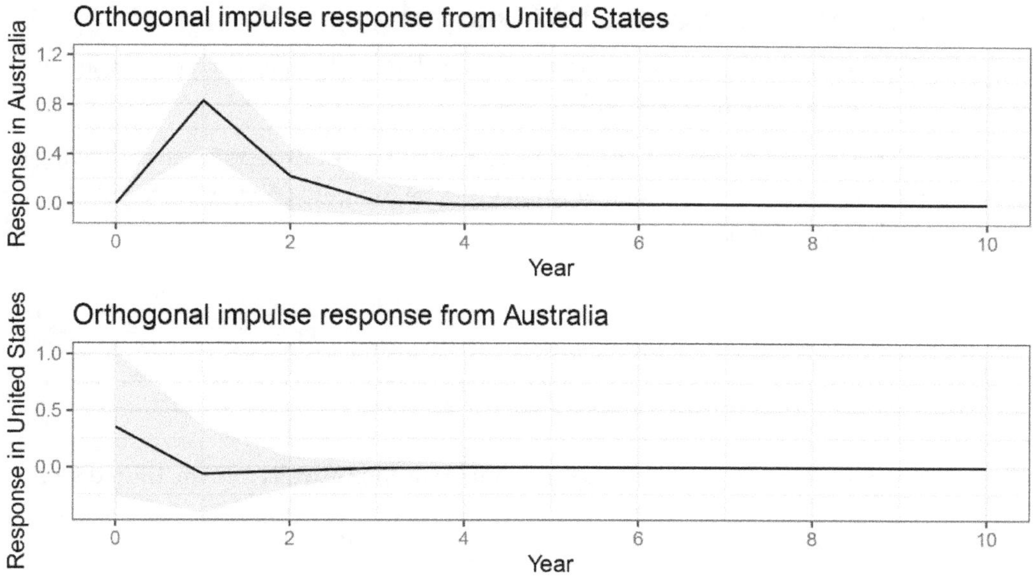


*Figure 14.5: Orthogonal impulse response by countries*

*Figure 14.5* clearly illustrates that an impulse (or shock) to GDP growth in the United States at time zero significantly impacts GDP growth in Australia during the first period. However, this effect diminishes over time, ultimately fading after two periods. The shaded region surrounding the impulse response indicates the 95% confidence interval estimates for the impact of the shock. Notably, this interval suggests that after two time periods, the response of the Australian GDP growth to the initial shock from the United States is no longer statistically significant.

Additionally, the response of GDP growth in the United States demonstrates resilience to shocks originating from changes in Australian GDP growth. This indicates that fluctuations in Australia do not significantly alter the growth trajectory of the USA economy.

The IRF not only enhances our understanding of the immediate and prolonged effects of economic shocks but also highlights the importance of considering international economic dynamics in forecasting and policy-making.

Those interested in the mathematical representation of the IRF, comprehensive details on VAR and related models can be found in Lütkepohl (2010) and Hill et al. (2018). We encourage readers who wish to dive deeper into the analysis of economic time series to consult these resources.

## 14.2.7 Forecasting from a VAR model

Forecasting from a VAR model follows the exact process that we used in other models; apply the `forecast()` function on the fitted model object.

Let's use the fitted VAR(1) model from *Section 14.2.3* and compute forecasts for the following 8 years with prediction intervals. The following lines of code compute point and interval forecasts, and then plot them:

```
var_fit |>
  select(var_aicc) |>
  forecast(h = 8) |>
  autoplot(GDP_growth_dt, level = 95, linewidth = 0.9)
```

*Figure 14.6* shows the forecasted values with 95% intervals for Australia and the United States:

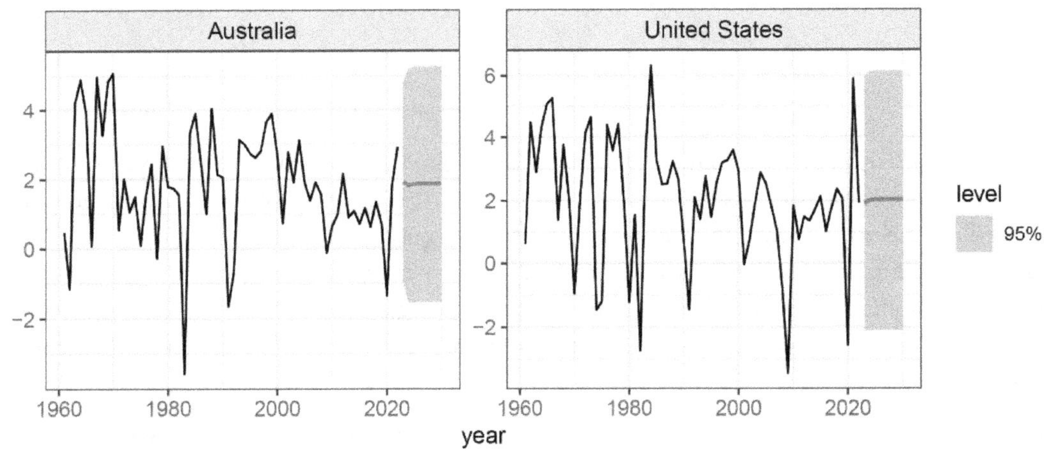


*Figure 14.6: Forecasts from the VAR(1) model fitted on GDP growth from Australia and the USA*

It is important to note that VAR is well-suited for short- to medium-term forecasting in systems where the dynamic relationships between the variables remain stable.

You should be aware of some important remarks when using the VAR model in analysis, such as the following:

- **Non-stationary data**: If your time series are non-stationary and cannot be made stationary through differencing, then a VAR model is inappropriate. In such cases, you might consider a **Vector Error Correction Model** (**VECM**) if the series are cointegrated.

- **Many time series with short history**: If you have too many time series or very few data points, VAR can become overly complex. This leads to overfitting and poor out-of-sample performance.
- **Highly correlated time series**: If your time series variables are highly collinear, VAR can produce unstable estimates. In such cases, regularization techniques, such as Ridge or Lasso, or dimensionality reduction methods, such as PCA, might be more appropriate.
- **Nonlinear relationships**: VAR assumes linear relationships between the time series. If you suspect nonlinear interactions, other models, such as **Vector Autoregressive Moving Average (VARMA)**, nonlinear autoregressive models, or machine learning techniques may be more suitable.
- **Long-term forecasting**: VAR typically works best for short-term forecasting. Over longer periods, the relationships between time series might drift, making VAR less reliable.

It is also worth noting that the VAR model does not assume any underlying theory in economics. Despite these facts, VAR has proven useful for forecasting a collection of related variables where no explicit interpretation is required.

When forecasting is required for multiple time series that are non-stationary and have complex correlation patterns, but do not necessarily exhibit bidirectional relationship, then we need more flexible approaches to handle such data structure. Hierarchical and grouped time series are ideal options for such a scenario, which we discuss in the following section.

# 14.3 Hierarchical and grouped time series

Many business applications face the challenge of managing multiple time series that exhibit a hierarchical structure. This hierarchy implies that time series data is aggregated across multiple categorical levels. For instance, the time series representing the total number of tourists visiting a country can be disaggregated into time series at the state level and further broken down by the purpose of travel (e.g., business, leisure). Each of these categories can then be further subdivided into more granular segments, such as adult versus child travelers.

Forecasting at different levels within this hierarchy is often required to support distinct business decisions tailored to various stakeholders. For example, national-level tourism forecasting is crucial for broad economic and policy planning, providing insights into overall trends. At the state level, more specific forecasting may be required to address localized needs. A state that attracts many business travelers may require a different set of services and infrastructure compared to a state that caters primarily to family vacations, where a higher proportion of children is expected. The ability to forecast demand at each hierarchical level is critical for ensuring that the appropriate resources and services are allocated effectively, reflecting the unique requirements of each segment of the tourism market.

In this example, we have a hierarchy with three levels denoted by $L1, L2, L3$. We illustrate this hypothetical hierarchy with two child series for each level down, to formalize the concept of the data structure schematically in *Figure 14.7*. The top level is generally denoted by index $L0$. For simplicity, we use a symmetric structure, although it is not a requirement for modeling.

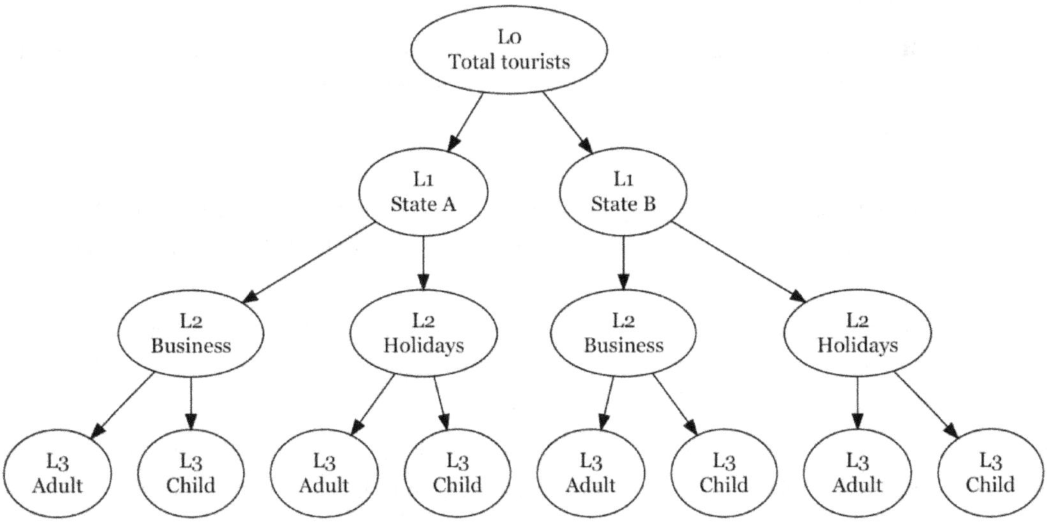


*Figure 14.7: Structure of hierarchical time series with three levels*

A grouped time series is a variation of hierarchical time series in which the aggregation structure arises from the crossing of categories rather than from a strict nesting hierarchy. In hierarchical time series, the relationships between series are defined by parent-child structures, where higher-level aggregates roll up from the lower-level series. However, in grouped time series, categories of interest intersect to create groupings that are not naturally nested.

For example, consider a large retail chain that sells various product categories such as clothing, electronics, and kitchenware. These categories could be further segmented by region, such as metro versus non-metro areas. In this context, the demand for products can be grouped in multiple ways. One approach could be to first group by product category (for example, electronics) and then by region (such as, metro). Alternatively, the demand could be first grouped by region, and then by product category. In a grouped time series, these different groupings exist side by side, and there is no inherent parent-child relationship, unlike in a hierarchical structure.

A fundamental characteristic of both hierarchical and grouped time series is that they must remain aggregate consistent. This means that the data from the lower, more granular levels must sum up correctly to the higher, aggregated levels. Ensuring this consistency is essential for accurate forecasting and planning at multiple levels.

To forecast these time series, specialized hierarchical forecasting methods are employed. These methods are designed to handle both hierarchical and grouped structures, maintaining aggregate consistency across all levels. In this book, for the sake of simplicity, we will refer to the forecasting of both hierarchical and grouped time series as hierarchical forecasting. This terminology simplifies the discussion, but the methods applied are equally valid for grouped time series.

## 14.3.1 Traditional hierarchical forecasting: top-down, bottom-up, and middle-out approaches

Hierarchical forecasting involves specialized methods for forecasting that can be used for both hierarchical and grouped time series. These methods have matured over time in the inventory planning and demand forecasting literature (Dangerfield et al., 1992) and (Fliedner, 2001). Some traditional and widely used methods are top-down, bottom-up, and middle-out approaches. The top-down approach generates an aggregate forecast and distributes it downward using historical or predicted proportions. The bottom-up approach aggregates granular-level forecasts into higher-level totals, while the middle-out method combines both by forecasting at an intermediate level, then reconciles upward via aggregation and downward via disaggregation. Each method has its own use cases, advantages, and disadvantages.

In hierarchical forecasting, the process starts by fitting a univariate time series forecast at the level defined by the approach used for reconciliation. That means, for the top-down approach, the base forecasting model will be fitted on the series at the highest level. In the bottom-up approach, the univariate forecasting models are fitted on each individual series at the bottom most-level. Irrespective of which level's data the models are applied to, you can choose the "base" univariate forecasting model from any available model classes, such as automated ARIMA, ETS (Hyndman et al., 2008), or effort-heavy DL or ML forecasting models. In further discussions, we use an ETS model as the "base" model across all approaches so that we can compare the forecasting performance across the reconciliation techniques.

For the same reason, we will use a single dataset, the `aus_retail` data from the `tsibbledata` package, in all examples in this section. The dataset has monthly retail turnover (millions of AUD) from the main industries in Australia. The retail turnover from January 2008 to December 2018 is used for the selected industries: (i) clothing retailing, (ii) food retailing, and (iii) other recreational goods retailing, for the states of Queensland and Victoria. We made this choice deliberately to mimic real business scenarios, for example, the difference in seasonality patterns in clothing retail turnover between the two states due to the fact that Queensland enjoys a tropical (warmer) climate compared to the temperate, cooler climate with a wet winter in Victoria. Hence, the example and discussion will equip you well, so that you will be able to apply these techniques in a business context.

Let's organize the data by creating a hierarchical structure specifying successive levels for the hierarchy and the aggregation function for the turnover. In the data structure, States is used at the top-level (level 1), Industry at the middle-level (level 2), and the individual retail series at the bottom-most level (level 3). The `aggregate_key()` function from the `fabletools` package allocates this structure appropriately.

The following code shows the complete operation:

```
retail_sel <- aus_retail |>
  filter(Month >= yearmonth("2008 Jan") ) |>
  filter(State %in% c("Queensland", "Victoria")) |>
  filter(Industry %in% c(
    "Clothing retailing", "Food retailing",
    "Other recreational goods retailing" )
  ) |>
  mutate(
    Industry  = case_when(
      Industry == "Clothing retailing" ~ "Clothing",
      Industry == "Food retailing" ~ "Food",
      Industry == "Other recreational goods retailing" ~ "Recreational",
      .default = Industry
    )
  ) |>
```

```
  rename(
    month = Month,
    state = State,
    industry = Industry,
    turnover = Turnover
  ) |>
  aggregate_key((state/industry), turnover = sum(turnover))

retail_sel
```

There will be a total of 9 series with state and industry aggregated as shown in the following output:

```
# A tsibble: 1,188 x 4 [1M]
# Key:       state, industry [9]
      month state        industry     turnover
      <mth> <chr*>       <chr*>          <dbl>
 1 2008 Jan <aggregated> <aggregated>    3799.
 2 2008 Feb <aggregated> <aggregated>    3509.
 3 2008 Mar <aggregated> <aggregated>    3778.
 4 2008 Apr <aggregated> <aggregated>    3649.
 5 2008 May <aggregated> <aggregated>    3762.
 6 2008 Jun <aggregated> <aggregated>    3546.
 7 2008 Jul <aggregated> <aggregated>    3843.
 8 2008 Aug <aggregated> <aggregated>    3891.
 9 2008 Sep <aggregated> <aggregated>    3812.
10 2008 Oct <aggregated> <aggregated>    4076.
# i 1,178 more rows
```

After hierarchy allocation, let's visualize each series to visually examine them. *Figure 14.8* presents time plots of all series in the hierarchy:

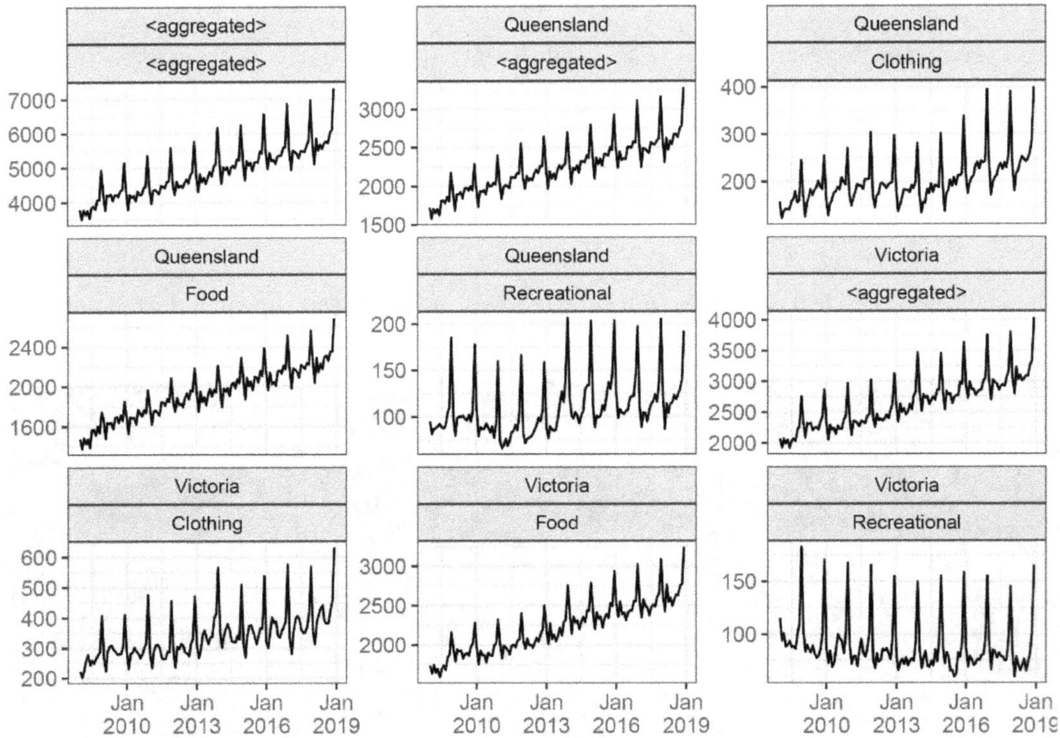


*Figure 14.8: Time plot of hierarchical retail turnover data*

The time plots in *Figure 14.8* show the variations in patterns by states and across industries. Victoria clothing retail declines in the middle of the year and in the winter. This pattern is not evident in the Queensland clothing industry. Turnover for food retailing also shows some difference in seasonality between states, despite both having upward trends. In Victoria, the trend in the other recreational goods retailing turnover has moved downward, whereas the trend in the same industry in Queensland stayed at a similar level.

Now, split the data into a training set (between January 2008 and December 2016) and a test set (between January 2017 and December 2018), as shown in the following code:

```
retail_sel_train <- retail_sel |>
  filter(month <= yearmonth("2016 Dec"))

retail_sel_test <- retail_sel |>
  filter(month >= yearmonth("2017 Jan"))
```

With a hierarchical data structure at hand, we will now discuss how to apply various forecast reconciliation techniques.

### 14.3.1.1 Top-down approach

In the top-down approach, the forecasting process begins by generating a forecast at the highest, most aggregated level of the hierarchy. The "base" model should be selected according to the characteristics of the overall, top-level aggregated time series. Once an optimal base model is fitted and total forecasts for the top-level series are obtained, then it is disaggregated down to the lower levels of the hierarchy.

Several methods can be used to perform this disaggregation; two of the most applied methods are as follows:

- **Historical proportion**: In this method, the top-level forecast is allocated to the lower-level series based on historical proportions (Gross et al., 1990). This means that each lower-level series receives a share of the total forecast corresponding to its historical contribution to the top-level series. The assumption is that the relationships between the levels of the hierarchy remain stable over time, with each lower-level series maintaining a consistent proportion of the total.
- **Forecast proportion**: This method disaggregates the top-level forecast using base forecasts generated for the lower-level series or the proportion series (Athanasopoulos et al., 2009). Here, base forecasts (which can also be derived from univariate forecasting models) are used to determine how the top-level forecast should be split among the lower levels. Unlike the historical proportion method, this approach allows for more flexibility by adjusting for expected changes in the lower-level series. The proportions can vary over time, reflecting dynamic relationships between the levels. Although conceptually better, this is a very effort demanding approach.

The top-down approach is straightforward and simple to implement, primarily because it only requires a single base model for the aggregate series. This simplicity makes it computationally efficient, especially for large hierarchies, as there is no need to develop separate models for each lower-level series. The reduced number of forecasts required also makes this approach more practical when dealing with limited data or computational resources.

However, the top-down approach has several limitations; the most alarming one is that it often fails to account for variability or distinct patterns that may exist at the lower levels of the hierarchy. Because the forecasts for the lower-level series are derived entirely from the top-level forecast, important information such as localized trends, seasonality, or irregular fluctuations at the disaggregated levels may be overlooked. As a result, the method can lead to inaccurate forecasts for individual lower-level series, particularly when there are significant differences in behavior between series at different levels of the hierarchy.

Furthermore, the dependence on the accuracy of the top-level forecast is a potential drawback. If the top-level forecast is inaccurate, it can propagate these errors down the hierarchy, negatively impacting the forecasts at all subsequent levels. Additionally, the method may not be well-suited for capturing demand fluctuations or evolving trends at lower levels, which can result in suboptimal forecasting performance for finer granularities, particularly in dynamic environments where relationships between levels are changing over time.

The top-down forecast using historical proportions can be derived by fitting a suitable univariate model to the aggregated turnover. We use automated ETS models for this purpose. After the top-level base forecast is generated, it is distributed to successive lower levels using historical proportions using the `average_proportions` method. From the `fabletools` package, the `reconcile()` function in combination with the `top_down()` function achieves that.

First, in the training, we fit automated ETS models for each series in the hierarchy for comparison purposes. Then we just take the fitted ETS model of the top-level aggregated series, and the forecasts from that model will be distributed to all levels according to the average proportions. This process avoids further re-fitting the model. Once forecast redistribution is performed for the hierarchy, then use the model object to generate forecasts for the next 2 years. The complete operations are performed using the following code:

```
hfm_base_fit <- retail_sel_train |>
  model(base = ETS(turnover))

hfm_td_fit <- hfm_base_fit |>
  reconcile(td_hp = top_down(base,
            method = c("average_proportions")))

hfm_td_fcst <- hfm_td_fit |>
  forecast(h = "2 years")

hfm_td_fit
```

The code will produce the model object, as shown in the following output (the `td_hp` column will be grayed out when printed on the R console to indicate that no further model fitting is performed):

```
# A mable: 9 x 4
# Key:     state, industry [9]
  state        industry                base td_hp
  <chr*>       <chr*>               <model> <model>
1 Queensland   Clothing     <ETS(M,Ad,M)> <ETS(M,Ad,M)>
```

```
2 Queensland   Food          <ETS(M,Ad,M)> <ETS(M,Ad,M)>
3 Queensland   Recreational  <ETS(A,N,A)> <ETS(A,N,A)>
4 Queensland   <aggregated> <ETS(M,Ad,M)> <ETS(M,Ad,M)>
5 Victoria     Clothing      <ETS(M,Ad,M)> <ETS(M,Ad,M)>
6 Victoria     Food          <ETS(M,Ad,M)> <ETS(M,Ad,M)>
7 Victoria     Recreational  <ETS(A,A,A)> <ETS(A,A,A)>
8 Victoria     <aggregated> <ETS(M,Ad,M)> <ETS(M,Ad,M)>
9 <aggregated> <aggregated> <ETS(M,Ad,M)> <ETS(M,Ad,M)>
```

An $ETS(M, A_d, M)$ model is selected for the top-level series (row 9 in the preceding output). That means a model with multiplicative error, additive damped trend, and multiplicative seasonality, deemed appropriate for the top-level aggregated series. Even though the `td_hp` column mimicked the base column model names, technically, no further model was fitted.

To aid visual inspection, observations in a shorter recent training window and test window, and the forecasts obtained from both the base and top-down approaches are plotted. The visualization is presented in *Figure 14.9*:

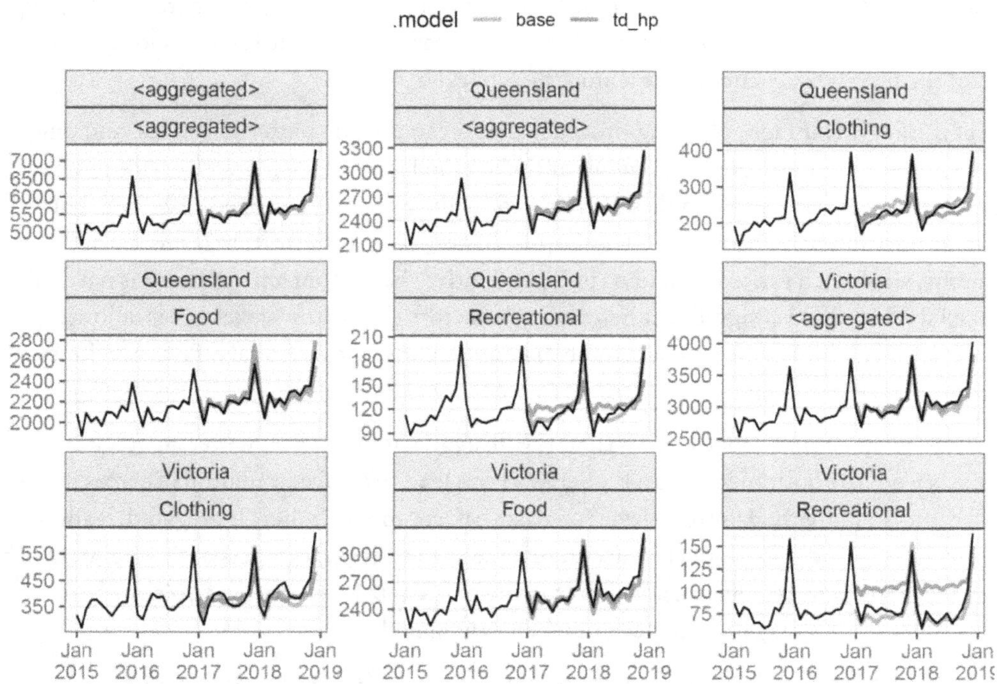


*Figure 14.9: Top-down forecasts of retail turnover*

*Figure 14.9* shows strong forecast performance of the top-down approach for the top-level aggregated data, as expected. The performance remains strong when distributed to the state level (level 2) and diminishes considerably at the bottom level (level 3). In the whole hierarchy, the forecast performance is strong for only those series that exhibit trend and seasonality patterns similar to the top-level series. Otherwise, the forecasts were way off the true values, and the top-down approach is worse than the base forecasts obtained for each series individually.

The top-down approach can be computationally efficient and simple to apply, but it also sacrifices accuracy for series at the lower levels due to its reliance on the top-level forecast and inability to capture detailed patterns that exist further down the hierarchy. The bottom-up approach is used to mitigate this particular issue.

#### 14.3.1.2 Bottom-up approach

The bottom-up approach to hierarchical forecasting begins by fitting individual forecasting models to every series at the bottom-level of the hierarchy. Depending on the data characteristics, a user can choose from a wide variety of forecasting model ranges, from simple naive to DL/ML. Once forecasts for all the bottom-level series have been generated, they are aggregated upwards by summing the individual forecasts to produce forecasts for the next level of the hierarchy, eventually reaching the top level.

One of the key advantages of this approach is that it can capture patterns, trends, and unique features present in individual time series at the detailed level of the hierarchy. This approach provides more flexibility to handle series-specific events, such as localized seasonality or irregular fluctuations. This can result in improved accuracy at the bottom levels of the hierarchy, since each series is modeled independently. The bottom-up approach is particularly beneficial when there is significant heterogeneity across the bottom-level series, allowing the forecasts to account for varying demand behaviors, trends, and seasonal effects in different segments of the hierarchy.

The bottom-up approach performs well at the disaggregated level. However, aggregating forecasts from the bottom level to a higher level can lead to an accumulation of errors. Even small errors in individual bottom-level forecasts can compound when aggregated, leading to higher variance and potentially less accurate forecasts at the upper levels of the hierarchy. This approach commonly results in under-forecasting of the top-level series, as noise or bias in individual forecasts may not balance out when aggregated.

The bottom-up approach requires more computational power compared to the top-down approach involving historical proportions. Each bottom-level series needs to be forecast separately, which means that for large hierarchies, hundreds or even thousands of individual forecasts might be needed. This can be computationally expensive in terms of resources, especially when large-scale models are applied. Moreover, it requires sufficient and reliable data for each bottom-level series. In many real-world applications, data at the bottom level may be sparse or noisy, which can make univariate forecasting more difficult.

We have already fitted ETS models to each individual series available in the bottom level of the hierarchy. The selected models are as follows(the same output is reproduced):

```
> hfm_base_fit

# A mable: 9 x 3
# Key:     state, industry [9]
  state        industry            base
  <chr*>       <chr*>           <model>
1 Queensland   Clothing     <ETS(M,Ad,M)>
2 Queensland   Food         <ETS(M,Ad,M)>
3 Queensland   Recreational  <ETS(A,N,A)>
4 Queensland   <aggregated> <ETS(M,Ad,M)>
5 Victoria     Clothing     <ETS(M,Ad,M)>
6 Victoria     Food         <ETS(M,Ad,M)>
7 Victoria     Recreational  <ETS(A,A,A)>
8 Victoria     <aggregated> <ETS(M,Ad,M)>
9 <aggregated> <aggregated> <ETS(M,Ad,M)>
```

As you can see, the automated ETS has selected the optimal model for each series by assessing their characteristics individually. Now, the forecasts from these base models will be aggregated to derive forecasts for successive higher levels. Bottom-up aggregation is performed via the `reconcile()` and `bottom_up()` functions.

We will use the previous ETS model fit and reconcile them according to the bottom-up approach. Finally, forecasts for the next 2-years are generated from the whole hierarchy. The following code shows the operations:

```
hfm_bu_fit <- hfm_base_fit |>
  reconcile(bu_hp = bottom_up(base))

hfm_bu_fcst <- hfm_bu_fit |>
   forecast(h = "2 years")
```

*Figure 14.10* presents the most recent 2 years training data, test set, and the forecasts from the base and bottom-up approach:

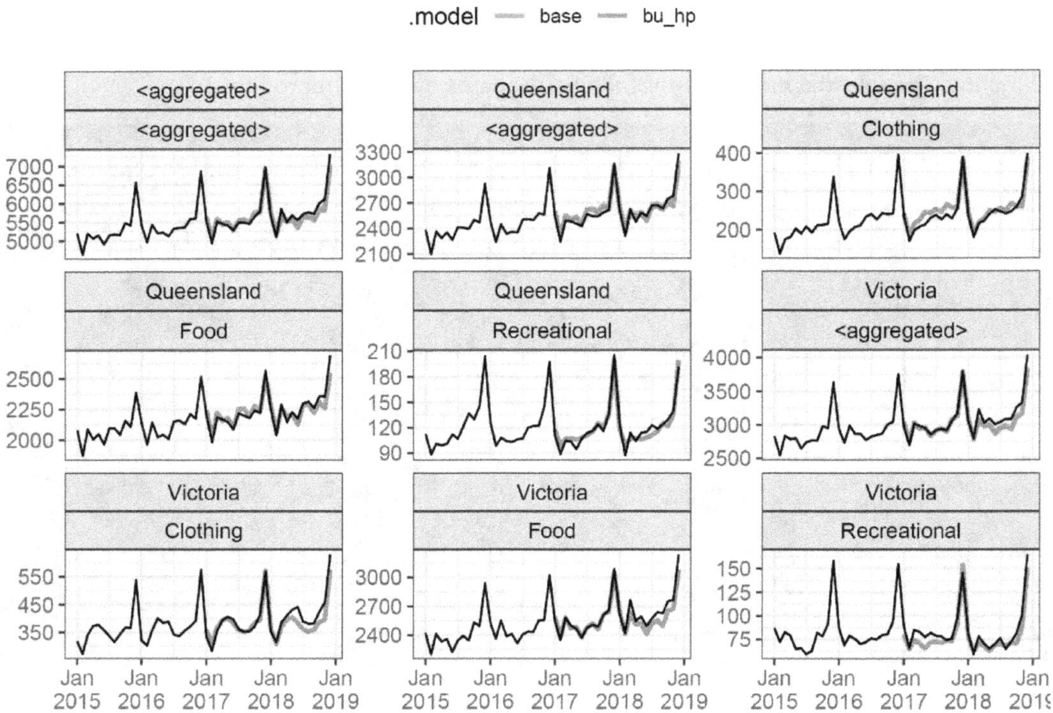


*Figure 14.10: Bottom-up forecasts of retail turnover*

Through visual inspection from *Figure 14.10*, we see the strong performance of the bottom-up approach for the bottom level series. The performance remains strong when aggregated to successive aggregation level. This better forecasting performance of the bottom-up approach is partly due to well-behaved bottom level data. In many business applications that might not be the case, and the bottom-up approach will get worse as one moves up further in the aggregation level.

While the bottom-up approach offers high accuracy and flexibility at the lowest level in a hierarchy, it can be resource-intensive and prone to error accumulation when aggregating forecasts to higher levels. The success of this method largely depends on the availability and quality of data at the bottom level, making it less suitable for cases where data is sparse or noisy.

Next, we'll discuss the middle-out approach which may help to mitigate the error accumulation issue.

### 14.3.1.3 Middle-out approach

The middle-out approach to hierarchical forecasting is a hybrid method that combines elements of both the top-down and bottom-up approaches. Instead of focusing on the highest (top-level) or the most disaggregated (bottom-level) series, forecasts are generated at an intermediate (middle) level of the hierarchy. Once forecasts are produced for the middle level, the approach then diverges into two paths:

- **For levels below the middle**: The top-down approach is applied to distribute forecasts from the middle level to the lower levels. This allows forecasts for the more detailed series to be derived from the intermediate-level forecasts.
- **For levels above the middle**: The bottom-up approach is used, where forecasts for higher levels are the aggregation of middle-level forecasts upwards.

The forecasting process begins by selecting a middle level in the hierarchy. This level should strike a balance between the level of detail needed and the computational resources available. Univariate time series forecasting models, such as ARIMA or ETS, are applied to each series at the middle level. These models are chosen based on the characteristics of the intermediate-level time series.

By focusing on an intermediate level, the middle-out approach balances the granularity of the bottom-up approach (which captures detailed patterns at lower levels) with the computational efficiency of the top-down approach (which requires fewer forecasts). This makes it a good compromise between the two methods. Nevertheless, in the middle out approach the number of forecasts is reduced compared to the bottom-up approach, as the method does not require individual forecasts for every bottom-level series. At the same time, it retains more detail than the top-down approach, which might overlook key patterns in the data at lower levels.

One of the primary challenges of the middle-out approach is determining the optimal middle level to generate forecasts. If the chosen middle level is not optimal, the benefits of the approach may be diminished.

A suboptimal choice could result in forecasts that are either too aggregated or too detailed, leading to less accurate results. Since different levels of the hierarchy are forecast using different methods (top-down for lower levels and bottom-up for higher levels), there is a risk of inconsistencies across levels if the two methods do not align well. Ensuring consistency across the hierarchy requires careful handling of the transition between the top-down and bottom-up components of the method.

Let's closely examine which ETS models were fitted for the middle-level (i.e., Industry) from the base model fit object:

```
> hfm_base_fit |> slice(c(4,8))

# A mable: 2 x 3
# Key:     state, industry [2]
  state      industry              base
  <chr*>     <chr*>             <model>
1 Queensland <aggregated> <ETS(M,Ad,M)>
2 Victoria   <aggregated> <ETS(M,Ad,M)>
```

The $ETS(M, A_d, M)$ model is fitted for the aggregated series at the industry level. In the middle-out approach, forecasts from these two estimated models will be aggregated for the single top-level series and distributed to the bottom six series. We will use the base model fit to perform the middle-out approach and generate forecasts for the next 2 years for the whole hierarchy. The following code snippet provides the complete operation:

```
hfm_mo_fit <- hfm_base_fit |>
  reconcile(mo_hp = middle_out(base, 1))

hfm_mo_fcst <- hfm_mo_fit |>
  forecast(h = "2 years")
```

As we did earlier, *Figure 14.11* presents plot of the middle-out forecasts for visual inspection:

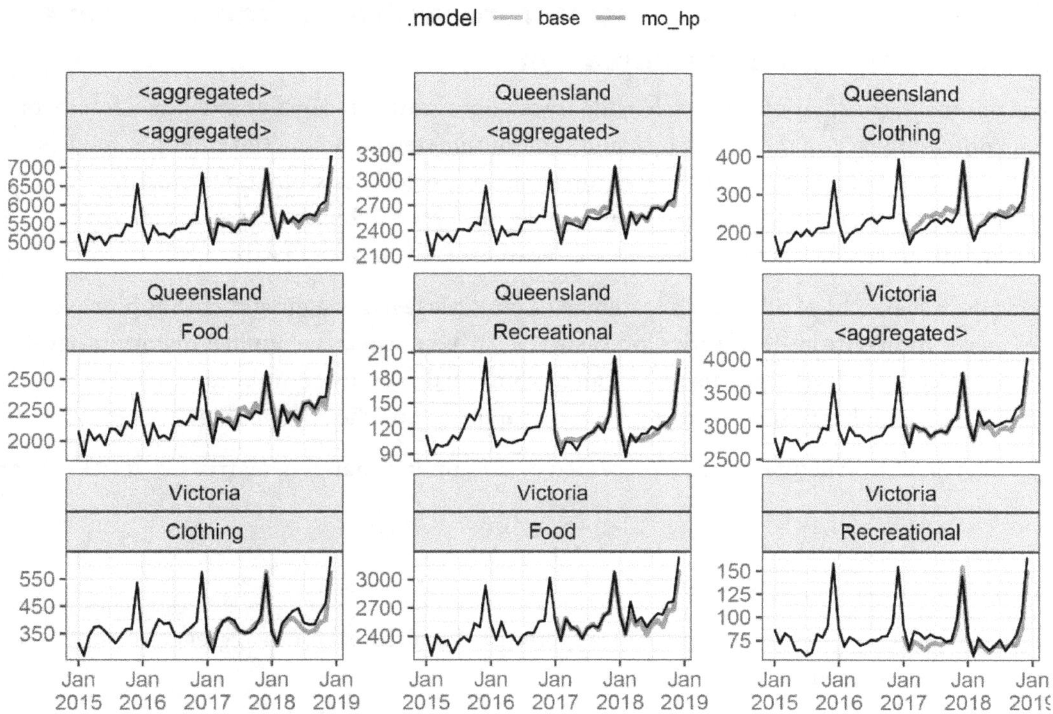


*Figure 14.11: Middle-out forecasts of retail turnover*

*Figure 14.11* shows a very strong performance of the middle out approach for state-level aggregated series. The performance remains strong when aggregated to the aggregation level or distributed to the industry level due to the simple nature of this hierarchy.

Despite the relative strengths of each traditional hierarchical forecasting approach, none of these methods captures all the dynamics in the data at different levels and between levels for forecasting all series in a hierarchy. Each approach has its own trade-offs in terms of accuracy, complexity, and computational demand.

So far, we have discussed three different forecast aggregation/distribution methods to apply to a hierarchical data structure. Next, we'll present an optimal forecast reconciliation method.

## 14.3.2 Optimal reconciliation forecast for hierarchical time series using trace minimization

The optimal reconciliation approach using trace minimization reconciles individual forecasts from different levels of the hierarchy using some automated modeling (for example, models discussed in Hyndman et al. (2008)), ensuring that these forecasts sum up to match aggregated totals, minimizing errors across all levels. This is achieved by solving an optimization problem that minimizes the forecast error covariance matrix.

Consider a hierarchical time series system with multiple levels of aggregation. Suppose we have a vector of bottom-level series forecasts $\hat{y}_b \in \mathbb{R}^m$, where $m$ is the number of bottom-level series. The higher-level series forecasts can be computed as sums of the bottom-level series. Let the total number of series across all levels be $n$, including the bottom level.

The aggregation structure can be represented by a summing/mapping matrix $S \in \mathbb{R}^{n \times m}$, where

$$\hat{y}_h = S\hat{y}_b$$

Here,

- $\hat{y}_h \in \mathbb{R}^n$ is the vector of h-step ahead forecasts for all levels (both aggregated and disaggregated).
- $S$ is the matrix that maps the bottom-level forecasts $\hat{y}_b$ to the higher level series forecasts by summing the relevant bottom-level components.

For instance, $S$ could include rows where some elements are set to 1 (to represent the summation of the relevant bottom-level series) and others set to 0 (where series do not contribute).

Thus, the total forecast vector for all levels can be written as:

$$\hat{y}_h = \begin{bmatrix} \hat{y}_{\text{top}} \\ \hat{y}_{\text{mid}} \\ \hat{y}_b \end{bmatrix}$$

where $\hat{y}_{\text{top}}$, $\hat{y}_{\text{mid}}$, and $\hat{y}_b$ represent the forecasts at the top level, intermediate levels, and the bottom level, respectively.

### 14.3.2.1 Reconciliation problem Setup

Let $\tilde{y}_h \in \mathbb{R}^n$ represent the base (unreconciled) forecasts from any forecasting method, where these forecasts are not necessarily aggregate consistent. Our goal is to reconcile these forecasts to ensure consistency across all levels.

The reconciled forecasts $\hat{y}_h^{\text{rec}}$ are obtained by adjusting the base forecasts $\tilde{y}_h$ as follows:

$$\hat{y}_h^{\text{rec}} = G\tilde{y}_h$$

where $G \in \mathbb{R}^{n \times n}$ is the reconciliation matrix. This matrix adjusts the base forecasts to ensure aggregate consistency.

The objective is to find $G$ that minimizes the forecast errors while maintaining consistency.

### 14.3.2.2 Optimal reconciliation via trace minimization

The trace minimization approach finds the optimal reconciliation matrix $G$ by minimizing the trace of the covariance matrix of the reconciled forecast errors. The objective is to ensure that the reconciled forecasts minimize the overall forecast error across all levels, subject to aggregate consistency.

The optimal reconciliation problem is formulated as follows:

$$\min_G \text{Tr}(G\Sigma_h(G)^\top)$$

subject to the constraint:

$$SG = S$$

This constraint ensures that the reconciled forecasts are aggregate consistent, meaning the reconciled bottom-level forecasts still sum to the higher-level forecasts.

Here:

- $\Sigma_h \in \mathbb{R}^{n \times n}$ is the covariance matrix of the base forecast errors.
- $\text{Tr}(\cdot)$ denotes the trace of a matrix (the sum of the diagonal elements).

The solution to this optimization problem is

$$G^\star = S^\top(S\Sigma_h S^\top)^{-1}S\Sigma_h$$

This formula gives the optimal reconciliation matrix $G^\star$, which, when applied to the base forecasts, provides the reconciled, aggregate-consistent forecasts.

To use this in practice, we need to estimate $\Sigma_h$ the forecast error variance of the h-step ahead base forecasts. The simplest choice of $\Sigma_h = k_h I$ for all $h$ and $k_h > 0$. This essentially says the matrix $G$ is independent of data, and the solution is generally referred to as an OLS-based solution (Hyndman et al., 2011). This provides substantial computational savings. The disadvantage, however, is that this specification does not account for the differences in scale between the levels of the structure, or for relationships between series. As an alternative, one could use _eqn_203.png for all _eqn_204.png and _eqn_205.png. This assumes that the error covariance matrices are proportional to each other, and we directly estimate the full one-step covariance matrix _eqn_206.png. The most obvious and simple way would be to use the sample covariance (Wickramasuriya et al., 2019).

The linear-algebra-heavy method can be implemented in R using a few lines of code. For the retail turnover example, both OLS and sample covariance-based optimal reconciliation forecasts can be derived by passing the method's name into the `min_trace()` function, as shown in the following code:

```
hfm_mint_fit <- retail_sel |>
  filter(month <= yearmonth("2016 Dec") ) |>
  model(base = ETS(turnover)) |>
  reconcile(ols = min_trace(base, method = "ols"),
            mint_cov = min_trace(base, method = "mint_cov"))

hfm_mint_fcst <- hfm_mint_fit |>
  forecast(h = "2 years")
```

*Figure 14.12* presents the base, OLS and MinT reconciliation forecasts for visual inspection:

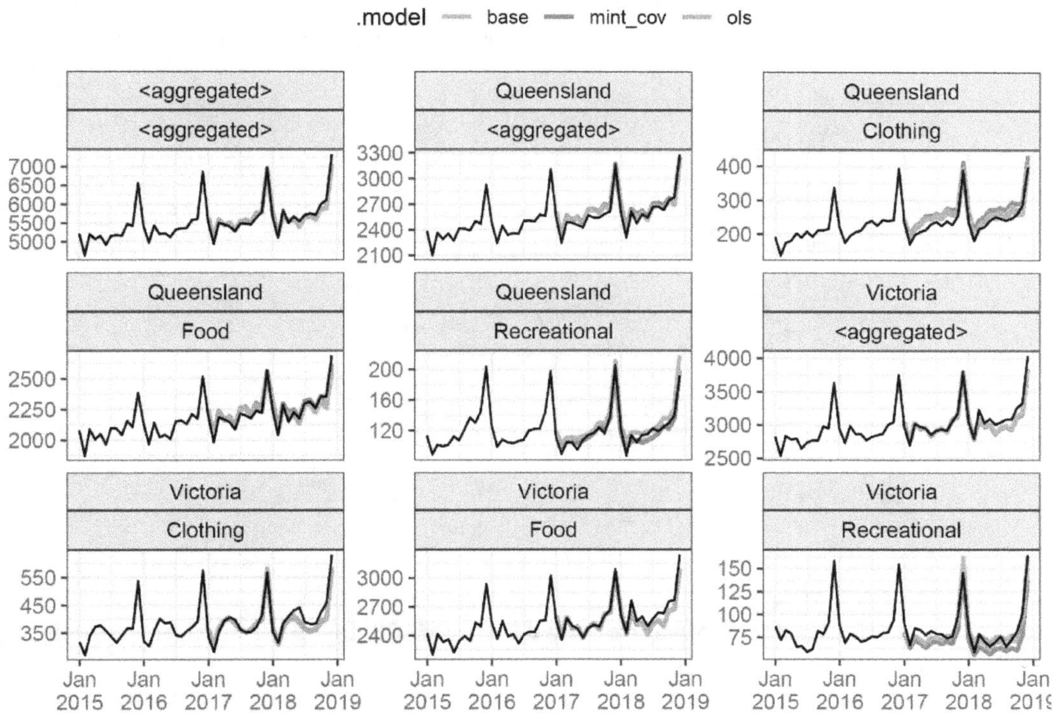


*Figure 14.12: Optimal reconciliation forecasts of retail turnover data*

The visual comparison in *Figure 14.12* shows the strong performance of the alternative implementation of optimal reconciliation approaches across all levels of the hierarchy. Both OLS and MinT -based forecast reconciliation give very close forecasts at aggregated levels.

So far, we used visual inspection to evaluate the performance of the several forecasting approaches of the hierarchical structure. In the following section we compare them via forecast accuracy metrics.

## 14.3.3 Forecast accuracy evaluation

In a hierarchical structure, visual inspection is challenging to distinguish the performance of alternative forecasting methods as the scale of the data is different in different panels and therefore, accuracy at a given series or level or across the hierarchy is not easy to estimate. An objective approach is to calculate accuracy of alternative forecasts in the test data.

To facilitate the forecasting performance comparison for all the methods, we refit them and organized in one object, generate forecasts and compute accuracy metrics as follows:

```
fit_hfm <- retail_sel_train |>
  model(base = ETS(turnover)) |>
  reconcile(
    td_hp =  top_down(base, method = c("average_proportions")),
    bu = bottom_up(base),
    mo = middle_out(base, 1),
    ols = min_trace(base, method = "ols"),
    mint_cov = min_trace(base, method = "mint_cov")
  )

acc_hfm <- fit_hfm |>
  forecast(h = "2 years") |>
  accuracy(data = retail_sel |>
             filter(month >= yearmonth("2012 Jan")))
```

*Table 14.3* organized the **Mean Absolute Scaled Error** (**MASE**) for all hierarchical forecasting methods discussed in this chapter:

| **State** | **Industry** | **Top-down** | **Bottom-up** | **Middle-out** | **OLS** | **Tr.min. (cov)** |
|---|---|---|---|---|---|---|
| Queensland | Clothing | 1.479 | 0.760 | 0.811 | 1.039 | 1.648 |
| Queensland | Food | 0.671 | 0.634 | 0.570 | 0.599 | 0.599 |
| Queensland | Recreational | 1.282 | 0.493 | 0.475 | 0.732 | 0.481 |
| Queensland | All industries | 0.596 | 0.642 | 0.623 | 0.625 | 0.752 |
| Victoria | Clothing | 1.232 | 0.671 | 0.645 | 0.591 | 0.587 |
| Victoria | Food | 0.460 | 0.508 | 0.451 | 0.481 | 0.504 |

| State | Industry | Top-down | Bottom-up | Middle-out | OLS | Tr.min. (cov) |
|---|---|---|---|---|---|---|
| Victoria | Recreational | 5.349 | 1.284 | 1.250 | 1.048 | 2.140 |
| Victoria | All industries | 0.521 | 0.603 | 0.539 | 0.527 | 0.606 |
| All states | All industries | 0.487 | 0.565 | 0.495 | 0.496 | 0.478 |

*Table 14.3: Accuracy of alternative forecast reconciliation methods for different levels of data hierarchy*

For this simple and well behaved dataset, except for the top-down approach, all other methods give very competitive results.

We have nearly completed our exploration of multiple time series analysis. To wrap up, we want to highlight the potential of applying machine learning to hierarchical forecasting, especially given its success as the winning method in the recent M5 forecasting recent competition Makridakis et al. (2022).

The next section offers a general overview of this application. We will skip the specific code here, as the foundational structure for machine learning forecasts was already detailed in the previous chapters.

## 14.4 Machine learning for multiple time series forecasting

In *Chapter 13*, you learned how to use machine learning for univariate time series forecasting and implemented it using the `forecastML` R package. This method was applied to the `AirPassengers` time series, which exhibits strong trends and seasonality, similar to retail turnover. Consequently, ML can also be utilized for forecasting multiple time series, regardless of whether individual series are stationary. The challenge of accounting for potential complex correlations between multiple time series can be addressed by empirically learning from the data. This means that when modeling one series using a machine learning model, the remaining series can be used as features to account for the impact of correlations.

As discussed in *Subsection 13.3.1*, you can include other time series as dynamic features within the `forecastML` workflow. You can also incorporate static features relevant to your models. With properly curated features, machine learning-based methods can produce highly accurate point forecasts, as evidenced by results from recent forecasting competitions. Despite their high accuracy, in real-life applications, you may need to weigh the trade-off between speed and accuracy. Many modern machine learning algorithms are fast when run at scale, but additional static and dynamic feature curation can be time-consuming and should be factored into the project's timeline.

Nevertheless, machine learning-based forecasts do not reconcile in the same way as hierarchical forecasting methods. If you require a single forecasting framework that serves multiple stakeholders at different levels of the hierarchy and requires aggregation consistency in reporting, the machine learning-based approach might not be the optimal choice, despite its favorable accuracy measures.

# Summary

In this chapter, you learned about two important classes of methods for analysis and forecasting multiple time series. You gained a solid understanding of how to model bidirectional relationships from multiple time series.

You developed a solid understanding of when the VAR model should or should not be used, and how to make such a decision by inspecting the data. You learned a wide range of methods for forecasting hierarchical and grouped time series that can be used for inventory management and demand forecasting. You now know how to use tidy R code to apply these advanced methods for multiple time series analysis and forecasting.

This chapter concludes our discussion on time analysis with a core focus on forecasting. In the next chapter, we will extend these concepts to business decision-making, particularly in scenarios where A/B testing tools are less relevant or not applicable.

# Further reading


- Athanasopoulos, G., Ahmed, R. A., & Hyndman, R. J. (2009). *Hierarchical forecasts for Australian domestic tourism. International Journal of Forecasting*, 25(1), 146–166.
- Dangerfield, B. J., & Morris, J. S. (1992). *Top-down or bottom-up: Aggregate versus disaggregate extrapolations. International Journal of Forecasting*, 8(2), 233–241.
- Fliedner, G. (2001). *Hierarchical forecasting: Issues and use guidelines. Industrial Management & Data Systems*, 101(1), 5–12.
- Granger, C. W. (1969). *Investigating causal relations by econometric models and cross-spectral methods. Econometrica: Journal of the Econometric Society*, 424–438.
- Gross, C. W., & Sohl, J. E. (1990). *Disaggregation methods to expedite product line forecasting. Journal of Forecasting*, 9(3), 233–254.

- Hill, R. C., Griffiths, W., & Lim, G. C. (2018). *Principles of econometrics. John Wiley & Sons.*
- Hyndman, R. J., Ahmed, R. A., Athanasopoulos, G., & Shang, H. L. (2011). *Optimal combination forecasts for hierarchical time series. Computational Statistics & Data Analysis*, 55(9), 2579–2589.
- Hyndman, R. J., & Khandakar, Y. (2008). *Automatic time series forecasting: The forecast package for R. Journal of Statistical Software*, 27(3), 1–22. `https://doi.org/10.18637/jss.v027.i03`
- `Lütkepohl`, H. (2010). *Impulse response function. Springer.*
- Makridakis, S., Spiliotis, E., & Assimakopoulos, V. (2022). *M5 accuracy competition: Results, findings, and conclusions. International Journal of Forecasting*, 38(4), 1346–1364.
- Sims, C. A. (1980). *Macroeconomics and reality. Econometrica: Journal of the Econometric Society*, 1–48.
- Sims, C. A. (1987). *Vector autoregressions and reality]: comment. Journal of Business & Economic Statistics*, 5(4), 443–449.
- Wickramasuriya, S. L., Athanasopoulos, G., & Hyndman, R. J. (2019). *Optimal forecast reconciliation for hierarchical and grouped time series through trace minimization. Journal of the American Statistical Association*, 114(526), 804–819.


# Join our community on Discord

Join our community's Discord space for discussions with the author and other readers:

`https://discord.gg/G4YmB6DQvf`

# Part 5

# Plot Twists and Turning Points: Measuring Impact and Detecting Anomalies

It is common to have auto-correlated data in many business scenarios. Standard causal impact measurement methodologies may become unreliable if such inter-dependencies are not properly handled. In this part, we address this problem and present time-series-based techniques for causal impact measurement. Anomalies and change point detection are two crucial phenomena that are observed frequently in business settings. We discuss some methodologies to detect anomalies or change points in the data and how to handle them. This part of the book is suitable for intermediate to advanced professionals.

*Part 5* comprises the following chapters:

- *Chapter 15, Causal Impact Estimation*
- *Chapter 16, Changepoint Detection*
- *Chapter 17, Anomaly Detection and Imputation*

# 15

# Causal Impact Estimation

Modern businesses operate in a rapidly evolving environment shaped by a multitude of dynamic factors. These include product and service innovations, internal system and process transformations, fluctuating market conditions, competitive actions, and changing regulatory landscapes. To thrive in such complexities, businesses must base their decisions ranging from strategic initiatives to operational updates not only on data but also on reliable evidence. A central challenge in this context is determining the true impact of a business decision or action.

Traditional data analysis methods have long sought to address this question through various approaches, with the quality of insights often tied to the sophistication of an organization's analytics capabilities. However, objective and confident decision-making requires more than surface-level analysis. It demands high-quality evidence that unveils deeper cause-and-effect relationships. For critical decisions, such as launching a new product, entering a new market, or implementing a strategic initiative, this evidence must inspire confidence and meet rigorous standards, underscoring the need for robust analytical approaches.

The idea of a hierarchy of evidence has been a cornerstone in clinical and public health research, serving as a framework for evaluating the effectiveness of interventions. This concept is now gaining traction in the business world, where organizations aim to anchor their decisions in reliable causal insights. The process of measuring and understanding the impact of business initiatives falls within the domain of causal inference. By leveraging causal inference principles, businesses can avoid common pitfalls of correlation-based analyses, enabling evidence-based decision-making that aligns with strategic goals and the demands of an ever-changing marketplace.

Given the broad scope of topics covered, the chapter is organized as follows:

- Section 15.1: What is causal inference?

- Section 15.2: Causal inference with experimentation
- Section 15.3: Causal inference with observational data
- Section 15.4: Equivalence of hypothesis tests and regression
- Section 15.5: Interrupted Time Series (ITS) analysis
- Section 15.6: ITS on stationary time series
- Section 15.7: ITS on time series with trends
- Section 15.8: ITS on time series with autocorrelation
- Section 15.9: ITS on time series with AR(1) and trends
- Section 15.10: Flexibilities of ITS-based inference
- Section 15.11: Monitoring guardrail metrics in ITS



# Technical requirements

Let's load the packages that are required in this chapter.

```
#install.packages("pacman")
library(pacman)
p_load("broom"
       ,"conflicted"
       ,"datasets"
       ,"desk"
       ,"fable"
       ,"fabletools"
       ,"forecast"
       ,"here"
       ,"janitor"
       ,"lmtest"
       ,"patchwork"
       ,"prais"
       ,"pwr"
```

```
        ,"sandwich"
        ,"tidyverse"
        ,"tsibble"
        ,"tsibbledata" )

conflicts_prefer(
  fabletools::accuracy
  ,fabletools::forecast
  ,dplyr::filter
  ,dplyr::select )
```

The code file is available in the GitHub repo at `https://github.com/PacktPublishing/Modern-Time-Series-Analysis-with-R.git` for you to download and practice.

# 15.1 What is causal inference?

Causal inference is the process of determining whether a cause-and-effect relationship exists between variables and quantifying the strength of that relationship. It goes beyond simply identifying correlation to answer fundamental questions: what causes what, and by how much? This process is central to understanding the underlying mechanisms that drive changes in outcomes, allowing decision-makers to pinpoint the true impact of one variable on another. Unlike correlation analysis, causal inference emphasizes direction and the pathways through which variables influence each other.

Correlations can offer insights into bivariate relationships, but they are prone to misinterpretation and may lead to flawed decisions if incorrect causality is assumed. For example, while a correlation analysis might reveal that increased advertising expenditure correlates with higher sales, such an association does not necessarily mean that the increased expenditure caused the sales to rise. Both variables could be influenced by a third factor, such as seasonal demand or economic conditions, commonly termed a *confounder*. Causal inference seeks to disentangle these complexities, determining whether advertising expenditure is the true driver of sales or merely coincidental. Acting solely on correlation-led analysis could result in investing resources in initiatives that do not deliver the desired outcomes or missing opportunities to address the actual drivers of change. Causation, on the other hand, provides the actionable insights necessary for strategic planning and operational execution.

Causal inference relies on a combination of theoretical principles, experimental designs, and statistical tools to isolate the causal effect of one variable on another. It is grounded in frameworks such as counterfactual reasoning, where the impact of a treatment or intervention is measured against what would have happened in its absence. Experimental methods are often considered the gold standard for causal inference because they effectively eliminate confounding factors in counterfactual reasoning through random assignment of treatment. However, in many real-world scenarios, treatment assignment is impractical or unethical, thus making the experiment unfeasible, and in such scenarios, observational data and advanced statistical techniques such as *matching, instrumental variables methods*, and *time series models* are employed to estimate the causal effect.

In the terminology of causal inference, a "treatment" refers to an intervention or initiative, such as a business decision or policy change, while an "outcome" or "response" denotes the measurable result or effect of that intervention. For businesses, this framework provides a systematic approach to evaluate the true impact of their actions, whether it involves launching a new product, implementing a marketing campaign, or optimizing internal processes.

Building upon a fundamental understanding of the core concepts and essential terminologies associated with causal inference, the subsequent section will formally introduce the foundational tools and methodologies necessary for conducting causal analysis.

## 15.2 Causal inference with experimentation

Over the centuries, several scientific methods have been developed to identify and estimate causal relationships, each tailored to specific scenarios and data contexts. The reliability of evidence derived from these methods varies, giving rise to the concept of a hierarchy of evidence. This concept has been extensively discussed and refined in fields, such as clinical and public health research (Evans, 2003). More recently, it has begun to play a pivotal role in business decision-making processes, as organizations recognize the need for robust and reliable insights to guide their strategies.

At the top of this hierarchy lies scientific experimentation, which provides the most reliable form of evidence for causal inference. Foundational texts, such as Montgomery (2017), offer a detailed exploration of experimental designs, including those specifically crafted to uncover cause-and-effect relationships. Among these, the most fundamental and powerful design is the randomized controlled trial (RCT). In an RCT, subjects are randomly assigned to either a treatment group or a control group, ensuring that, on average, the two groups are comparable in all respects except for the treatment being studied. This randomization process minimizes the influence of confounding variables (external factors that could affect both the treatment and the outcome), thereby enabling a direct attribution of observed differences to the treatment itself.

RCTs have gained significant traction in the technology industry in recent times and are often referred to as experiments or A/B testing. At Microsoft, A/B testing was employed to substantiate claims that the choice of color for presenting search results in the Bing search engine could generate tens of millions in annual revenue (Kohavi et al., 2017). For businesses with millions or billions of users, A/B testing on the individual response serves as a reliable method for making objective decisions.

However, conducting proper A/B testing in a business scenario faces practical challenges that are highlighted in the following sub-section.

## 15.2.1 Practical challenges of conducting experiments

While experimental tools such as A/B testing are often regarded as the gold standard for establishing causal relationships, there are many scenarios where conducting such experiments is either impractical or unfeasible. One of the common challenges is obtaining the required sample size, particularly when businesses have a limited number of customers to study. This issue is further compounded when the effect size of the intervention is small. For instance, detecting a 0.5% increase in click-through rate (CTR) demands a significantly larger sample size compared to detecting a 5% increase.

The challenge becomes even greater when businesses aim for a higher degree of confidence in their results. Achieving a higher level of significance of a statistical test reduces the likelihood of false positive results but necessitates a larger sample size. Similarly, when decision-makers require greater statistical power ensuring a lower likelihood of false negatives, then the sample size requirement grows further.

Variability in individual responses to an intervention also complicates matters, as higher variance inflates the sample size needed to achieve reliable results. While advanced experimental designs can sometimes address these issues by reducing the sample size and stabilizing variance, implementing such strategies requires deep insights into the population under study. In practice, many businesses do not have such an advantage. Consequently, for medium-sized or even some large businesses, conducting experiments might not be practical due to these resource constraints.

In addition to the limitations posed by sample size, not all experiments in a business setting involve individual customers or users. Often, experiments need to be designed around larger units, such as geographic regions or retail outlets. The number of such units is typically far smaller than the number of individual users, making the sample size requirement unrealistic.

Beyond these, practical constraints in several scenarios where A/B testing is inherently unsuitable are as follows (but not limited to these):

- **Post implementation analysis**: When an intervention has already been implemented, it becomes impossible to design a randomized experiment retrospectively.
- **Resistance to randomization**: Stakeholders or participants may resist random assignment due to concerns about the potential adverse effects of the intervention on key business metrics.
- **Ethical considerations**: Testing certain interventions, such as subsidies or aids targeted at sub-populations, may raise equity or ethical concerns, especially if withholding treatment is perceived as unfair.
- **Long-term follow-up**: Experiments requiring prolonged monitoring, such as studying the impact of an intervention on customer complaints over several months may be impractical due to resource or time constraints.
- **Low incidence events**: Measuring the effect of interventions on rare but critical events, such as system failures or severe customer complaints, is challenging due to the scarcity of such events.

Due to these practical limitations, businesses often need to look for analytical tools beyond A/B testing to assess the impact of their interventions. These constraints also highlight the importance of alternative approaches, such as observational data analysis, which can provide valuable causal insights when experimentation is not feasible. We discuss the method in the following section.

# 15.3 Causal inference with observational data

Observational studies provide a practical and reliable means of investigating causal relationships in real-world contexts where experiments are not feasible. Unlike experimental studies, observational research involves the collection of data on subjects without applying any interventions or manipulating variables in a randomized way. Instead, subjects are observed for outcomes in their natural environments, allowing researchers to analyze outcomes without directly influencing them. Depending on their design and objectives, observational studies can be categorized as either *descriptive* (focusing on characterizing phenomena) or *analytical* (aimed at exploring relationships between variables).

Though descriptive studies provide powerful insights to uncover hidden patterns, it is not the focus of this chapter. We'll focus on analytical studies leveraging statistical modeling and inference. Two common types of analytical observational studies are as follows:

- **Cohort studies**: These studies track a group of subjects over time to examine the relationship between exposure to specific factors and the development of an outcome.

Cohort studies may be prospective, observing events as they unfold, or retrospective, analyzing past data to uncover patterns.

- **Case-control studies**: These studies compare subjects who have a particular outcome (cases) with those who do not (controls) to identify factors that may have contributed to the outcome. Case-control studies can also capture how the outcome evolves over time.

The main focus of this book is on the application of time-series-based methodologies in a business context. Hence, we limit the discussions of cohort studies using time-series-based methodologies.

Experimental data typically focuses on unit-level responses, such as whether an individual responds positively to a treatment or not. On the other hand, observational studies can analyze both unit-level responses and macro-level outcomes. For instance, a macro-level response might measure the total number of customers contacting a mobile phone service provider about connection problems due to a cellular network fault in different geographic regions.

Establishing causality in observational studies, however, is inherently more challenging than in experimental designs due to the lack of direct control over the treatment variable. Observational studies are particularly susceptible to confounding factors—variables that influence both the independent and dependent variables, potentially biasing the results. To address these challenges, observational studies often adopt strategies to mimic experimental designs, such as quasi-experiments (e.g., propensity score matching) for observable confounders, or rely on sophisticated statistical methods to control for unobservable confounders (e.g., Heckman correction in regression) during the analysis. These approaches are extensively discussed in existing literature and are integral to ensuring valid inferences.

We will not cover discussions on the broader methodologies for establishing causality from observational data. Instead, our goal is to demonstrate how observational study methods can be utilized in scenarios where a known cause-and-effect mechanism already exists, and the objective is to estimate the impact of the causal factor on the response.

The reliability of causal effect estimates depends on the adequacy of the analytical tools employed. By applying advanced time series methods to observational data, we aim to highlight the improved robustness of these tools, ensuring that the resulting quantitative evidence is actionable, and can reliably estimate the effects of interventions or treatments. As established in earlier chapters, many time series models are fundamentally rooted in regression techniques. Therefore, prior to leveraging time series regression for causal impact estimation, it is essential to formally understand the inherent connection between hypothesis testing and regression analysis, which will be the focus of the next section.

## 15.4 Equivalence of hypothesis tests and regression

Over the centuries, a wide array of statistical methods and models have been developed in both academic and applied research to analyze data derived from experiments and observational studies. For data collected from scientific experiments, statistical hypothesis testing has been the cornerstone of analysis. Comparing group means, for example, often relies on methods such as t-tests, z-tests, and chi-squared tests. When the comparison involves multiple groups, researchers frequently turn to F-tests or Analysis of Variance (ANOVA). These methods are deeply rooted in specific assumptions: the samples must be random, independent and identically distributed (i.i.d.); and they should exhibit approximately normal distributions. The validity of any conclusions drawn relies on the fulfillment of these fundamental assumptions.

In contrast, regression analysis offers an alternative approach, particularly useful for observational studies. The Time Series Linear Models (TSLM) presented in Chapter 10, Section 10.7, and advanced time series regression techniques, such as ARIMAX in Chapter 12, Section 12.6, provide the flexibility to incorporate multiple factors or predictors into the analysis, enabling a more nuanced understanding of the relationships within the data. However, the validity of regression-based inference also depends on satisfying critical assumptions: the relationship between predictors and the outcome must be linear, predictors must be independent of one another, residuals should have a mean of zero with constant variance, and residuals must remain independent of both the predictors and each other.

Despite their distinct origins and applications, hypothesis testing and regression analysis share a fundamental equivalence when the required assumptions for analysis are met. Indeed, many common statistical tests, such as t-tests and ANOVA, can be viewed as specific cases of inference about parameters using linear regression models. This equivalence underscores the power of regression not only for prediction but also for causal inference, provided that confounding factors are adequately controlled.

To illustrate this inter-connectedness, we turn to a classic dataset, the randomized experiment on sleep conducted by Student, the pseudonym of William S. Gosset, who developed the t-distribution. The dataset measures the effect of two soporific drugs on `10` patients, recording the increase in hours of sleep. The paired design of this randomized controlled trial (RCT) ensures that confounders are minimized, as each individual receives both treatments, allowing for direct comparisons. The `Student` dataset, available in the `datasets` package in R, provides an excellent example for exploring the connections between traditional hypothesis testing and regression analysis.

Let's have a look at the data in the following output:

```
> str(sleep)

'data.frame':    20 obs. of  3 variables:
 $ extra: num  0.7 -1.6 -0.2 -1.2 -0.1 3.4 3.7 0.8 0 2 ...
 $ group: Factor w/ 2 levels "1","2": 1 1 1 1 1 1 1 1 1 1 .
 $ ID   : Factor w/ 10 levels "1","2","3","4",..: 1 2 3 4
```

The preceding output presents the structure of the `sleep` dataset showing variables and their types. The group column indicates the treatment allocation, and the extra column records the response, extra amount of sleep. In our analysis, we will organize the data into a tidy format and redefine the group variable to make it more intuitive. Through the application of statistical tools, we will demonstrate how hypothesis testing and regression analysis yield consistent results when assumptions are satisfied.

First, we will illustrate equivalence of hypothesis testing and linear model in a paired sampling design in the following subsection.

## 15.4.1 Paired sample

Consider a scenario where two treatments are compared in terms of their effect on sleep as a paired design. To begin, calculate the mean and variance for two groups using the following code:

```
slpd <- sleep |>
  tibble() |>
  clean_names() |>
  mutate(
    group = case_when(
      group == 1 ~ "Trt_1",
      .default = "Trt_2"
    ) |>
    factor() |>
    fct_relevel(c("Trt_1", "Trt_2"))
  )

sleep_summary <- slpd |>
  group_by(group) |>
  summarise(
    mean = mean(extra),
    variance = var(extra)
```

```
)

sleep_summary
```

The summary statistics for treatment groups are printed in the following output:

```
# A tibble: 2 × 3
  group  mean variance
  <fct> <dbl>    <dbl>
1 Trt_1  0.75     3.20
2 Trt_2  2.33     4.01
```

The summary suggests, on average, treatment 1 resulted in an additional 0.75 hours of sleep, while treatment 2 led to 2.33 extra hours of sleep. This difference indicates that treatment 1 resulted in 1.58 fewer hours of extra sleep compared to treatment 2. But how consistent will this difference be if you perform the same study with other response cohorts? This is the exact reason we conduct hypothesis testing—to estimate confidence and measure the consistency of obtaining repetitive results observed in a study.

Now, perform a paired t-test using the `t.test()` function with the option `paired = TRUE`. The results will provide p-value and 95% confidence intervals that can be used to evaluate the difference between treatments. Run the following code:

```
pair_ttest <- t.test(
  extra ~ group, data = slpd,
  paired = TRUE, conf.level = 0.95)

pair_ttest_tidy <- tidy(pair_ttest) |>
  select(method, alternative, estimate, p.value,
         conf.low, conf.high)

pair_ttest_tidy
```

The following tidied output shows the results from the paired t-test:

```
# A tibble: 1 × 6
  method          alternative estimate p.value conf.low conf.high
  <chr>           <chr>          <dbl>   <dbl>    <dbl>     <dbl>
1 Paired t-test two.sided      -1.58 0.00283    -2.46    -0.700
```

The paired t-test shows p-value $< 0.05$, indicating a statistically significant difference between the two treatments at a $5\%$ significance level. To interpret the p-value, recall that it represents the probability of observing a difference as extreme as $-1.58$ under the null hypothesis that assumes no difference between the means. With such a small p-value, we confidently reject the null hypothesis. The $95\%$ confidence interval for the mean difference ranges from $-2.46$ to $-0.70$, further supporting the conclusion that treatment $1$ causes significantly less extra sleep compared to treatment $2$.

While a paired t-test can technically be performed without explicitly computing the difference between measurements obtained under alternative treatments, the underlying statistical calculation inherently operates on difference. Difference in measurement is also needed to demonstrate the use of a linear model on this data. To validate the consistency between approaches, we will first run a one-sample t-test on the calculated difference data to ensure that the results align with those derived from the paired t-test applied earlier. The following code computes the difference between two treatments and then performs a one-sample t-test:

```
slpdiff <- slpd |>
  pivot_wider(names_from = group,
              values_from = extra) |>
  mutate(diff_group =  Trt_1 - Trt_2)

onesamp_diff_ttest <- t.test(
  slpdiff$diff_group,
  conf.level = 0.95)

onesamp_diff_ttest_tidy <- tidy(onesamp_diff_ttest) |>
  select(
    method, alternative, estimate,
    p.value, conf.low, conf.high)

onesamp_diff_ttest_tidy
```

The following output confirms that the one-sample t-test on the difference between two treatments provides results identical to the results obtained from the paired t-test:

```
# A tibble: 1 × 6
  method            alternative estimate p.value conf.low conf.high
  <chr>             <chr>          <dbl>   <dbl>    <dbl>     <dbl>
1 One Sample t-test two.sided      -1.58 0.00283    -2.46    -0.700
```

To achieve the same results using linear regression, we fit a linear model on the `diff_group` column with no independent variable. Coefficient estimates from this model will essentially make inferences about the mean of `diff_group` (difference in treatment responses). You can fit the linear model using the `lm()` function, tidy the model object, and compute confidence intervals using the `confint()` function as shown in the following code:

```
pair_diff_lm <- lm(diff_group ~ 1, data = slpdiff)
tidy(pair_diff_lm)
```

The results are presented in the following output:

```
# A tibble: 1 × 5
  term        estimate std.error statistic p.value
  <chr>          <dbl>     <dbl>     <dbl>   <dbl>
1 (Intercept)    -1.58     0.389     -4.06 0.00283


> round(confint(pair_diff_lm), 4)

              2.5 %  97.5 %
(Intercept) -2.4599 -0.7001
```

As you will have noticed, the regression results mirrored those of the paired t-test, including identical estimates, p-values, and confidence intervals. This demonstrates that a one-sample t-test and a linear regression model for the mean difference are mathematically equivalent.

In the following section, we look at the equivalence between hypothesis testing and regression-based method in a two-sample study design.

## 15.4.2 Two independent samples

Now, let us shift the focus to the scenario of two independent response groups. Suppose the same data is reinterpreted as two independent samples rather than paired. We conduct a two-sample t-test, assuming equal variance. To maintain simplicity, we have opted not to perform formal statistical testing for the equality of variance. Instead, we rely on the established guidance that a two-sample t-test assuming equal variance remains robust even when sample variances differ by as much as a fourfold factor. In our data, the ratio of large and small sample variances, $\frac{4.01}{3.20} = 1.25$, is indeed less than 4. Hence, the equal variance assumption holds.

To perform a two-independent-samples t-test, pass the formula specification `extra ~ group` into the `t.test()` function with the option `var.equal = TRUE`. The following code shows you how to perform the test and organize the results:

```
twosamp_ttest <- t.test(extra ~ group, data = slpd,
                        var.equal = TRUE, conf.level = 0.95)

twosamp_test_tidy <- tidy(twosamp_ttest) |>
  select(method, alternative, estimate,
         p.value, conf.low, conf.high)

twosamp_test_tidy
```

The following results are observed:

```
# A tibble: 1 × 6
  method            alternative estimate p.value conf.low conf.high
  <chr>             <chr>          <dbl>   <dbl>    <dbl>     <dbl>
1 Two Sample t-t… two.sided        -1.58  0.0792    -3.36     0.204
```

The p-value $= 0.079$ is higher than $0.05$, and showed no statistically significant difference between the two treatments at a $5\%$ level of significance. The confidence interval ranges from $-3.36$ to $0.204$, which includes the null value zero, reflecting greater uncertainty in the estimated difference. This discrepancy arises because the standard error is inflated under the assumption of independence, as opposed to the paired design. This distinction highlights the importance of adopting an alternative experimental design in impact measurement; a paired-sample design has better control for the confounders than a two-independent-samples design.

So far, we have shown the equivalence of hypothesis testing and regression methods in on-sample and two-sample scenarios. Next, we will check the equivalence in a more general scenario, such as having three or more response groups.

While a t-test evaluates intervention on a single group or exactly two groups, the F-test serves as a general test that can assess multiple treatments or factors simultaneously to detect any statistically significant differences. The F-test is primarily employed in Analysis of Variance (ANOVA) to compare means across three or more treatments, and in regression to assess if a complex model fits the data significantly better than a simpler, intercept-only model.

We check the equivalence by fitting a linear model, `extra ~ 1 + group`, and perform an F-test as part of a two-sample ANOVA analysis. Like previous examples, the `Trt_2` is considered as the reference treatment. Let's perform the ANOVA analysis using the `aov()` function and extract results, as shown in the following code:

```
twosamp_anova <- aov(
  extra ~ 1 + group,
  data = slpd |>
    mutate(
      group = relevel(group,ref = "Trt_2")))

tidy(twosamp_anova)
```

The results presented in the following output show that the p-value from the F-test is identical to what was obtained from the two-sample t-test earlier:

```
# A tibble: 2 × 6
  term         df sumsq meansq statistic p.value
  <chr>     <dbl> <dbl>  <dbl>     <dbl>   <dbl>
1 group         1  12.5  12.5       3.46  0.0792
2 Residuals    18  64.9   3.60     NA     NA
```

When two or more treatments are compared, a pairwise comparison, such as Tukey's Honest Significant Difference (HSD) can be applied to identify which two treatments have different means. Even though there are only two treatments in the `sleep` data, we can still apply the HSD for illustration to identify which treatment group is significantly different. Before applying Tukey's HSD, one treatment should be kept as a reference. Throughout the prior examples, we used treatment 2 as a reference. The `TukeyHSD()` function requires an ANOVA output to test pair-wise differences. Pass the `twosamp_anova` object, as shown in the following code:

```
tukeyhsd_tidy <- tidy(TukeyHSD(twosamp_anova)) |>
  select(contrast,estimate, adj.p.value,
         conf.low, conf.high )
```

```
tukeyhsd_tidy
```

The result is presented in the following output:

```
# A tibble: 1 × 5
  contrast    estimate adj.p.value conf.low conf.high
  <chr>          <dbl>       <dbl>    <dbl>     <dbl>
1 Trt_1-Trt_2    -1.58      0.0792    -3.36     0.204
```

With two treatments, we can have only one pairwise comparison, which appears in the contrast column in the output. Estimated coefficients and confidence intervals from TukeyHSD matched those from the two-independent-samples t-test.

Finally, we check the equivalence of a regression model with an F-test in a scenario with three or more treatments. Run a linear model to perform a pairwise comparison for two treatments, keeping treatment 2 as a reference group, using the following code snippet:

```
twogrp_lm <- lm(
  extra ~ 1 + group,
  data = slpd |> mutate(
    group = relevel(group,ref = "Trt_2")))

tidy(twogrp_lm)
```

The coefficient estimates from the linear model and the 95% confidence intervals for the estimated coefficients are presented in the following output:

```
# A tibble: 2 × 5
  term        estimate std.error statistic p.value
  <chr>          <dbl>     <dbl>     <dbl>   <dbl>
1 (Intercept)     2.33     0.600      3.88 0.00110
2 groupTrt_1     -1.58     0.849     -1.86 0.0792


> round(confint(twogrp_lm), 4)


              2.5 %  97.5 %
(Intercept)  1.0686  3.5914
groupTrt_1  -3.3639  0.2039
```

As expected, the results obtained through linear regression models were consistent with those from t-tests and F-tests. Coefficients, p-values, and confidence intervals calculated using regression were identical, confirming that linear models offer a flexible framework for hypothesis testing in a general context.

Up to this point, we have utilized linear models to explain continuous outcome variables through categorical predictors, establishing the fundamental equivalence between various methods of impact measurement. It is critical to note that these equivalences remain valid only when the underlying assumptions of hypothesis testing are met and the regression model demonstrates an adequate fit, as verified through rigorous residual analysis. In scenarios where an outcome is influenced by a combination of continuous and categorical variables, regression models naturally extend into Analysis of Covariance (ANCOVA). However, to maintain the focus of our current progression, a detailed exploration of ANCOVA is intentionally omitted from this discussion.

The utility of simple regression extends beyond the analysis of experimental data in the context of time-series analysis. In earlier chapters, we discussed regression-based approaches for univariate, multivariate, and multiple time-series data. While Vector Autoregression (VAR) models were interpreted in terms of their coefficients, other methods, such as smoothing, filtering, and forecasting were emphasized for their predictive power rather than interpretability. It is worth noting that coefficients derived from regression models involving lagged outcome variables often lack natural interpretations. However, the regression framework remains central to understanding the relationships among variables, making it a foundational tool for advanced statistical inference.

This section has demonstrated the versatility of regression models as a powerful, unifying framework for various statistical analyses, including comparing treatments and analyzing time-series data. The equivalence of results obtained through methods, such as t-tests, ANOVA, and ANCOVA highlights the inherent flexibility of regression in handling diverse research questions. In the next section, you will find how to apply regression model-based concepts to measure impacts from time-series data.

# 15.5 Interrupted Time Series (ITS) analysis

Interrupted Time Series (ITS) analysis is a robust quasi-experimental method used to evaluate the impact of an intervention or policy change by comparing outcome metrics across distinct time segments. Unlike multi-group comparisons, a standard ITS focuses on a single population, analyzing time-series data to determine whether an intervention "interrupted" the pre-existing level or trend, or both. By examining shifts in both level (an immediate change) and trend (a change in the rate of increase or decrease), ITS provides a statistically rigorous way to attribute observed variations to a specific intervention.

A recent large-scale review of 200 academic studies involving 230 time series, conducted by Turner et al. (2021), summarizes the methods commonly used for ITS analysis. The most widely adopted approaches are regression-based methods estimated using Ordinary Least Squares (OLS), either with or without adjustments. Adjustments are critical when time series properties, such as autocorrelation, are present, as these properties violate key OLS assumptions. Autocorrelation, for instance, can bias regression coefficient estimates and render conventional statistical tests invalid due to incorrect standard error estimates.

Several methods are available to address the challenges posed by autocorrelation in ITS analysis, each offering varying degrees of sophistication. They are as follows:

- **OLS without adjustments**: This approach assumes no autocorrelation in the data, which can lead to biased estimates and unreliable statistical tests if the assumption is violated.
- **OLS with the Cochrane-Orcutt method**: This technique iteratively estimates the autocorrelation parameter and transforms the regression model to remove first-order autocorrelation (AR(1)) effects. The process continues until convergence is achieved.
- **Generalized Least Squares (GLS) using the Prais-Winsten method**: This method improves upon Cochrane-Orcutt by retaining the first observation in the dataset, resulting in more precise estimates for models with first-order autocorrelation (AR(1)).
- **OLS with Newey-West standard errors**: This adjustment accounts for both heteroskedasticity (non-constant variance) and autocorrelation in the error terms, ensuring more reliable statistical inference for time-series data.
- **Time-series regression models** (e.g., ARIMA with exogenous variables): These models explicitly account for the time series structure of the data, making them suitable for more complex scenarios involving autocorrelation and non-stationarity.

While a comprehensive discussion of these methods and their mathematical foundations is beyond the scope of this book, readers interested in the theoretical underpinnings of OLS adjustments can refer to the seminal work by Judge et al. (1991). This chapter will instead focus on the practical application of these methods to compare results, demonstrating how the concepts introduced in earlier chapters can be used to measure the impact of initiatives through ITS analysis.

By applying ITS methods, this chapter will also highlight the quality of evidence that can be expected when using different approaches to analyze observational data. The core objective is to equip you with the essential tools and insights necessary for the rigorous evaluation of intervention impacts. This includes understanding the inherent strengths and limitations of ITS methods within the context of real-world time series data.

To illustrate the concept of ITS, we will use simulated data throughout this chapter. Simulations are particularly useful for demonstrating ITS methods, as real-world business data used for measuring the impact of an initiative often remains confidential due to its sensitive nature. Businesses are generally hesitant to disclose the specifics of their initiatives, which are often designed to solve industry-wide problems or gain a competitive edge. Simulated data also allows us to benchmark the estimated benefits, as the true impact is known in advance. For this chapter, we will leverage the forecast package, which provides functionality for simulating time-series data under statistical model assumptions. While countless scenarios exist for simulating data, this chapter will focus on common and straightforward scenarios to serve as foundational building blocks for tackling more complex business cases.

# 15.6 ITS on stationary time series

The stationarity in the response metric is one of the most commonly assumed properties when analyzing data using experimental tools. When there is an interruption in the time series resulting in a level shift attributable to a business intervention, the resulting series covering the pre-intervention and post-intervention periods is no longer stationary. However, the metric can be stationary in the pre- and post-intervention periods separately. Consequently, a before-and-after comparison of the metric using experimental tools would violate the identically distributed assumption necessary for conducting statistical tests of significance.

To avoid this violation of assumption and to control bias due to confounders, experiment-based methods rely on comparisons between treatment and control groups to measure the impact of the intervention. Thus, in our data generation process, we adhere to these principles to ensure that a valid comparison can be made.

Let's generate simulated intervention data under the following scenario:

- `n_obs <- 120`: Total number of observations, 120
- `arima_ord <- c(0L, 0L, 0L)`: The $ARIMA(0,0,0)$ model represents a white noise or stationary process, which we use to simulate metrics for both the treatment and control groups with an error variance of 3. By simulating stationary data in this manner, we can effectively compare the treatment and control groups while maintaining the assumptions necessary for valid statistical inference. This will ensure that our analyses yield reliable and meaningful results.
- `n_trt <- 30`: Number of observations post-intervention
- `base_level <- 100`: Underlying level in the simulated data
- `level_increase <- 0.05`: A 5% increase in level was caused by intervention

Suppose we are measuring the number of customers who responded positively to an offer, such as clicking on a link in a marketing email. This is our business metric. In a design involving control and treatment groups, the intervention could be the position/font/color of the link in the email. The metric is measured for 120 equidistant time points. The time unit can be chosen to suit the size of the customer base for the business, thereby avoiding high variation in the response. For businesses with a large user base, the time unit could be in minutes (for example, 5 or 10 minutes), whereas for a smaller base, the time unit might need to be extended to hours or even days. Incorporating the minimum sample size required under a specified effect size could determine how long such an experiment should run.

The response metric is assumed to increase by 5% from the pre-intervention level once the intervention is applied. We consider 30 recent time points as post-intervention. It is worth noting that for treatment-control comparisons using experiments, pre-intervention data is not strictly necessary; however, for many business cases, it is possible to derive the pre-intervention metric for both groups if desired.

Now, generate the simulated data with the pre-defined parameters and organize them using the following code:

```
set.seed(1)
n_obs <- 120
arima_ord <- c(0L, 0L, 0L)
n_trt <- 30
base_level <- 100
level_increase <- 0.05

# Simulate two sets of data from ARIMA(0,0,0) model
a_mod2sim <- Arima(ts(rnorm(n_obs, sd = 3)), order = arima_ord)

sim_set1 <- simulate(a_mod2sim) + base_level
sim_set2 <- simulate(a_mod2sim) + base_level

# True impact size
# 5% increase in underlying level
impact_size <-base_level*level_increase

# Pre and post intervention metric
y_pre <- sim_set2[1:(n_obs - n_trt)]
y_post <- c(tail(sim_set2, n_trt)) + impact_size

# Working data as tibble
```

```
wd <- tibble(y_ctrl = sim_set1,
             y_trt =  c(y_pre, y_post)) |>
  mutate(tidx = 1:n()) |>
  mutate(period = c(rep("_pre", (n_obs - n_trt)),
                    rep("_post", n_trt)) |> factor() |>
           fct_relevel("_pre"))

wd
```

The first 10 rows of the simulated data are printed in the following output:

```
# A tibble: 120 × 4
   y_ctrl y_trt  tidx period
    <dbl> <dbl> <int> <fct>
 1   99.0 102.      1 _pre
 2  104.  103.      2 _pre
 3   99.8 101.      3 _pre
 4   99.9  98.0     4 _pre
 5  100.  103.      5 _pre
 6  102.   95.0     6 _pre
 7  100.   98.9     7 _pre
 8  100.   99.7     8 _pre
 9   98.5  99.9     9 _pre
10   99.5 103.     10 _pre
# i 110 more rows
```

You can plot the simulated data for the control and treatment groups, both pre- and post-intervention, using the following code:

```
wd |>
  pivot_longer(cols = c(y_ctrl, y_trt),
               names_to = "metric") |>
  ggplot(aes(x = tidx, y = value)) +
  geom_line(aes(linetype = metric), linewidth = 0.6) +
  scale_linetype_manual(values = c(4,1) ) +
  scale_x_continuous(limits = c(0,121), breaks = seq(0,121,30)) +
  geom_vline(xintercept = 90) +
  labs(x = "Time index", y = "Response metric") +
  theme(legend.position = "top")
```

The plot is presented in *Figure 15.1*:

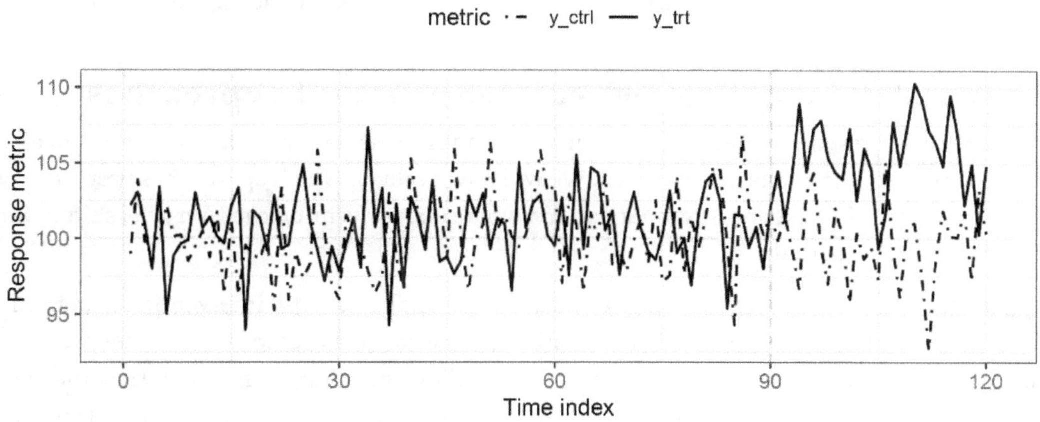


*Figure 15.1: Time plot of a stationary metric from the treatment and control groups*

From the plot, you can visually identify that the intervention caused an immediate shift in the metric for the treated group. As expected, there is a level shift in the post-intervention period, but no trend is introduced. The pre- and post-intervention summary is computed using the following code:

```
stationary_summary <- wd |>
  select(period, y_ctrl, y_trt) |>
  group_by(period) |>
  summarise_all(list(mean = mean, sd = sd))

stationary_summary
```

The summary is presented in the following output:

```
# A tibble: 2 × 5
  period y_ctrl_mean y_trt_mean y_ctrl_sd y_trt_sd
  <fct>        <dbl>      <dbl>     <dbl>    <dbl>
1 _pre          100.       101.      2.75     2.54
2 _post          99.7      105.      2.68     2.75
```

The summary confirms that the mean and standard deviation for the metric during the pre-intervention period were similar across both groups (99.7 versus 100). Following the intervention, the mean for the treatment group experienced a noticeable shift (increased to 105 from 101), while the standard deviation remained unchanged. The summary confirms that there is no change in variance and no trend introduced due to intervention, only the average level shifts upward. Hence, the response series after the intervention holds the stationary characteristics. Next, we will discuss methods to measure the impact of such intervention data

## 15.6.1 Stationary data impact measurement approaches

We previously presented three intervention data analysis methods: hypothesis testing, linear regression-based, and time-series-based. We will now evaluate which approaches best measure the true impact on a stationary metric where an intervention causes a level shift in the post-intervention period.

**Approach 1**: A two-independent-samples t-test can be applied to post-intervention data from the treatment and control groups. This statistical test allows us to determine whether there is a significant difference between the two groups. To facilitate meaningful interpretation, the data is re-leveled using the `fct_relevel()` function to make the control group a reference. This ensures that the measured change in the treatment group is attributable to the intervention.

After the data organization, you can perform the two-sample t-test with an equal variance assumption by running the following code:

```
stationary_ttest <- wd |>
  filter(period == "_post") |>
  select(y_ctrl, y_trt) |>
  pivot_longer(
    cols = c(y_ctrl, y_trt),
    names_to = "metric"
  ) |>
  mutate(
    metric = factor(metric) |>
    fct_relevel(c("y_trt", "y_ctrl"))
  ) %>%
  t.test(
    value ~ metric, data = .,
    var.equal = TRUE, conf.level = 0.95)

stationary_ttest_tidy <- tidy(stationary_ttest ) |>
  select(
    estimate, p.value, conf.low,
    conf.high, alternative)

stationary_ttest_tidy
```

The preceding code will produce the following output:

```
# A tibble: 1 × 5
  estimate  p.value conf.low conf.high alternative
     <dbl>    <dbl>    <dbl>     <dbl> <chr>
1     5.63 5.39e-11     4.23      7.03 two.sided
```

The results of the t-test indicate that the difference between the treatment and control groups is statistically significant, as evidenced by the very small p-value. The estimated impact of the intervention is calculated to be 5.63. Furthermore, the 95% confidence interval for this estimate ranges from 4.23 to 7.03. Importantly, this interval includes the true impact of the intervention, which is known to be 5. This alignment reinforces the reliability of the estimated impact and underscores the effectiveness of the intervention.

**Approach 2**: Alternatively, you could apply a linear model to measure the impact of the initiative within the treated group, using a pre- versus post-intervention comparison. This approach is particularly suitable given that the response data is stationary. By incorporating a dummy variable (period) to represent the intervention period, the regression model applied using the following code captures the impact of the post-intervention phase. The linear model allows testing the estimated parameter for significance and constructing confidence intervals for the estimated impact, which is also implemented in the following code:

```
stationary_lm <- lm(y_trt ~ period, dat = wd)
tidy(stationary_lm)
```

Here's the tidied model output:

```
# A tibble: 2 × 5
  term        estimate std.error statistic   p.value
  <chr>          <dbl>     <dbl>     <dbl>     <dbl>
1 (Intercept)   101.       0.274    368.   2.09e-182
2 period_post     4.67     0.547      8.53 5.78e- 14


> round(confint(stationary_lm),4)


               2.5 %   97.5 %
(Intercept) 100.1214 101.2049
period_post   3.5835   5.7506
```

Interestingly, the results from the linear model differ from what was obtained from the two-samples t-test. According to the linear model, the statistically significant estimated impact of the intervention is 4.67. The 95% confidence interval for this estimate ranges from 3.58 to 5.75, comfortably encompassing the true impact. Notably, this interval is narrower than the one obtained from the t-test, reflecting a higher degree of precision in the linear model's estimate.

To validate the fitted linear model's goodness of fit, evaluate the two diagnostic plots; residuals vs fitted and Q-Q distribution of the residuals are presented in *Figure 15.2*:

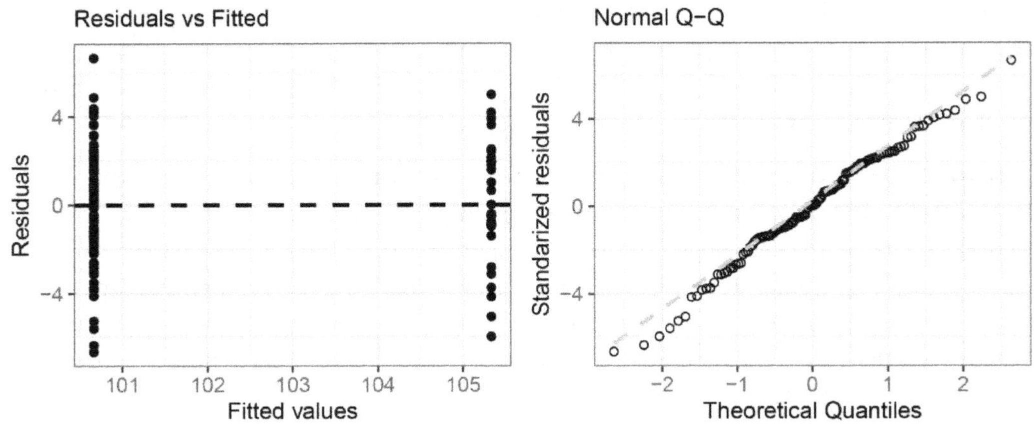


*Figure 15.2: Adequacy check for fitted linear model*

Visual inspection of *Figure 15.2* confirms that the residuals from the linear model are centered around zero when plotted against the fitted values. Additionally, the Q-Q plot demonstrates that the distribution of residuals is approximately normal. This further suggests that the impact measured by the regression method is valid and actionable.

A third approach could involve a time-series model-based approach. However, it is important to recognize that the pre- and post-intervention metrics are stationary. It does not exhibit any time-series patterns, such as trend, seasonality, or serial dependency. Hence, a time-series-based model will not be able to leverage additional time dynamics from data compared to what was already implemented in the prior two approaches.

However, how do we identify serial correlation, such as an $AR(1)$ process, in practical intervention data? The following section outlines a diagnostic procedure to address this question.

## 15.6.2 Durbin-Watson test to identify an AR(1) process

When conducting an experiment in real business scenarios, you might need to assess serial-dependency (autocorrelation structure) in the residuals from the linear regression model used for impact measurement. Formal statistical tests, such as the Durbin-Watson test, can be used to check for first-order autocorrelation. The `dwtest()` function from the `lmtest` package provides an efficient way to check it. Use the following code to test whether an $AR(1)$ process exists in the residuals obtained from the linear model used in the preceding section:

```
stationary_lm_ar1 <- dwtest(
  stationary_lm, alternative = "two.sided")
```

The test will provide the following output:

```
tidy(stationary_lm_ar1) |>
  select(statistic, p.value, alternative)

# # A tibble: 1 × 3
#   statistic p.value alternative
#       <dbl>   <dbl> <chr>
# 1      2.15   0.454 true autocorrelation is not 0
```

The p-value from this test is large ($0.454$), indicating that the first-order autocorrelation in the model residuals is not significantly different from zero. The test result further supports that the linear model is adequate to measure impact when data following an intervention is stationary.

Next, we provide a brief overview of power analysis as an essential tool for experimental design, then discuss how these principles extend to impact estimation via regression modeling.

## 15.6.3 Statistical power analysis in regression

Power analysis is a vital statistical technique used to determine the minimum sample size required for an experiment. By balancing the risk of making a false positive error (significance level) against the need for sufficient strength to detect a real effect (statistical power) and the expected effect size, it ensures the study includes enough data to detect a true effect. This method helps confirm that the results of an experiment or survey are statistically significant and not simply due to random chance. For standard experimental designs, such as A/B testing, it is critical to calculate the required sample size upfront.

The concept of statistical power calculation extends to observational data modeling involving regression. In observational data modeling, generally, we don't have any flexibility in collecting additional data besides what is already collected. This leaves only the option to check whether the minimal sample size requirement was met for regression under the hypothesized effect. A higher effect size indicates that the mean difference is large relative to the overall variability.

The pwr package provides functions for performing power analysis for regression. This package allows researchers to calculate sample sizes for hypothesized effects at a specified significance level and power. In a regression setting, the effect size is defined as a ratio of between and within the sum of squares, computed as $f^2 = R^2/(1 - R^2)$, where $R^2$ is the coefficient of determination of the regression model. Cohen's effect size thresholds (Cohen, 2013) are $0.02$, $0.15$, and $0.35$, corresponding to small, medium, and large effects respectively, for regression. Now, compute the minimum sample size required for a linear regression model to confidently measure the observed effect of $f^2 = 0.35$ with $80\%$ power at a $5\%$ significance level. The value of $u = 1$ indicates degrees of freedom or the number of independent variables excluding the intercept in the model:

```
pwr_analysis_lm <- pwr.f2.test(
  u = 1, f2 = 0.35,
  sig.level = 0.05,
  power = 0.80)

min_sample_size <- ceiling(
  pwr_analysis_lm$u + pwr_analysis_lm$v + 1)
```

The result can be used to determine the minimum sample size required for a regression with one predictor, as shown in the following output:

```
min_sample_size
# [1] 25
```

In the formula for calculating the minimum sample size, $v$ represents the residual degrees of freedom calculated by the power test. It accounts for the sample data leftover after all parameters are estimated. The additional $+ \ 1$ term accounts for the intercept (i.e., the constant term) in the regression equation.

In the simulation, we kept $30$ observations after intervention for measuring impact. This is sufficient as it is higher than the recommended $(25)$ to observe a large effect size $(0.35)$ with $80\%$ power and $5\%$ level of significance.

Computing the achieved effect size is crucial because it quantifies the measured effect from the observed data. This also allows businesses to determine whether the regression model is practically meaningful, rather than the coefficients being just statistically significant. It is expected that the effect size should be at least as large as the effect size assumed in the power analysis. If the achieved effect size is smaller than the pre-specified effect size, then it is an indication that the model's estimated impact might not be reliable.

Let's compute the achieved effect size from the fitted linear model to the stationary data (`stationary_lm`) using the following code:

```
# Extract R-squared
rsq <- summary(stationary_lm)$r.squared
achieved_effect <- rsq/(1-rsq)
```

The result is shown in the following output:

```
achieved_effect
# [1] 0.6165695
```

The achieved effect size is 0.62 obtained under a study designed to detect a $0.35$ change (expected effect size for a regression model).

In business perspective, the intervention results, such as $1.7 \times$ the anticipated value or the $77\%$ exceeds from the performance benchmark—confirms the impact measured by the regression-based model is a robust, statistically sound response that can be used for broader intervention implementation with confidence.

A stationary time series is the simplest form of temporal data, conceptually, it is similar to cross-sectional data with an added time stamp. While non-stationary intervention data can be transformed into a stationary format to apply regression-based model, we aim to measure intervention effect directly by avoiding these cumbersome pre-processing transformations.

In the following sections, we will explore how Interrupted Time Series (ITS) analysis is applied to various non-stationary intervention data, beginning with series that exhibit trends.

# 15.7 ITS on time series with a trend

Time-series data is rarely stationary, as metrics often exhibit natural trends over time. These trends may arise from various factors, such as organic growth in the customer base, offering additional products and services over time, slow but steady growth of the brand in the market, and many more. Businesses are often interested in rolling out initiatives to improve metrics that inherently display such trends.

To illustrate such data, we will simulate a metric by incorporating a trend coefficient of 0.1 for each unit increase in the time index. The data was generated using the following code:

```
trend_size <- 0.1 # Effect of trend
wd <- wd |>
  mutate(y_c_trnd = y_ctrl + trend_size*tidx,
         y_t_trnd =  y_trt + trend_size*tidx)
```

The resulting metric, reflecting the effects of both the trend and the intervention, is presented in *Figure 15.3*:

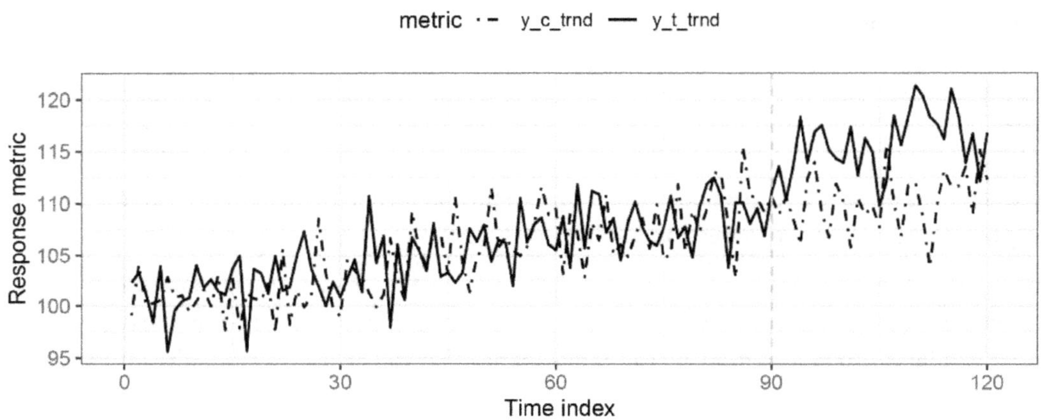


*Figure 15.3: Time plot of a metric exhibiting upward trend in the treatment and control group*

The time plots for the treatment and control groups in *Figure 15.3* suggest that the intervention caused an immediate upward shift in the treated group. Despite the presence of the intervention effect, the underlying trend remains evident. Similar to a stationary scenario, the treatment and control groups exhibit parallel trends over time, allowing us to distinguish the intervention's impact from the natural progression of the metric.

The plot of the simulated data demonstrates how interventions can interact with existing trends, offering insights into their potential effects in real-world scenarios.

Compute a descriptive summary from the control and treatment groups in pre- and post-intervention using the following code:

```
trend_summary <- wd |>
  select(period, y_c_trnd, y_t_trnd) |>
  group_by(period) |>
  summarise_all(list(mean = mean, sd = sd))
```

```
trend_summary
```

The summary table is provided in the following output:

```
# A tibble: 2 × 5
  period y_c_trnd_mean y_t_trnd_mean y_c_trnd_sd y_t_trnd_sd
  <fct>          <dbl>         <dbl>       <dbl>       <dbl>
1 _pre            105.          105.        4.14        3.74
2 _post           110.          116.        2.79        2.9
```

The summary of the experiment data suggests that the post-experiment mean has increased in both the control and treatment groups. Even though the standard deviation is lower in the post-experiment duration, that is due to the fact that this measure is scale dependent and it was measured around the higher mean value.

## 15.7.1 Trend data impact measurement approaches

Let's analyze the intervention data that exhibits a trend with hypothesis testing and a regression-based technique. Then we will explore the suitability of a time-series-based approach.

Like earlier, a two-sample t-test evaluates the intervention impact, keeping the control group as a reference. You may use the following code to perform the operation:

```
trend_ttest <- wd |>
  filter(period == "_post") |>
  select(y_c_trnd, y_t_trnd) |>
  pivot_longer(cols = c(y_c_trnd, y_t_trnd),
               names_to = "metric") |>
  mutate(metric = factor(metric) |>
         fct_relevel(c("y_t_trnd", "y_c_trnd"))) %>%
  t.test(value ~ metric, data = .,
        var.equal = TRUE, conf.level = 0.95)


trend_ttest_tidy <- tidy(trend_ttest) |>
  select(estimate,p.value, conf.low, conf.high, alternative)

trend_ttest_tidy
```

The result from the t-test is presented in the following output:

```
# A tibble: 1 × 5
  estimate  p.value conf.low conf.high alternative
     <dbl>    <dbl>    <dbl>     <dbl> <chr>
1     5.63 2.54e-10     4.15      7.11 two.sided
```

The t-test results indicate a statistically significant difference between the two groups, as evident by the very small p-value. The estimated impact of the intervention is 5.63, with a 95% confidence interval ranging from 4.15 to 7.11, comfortably encompassing the true impact of 5. However the interval is wide compared to the stationary data scenario which is attributed to the inflated variance.

While applying a linear model to measure intervention impact, you need to consider the existing trend in both control and treatment groups before the intervention was introduced. In this scenario, a regression model with a dummy variable for the intervention period should be considered to measure intervention impact. Since our data exhibits a time trend, we also include the time index variable as a predictor in the model to account for this trend effect. This is usually referred to as measuring the intervention impact adjusting for the trend.

Fit a linear model and generate the result as follows:

```
trnd_trt_lm <- lm(y_t_trnd ~ period + tidx, dat = wd)
tidy(trnd_trt_lm)
```

The model coefficients and confidence intervals are shown in the following output:

```
# A tibble: 3 × 5
  term        estimate std.error statistic   p.value
  <chr>          <dbl>     <dbl>     <dbl>     <dbl>
1 (Intercept)  100.        0.546     184.  6.39e-146
2 period_post    4.34      0.830       5.23 7.68e-  7
3 tidx           0.106     0.0104     10.2  8.47e- 18


> round(confint(trnd_trt_lm),4)
              2.5 %   97.5 %
(Intercept) 99.3307 101.4931
period_post  2.6925   5.9792
tidx         0.0850   0.1261
```

The linear model, adjusted for the time trend, estimates the intervention's impact at 4.34, with a 95% confidence interval ranging from 2.69 to 5.98. This interval comfortably contains the true impact and indicates a significant effect, as shown by the p-value < 0.001. Compared to the t-test, the confidence interval for the linear model is wider, reflecting a lack of precision while estimating more parameters from the same amount of data. Additionally, the model estimates a significant time trend effect, with the magnitude 0.106 closely matching the true value 0.1 used in simulation.

A third way to estimate impact from this data would be to use a time-series-based approach. Conceptually, a linear model for time-series data does not require further adjustments if a model already accounts for the time trend. In such cases, the linear model is expected to provide an adequate fit. This can be validated by analyzing the residuals. *Figure 15.4* presents the residual diagnostic plots from the fitted `trnd_trt_lm` linear model:

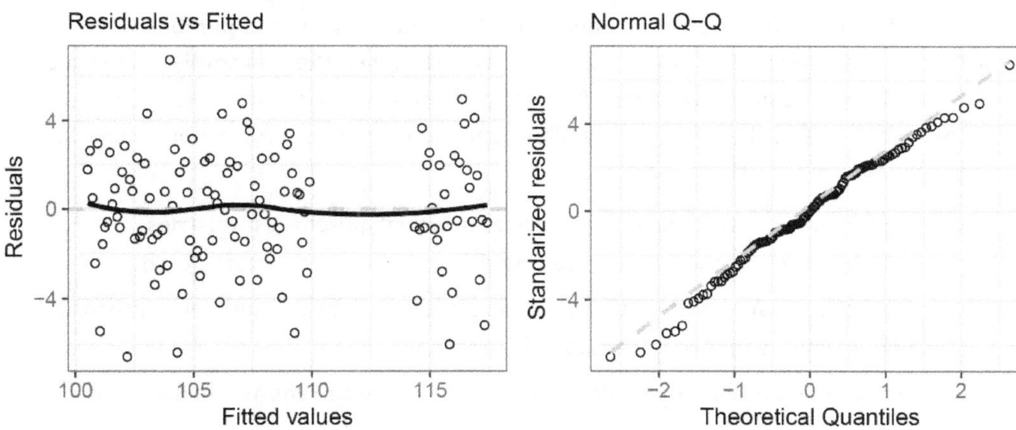


*Figure 15.4: Residuals analysis for checking the adequacy of the fitted linear model for measuring impact for stationary data*

In the preceding plots, residuals are centered around zero when plotted against the fitted values. Moreover, the Q-Q plot indicates that the residuals are approximately normally distributed.

To further evaluate the linear model, a formal test for first-order autocorrelation was conducted on the residuals, yielding a p-value of 0.499. This result suggests no evidence of autocorrelation in the linear model's residual, supporting the validity of the model fit.

From the data summary, we know that the variance in the treatment group changed after the intervention. While this could potentially affect the reliability of the coefficient tests in the linear models, it is important to recognize that the ratio of variances between the pre- and post-intervention data is 1.64. It is well known that t-tests are robust to variance differences of up to a four-fold magnitude, meaning this level of variance change is unlikely to compromise the results and that is the rationale behind using `var.equal = TRUE` in the t-test. Furthermore, had we tested the coefficient using a robust standard error estimate, such as the Newey-West standard error (discussed in Section 15.8.2), the p-value would still have been $< 0.001$. The resulting confidence interval would range from $2.72$ to $5.95$, representing only a negligible change. We leave this calculation as an exercise for you to try after reviewing the relevant section.

Next, we will move on to estimating impact from serially correlated data where dependency is not induced by a long-term time trend but due to autocorrelation.

# 15.8 ITS on time series with autocorrelation

In earlier sections, we emphasized that when data is time-stamped but lacks strong time series properties such as serial dependency, a linear model produces results comparable to traditional experimental analysis tools, such as t-tests. However, those examples represent only a small subset of business problems. In practice, business metrics are often serially dependent, influenced by their past values. This dependency can arise for various reasons, such as the following:

- Customer behavior: In subscription services or repeat-purchase industries, a customer's past purchase behavior can significantly influence future purchases.
- Website traffic: Past website traffic can affect current traffic through mechanisms, such as user engagement, repeat visits, and viral trends.
- Production dynamics: Past production levels can shape current production due to factors such as capacity planning and supply chain constraints.

In this section, we explore options for ITS analysis using data with serial correlation to address such scenarios effectively. To illustrate, we simulate data from an ARIMA(1,0,0) process to ensure first-order autocorrelation with a moderate value of the AR parameter, $\phi = 0.5$. The intervention impact size is kept unchanged. The data was generated using the following code:

```
set.seed(2)
arima_ord <- c(1L, 0L, 0L)

# Simulate two sets of data from ARIMA(1,0,0) model
arima2sim <- Arima(ts(rnorm(n_obs, sd = 3)),
                   order = arima_ord,
                   fixed = c(0.5, NA))
sim_set1_ar <- simulate(arima2sim) + 100
sim_set2_ar <- simulate(arima2sim) + 100
```

```
# Pre and post intervention metric
y_pre <- sim_set2_ar[1:(n_obs - n_trt)]
y_post <- c(tail(sim_set2_ar, n_trt)) + impact_size

# Working data as tibble
wd_ar <- tibble(
  y_ar1_ctrl = sim_set1_ar,
  y_ar1_trt =  c(y_pre, y_post)
) |>
  mutate(tidx = 1:n()) |>
  mutate(
    period = c(
      rep("_pre", (n_obs - n_trt)),
      rep("_post", n_trt)
    ) |>
    factor() |>
    fct_relevel("_pre")
  )
```

Let's visualize data for control and treatment groups in *Figure 15.5*:

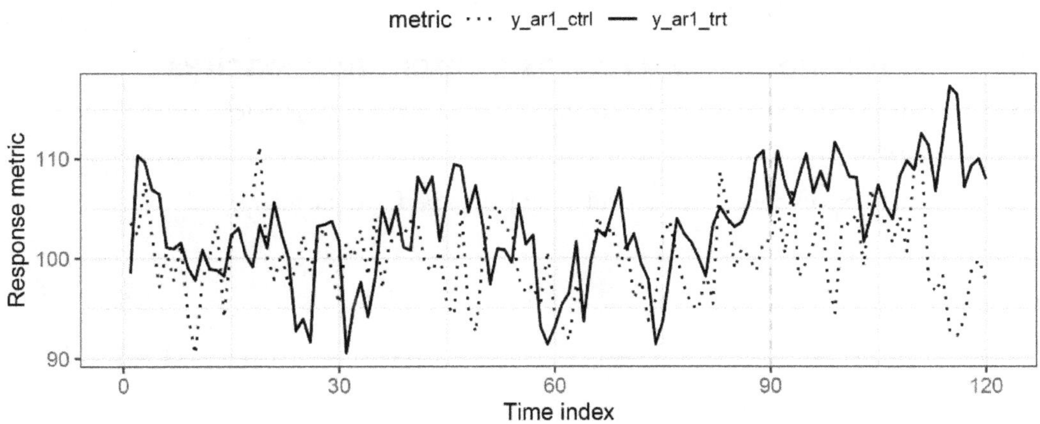


*Figure 15.5: Time plot of an AR(1) metric by treatment and control group*

The plot shows that the intervention caused a shift in the metric for the treated group immediately after its application. However, unlike previous scenarios, the visual identification of the intervention's impact is more challenging due to the serial correlation in the data, although the means for the pre- and post-intervention are clearly different. You can compute the summary of this data as follows:

```
ar_summary <- wd_ar |>
  group_by(period) |>
  summarise(mean = mean(y_ar1_trt),
            sd = sd(y_ar1_trt))

ar_summary
```

The mean and variance from both groups before and after intervention are as follows:

```
# A tibble: 2 × 3
  period  mean    sd
  <fct>  <dbl> <dbl>
1 _pre    101.  4.72
2 _post   109.  3.32
```

We observe a larger difference in the pre- and post-intervention means compared to what was used during the simulation. This has over-inflated the difference due to the serial dependency. This suggests that a method that can handle serial correlation data would be better suited to measure intervention impact.

### 15.8.1 AR(1) data impact measurement approaches

We will follow the same sequence for measuring the intervention's impact from data with serial autocorrelation.

The code for a two-sample t-test is provided for you to perform the test:

```
ar_ttest <- wd_ar |>
  filter(period == "_post") |>
  select(y_ar1_ctrl, y_ar1_trt) |>
  pivot_longer(cols = c(y_ar1_ctrl, y_ar1_trt),
               names_to = "metric") |>
  mutate(metric = factor(metric) |>
    fct_relevel(c("y_ar1_trt", "y_ar1_ctrl"))) %>%
  t.test(value ~ metric, data = .,
         var.equal = TRUE, conf.level = 0.95)
```

The following test results show the estimated impact of 7.82, which is significant, with a small p-value, and the 95% confidence interval, 5.73 to 9.91:

```
tidy(ar_ttest) |>
  select(estimate,p.value,
         conf.low, conf.high, alternative)
```

Following are the results:

```
# A tibble: 1 × 5
  estimate  p.value conf.low conf.high alternative
     <dbl>    <dbl>    <dbl>     <dbl> <chr>
1     7.82 4.41e-10     5.73      9.91 two.sided
```

The t-test result shows that the confidence interval does not contain the true intervention effect (5) used in the simulation. The hypothesis test-based approach overestimated the impact when there is autocorrelation in the intervention data.

In the linear regression model approach, the response variable (`y_ar1_trt`) should be regressed on the dummy variable period to capture the pre- versus post-intervention effect as shown in the following code:

```
ar_lm <- lm(y_ar1_trt ~ period, dat = wd_ar)
tidy(ar_lm)
```

The results in the following output show that the linear model overestimated intervention impact, 7.41 ($p$-value $< 0.05$), and the true intervention impact, 5, is outside of the 95% confidence interval (5.56,9.25):

```
# A tibble: 2 × 5
  term        estimate std.error statistic   p.value
  <chr>          <dbl>     <dbl>     <dbl>     <dbl>
1 (Intercept)   101.       0.465    218.   1.40e-155
2 period_post     7.41     0.931      7.96 1.18e- 12


> round(confint(ar_lm),4)

               2.5 %   97.5 %
(Intercept) 100.4325 102.2753
period_post   5.5657   9.2513
```

If you compare the results of the t-test and linear model, they are not drastically different, but both of them have overestimated the true intervention impact.

We should also assess the residual diagnostic plots to evaluate the linear model's validity to the serially correlated data. *Figure 15.6* presents the ACF and PACF for the residuals from the `ar_lm` model:

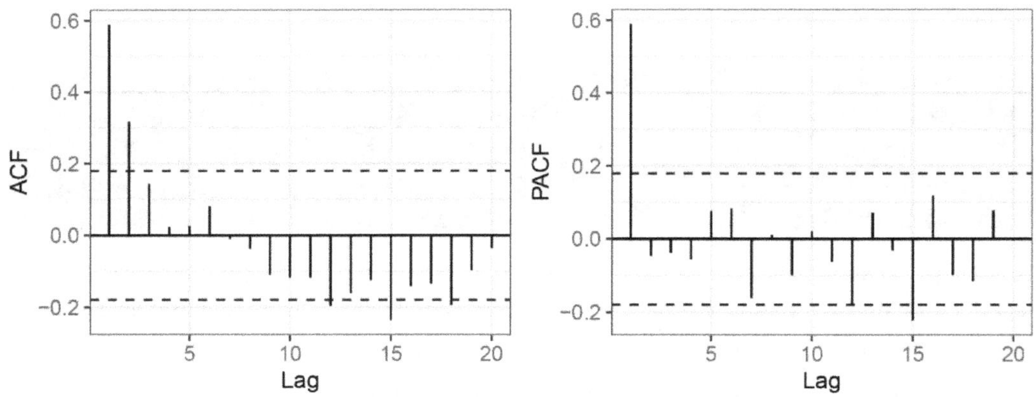


*Figure 15.6: Autocorrelation function of residuals from linear model fit on AR(1) time series*

The plot reveals strong first-order autocorrelation in the residuals. This is further supported by the Durbin-Watson test, which yields a p-value of 0, indicating significant first-order autocorrelation. As a result, the test for the coefficients in the linear model is not reliable, and the model fails to adequately account for the autocorrelation present in the intervention data.

It is a common phenomenon that t-test and regression-based-models are not adequate and overestimate the impact of intervention data that exhibits serial correlation. One way to improve the regression-based approach is applying proper adjustments on the estimated coefficients. In the next section, we will discuss three of such adjustments.

## 15.8.2 Adjustments in the linear model for AR(1) process

Theory of regression analysis tells us that the presence of $AR(1)$ in the residuals provides invalid inference unless the effect of autocorrelation is adjusted properly. In particular, autocorrelation can result in an inefficient estimate of model parameters by underestimating standard errors, which leads to false significance of the estimated coefficients.

To address this issue, three approaches are widely used:

- the Cochrane-Orcutt method
- the Prais-Winsten method
- the Newey-West adjustment.

These methods help to isolate the effect of autocorrelation and bring validity to the statistical tests. In R, packages such as `desk`, `prais`, and `lmtest` (in combination with sandwich) can be used to implement these adjustments.

Now, we will move on to the time-series-based approaches for measuring impact from intervention data as they exhibit time series characteristics, such as autocorrelation. Let's perform the adjustments and examine the results using these methods.

We will apply the Cochrane-Orcutt method to start with. The `cochorc()` function from the `desk` package is used to perform the adjustment on the linear model object. After that the 95% intervals are extracted as shown in the following code:

```
# Adjustment by Cochran-Orcutt
lm_CO_adj <- desk::cochorc(ar_lm, iter = 100)

# Derive 95% confidence interval
lm_CO_adj_int <- tibble(
  bind_rows(
    lm_CO_adj$results[,"coef"] -
    qnorm(0.975)*lm_CO_adj$results[,"std.err."],
    lm_CO_adj$results[,"coef"] +
    qnorm(0.975)*lm_CO_adj$results[,"std.err."]
  )
) |>
  t() |>
  `colnames<-`(c("2.5 %",    "97.5 %"))

lm_CO_adj
```

The model results are presented in the following output:

```
Cochrane-Orcutt estimation given AR(1)-autocorrelated errors
------------------------------------------------------------

                  coef  std.err.   t.value  p.value
(Intercept)   101.3394    0.8928  113.5036  < 1e-04
period_post     7.3009    1.6742    4.3607  < 1e-04


Number of iterations performed:        3
Final rho value:                  0.5898
```

The confidence intervals are as follows:

```
> lm_CO_adj_int

                2.5 %    97.5 %
(Intercept) 99.58950 103.08933
period_post  4.01946  10.58239
```

The estimated impact of the intervention using the Cochrane-Orcutt adjustment is 7.3, with a p-value < 0.001, indicating a statistically significant effect. The 95% confidence interval ranges from 4.02 to 10.58, which includes the true impact of 5. Additionally, the estimated magnitude of the autocorrelation coefficient is 0.589, closely aligning with the first-order ACF value observed in *Figure 15.6* and the true value used in the simulation.

The Prais-Winsten method uses the same model formula as the linear model with an additional argument for the time index. The model formula is passed directly to the `prai_winsen()` function from the `prais` package. The model fit, computation of intervals, and how to extract other information are shown in the following code:

```
# Adjustment by Cochran-Orcutt
lm_PW_adj <- prais::prais_winsten(
  y_ar1_trt ~ period,
   dat = wd_ar, index = 'tidx')

# Estimated coefficient and significance test
lm_PW_adj_sum <- lm_PW_adj |> summary()

# Derive 95% confidence interval
lm_PW_adj_int <- tibble(
  bind_rows(lm_PW_adj_sum$coefficients[,"Estimate"] -
  qnorm(0.975)*lm_PW_adj_sum$coefficients[,"Std. Error"],
  lm_PW_adj_sum$coefficients[,"Estimate"] +
  qnorm(0.975)*lm_PW_adj_sum$coefficients[,"Std. Error"])) |>
  t() |>
  `colnames<-`(c("2.5 %",    "97.5 %"))
```

Let's look at the results provided after the Prais-Winsten adjustment. The estimated correlation coefficient is as follows:

```
lm_PW_adj$rho |> last() |> c()
# [1] 0.5898135
```

The Prais-Winsten adjustment provides the following coefficients estimate:

```
> coef(lm_PW_adj_sum) |> round(3)

            Estimate Std. Error t value Pr(>|t|)
(Intercept)  101.339      0.893 113.504        0
period_post    7.301      1.674   4.361        0
```

The confidence intervals are as follows:

```
> round(lm_PW_adj_int,4)

              2.5 %   97.5 %
(Intercept) 99.5895 103.0893
period_post  4.0195  10.5824
```

The Prais-Winsten adjustments produce nearly identical results to the Cochrane-Orcutt method, reinforcing the need for time series-based adjustment when intervention generates an autocorrelation structure in responses.

Finally, we apply the Newey-West adjustment to the standard errors of the linear model coefficients as shown in the following code chunk using the `coeftest()` function from the `lmtest` package:

```
lm_NW_adj <- lmtest::coeftest(
  ar_lm,
  vcov = NeweyWest(ar_lm))

tidy(lm_NW_adj)
```

The results from the Newey-West are as follows:

```
# A tibble: 2 × 5
  term        estimate std.error statistic   p.value
  <chr>          <dbl>     <dbl>     <dbl>     <dbl>
1 (Intercept)   101.        1.00     101.  1.68e-116
2 period_post     7.41      1.38       5.37 4.04e-  7


> round(confint(lm_NW_adj),4)
```

```
              2.5 %   97.5 %
(Intercept) 99.3698 103.3380
period_post  4.6755  10.1415
```

The intervention impact estimated after the Newey-West adjustment (7.41) is the same as what was measured by the unadjusted linear model. However, the standard error changes substantially due to the use of a robust covariance matrix. The adjusted standard error increases to 1.38 (Newey-West adjustment) from 0.931 (unadjusted). Consequently, the 95% confidence interval expands to range from 4.67 to 10.14, which now contains the true impact of the intervention.

So far, the results demonstrated that when intervention data is autocorrelated, standard hypothesis tests and linear regression models overestimated impact and become unreliable. Because autocorrelation is a fundamental time-series component, it is best addressed through integrated time-series methodologies rather than a two-step 'regress-then-adjust' process. By applying robust time-series-based method directly on autocorrelated intervention data, we can obtain valid and reliable impact measures that account for the data's temporal structure.

Next, we will discuss pure time-series-based approach for measuring intervention impact.

## 15.8.3 ARIMAX for measuring the impact from an AR(1) process

Time series models are particularly well suited for autocorrelated data, as they explicitly account for lag dependencies. When an intervention is incorporated as an exogenous regressor in such models, its estimated impact is naturally adjusted for autocorrelation. To achieve this, a dummy variable can be used to distinguish between the pre- and post-intervention periods. In this discussion, we focus exclusively on ARIMA models discussed in *Chapter 12*, due to their ability to include exogenous regressors.

You might wonder why this topic appears later in the chapter, given that the book primarily focuses on time-series analysis in R. While the sequence may seem unconventional, it is intentional. It highlights the unique value that time-series models bring to scenarios involving autocorrelated data. By now, you should feel confident that linear models, when applied to observational data, serve as a robust alternative to hypothesis testing in experimental settings. Moreover, you have seen how linear models provide flexibility to incorporate adjustments, enabling improved inferences when assumptions are violated. To maintain a coherent narrative, all comparisons in this chapter were conducted on time series data with a known data generation process (the $ARIMA(0,0,0)$ process). In real-world applications, however, this level of clarity is rare, and determining the correct lag order becomes essential to efficiently adjust for autocorrelation.

In the literature, ARIMAX is often termed regression with ARIMA errors or dynamic regression. Theoretically, this is because the model combines a linear regression component (the exogenous variables, $X$) that captures time dynamics in the autocorrelated error terms through an ARIMA structure.

We will leverage the automated model selection capability of the ARIMA model and let it identify the correct autocorrelation structure. The `ARIMA()` function from the `fable` package by default uses `stepwise=TRUE`, which is optimized for forecasting model selection. On the contrary, we want ARIMA to identify an optimal model for the attribution perspective. That means the optimal model should have the lowest possible information criteria (either $AIC$, $AIC_c$, or $BIC$) among all the plausible models. That is why the `stepwise=FALSE` option is used in automated model selection. Now fit the model, extract its coefficients, and compute the confidence intervals of the coefficients. The code progresses as follows:

```
aa_trt <- wd_ar |>
  as_tsibble(index = tidx) |>
  model(arima = ARIMA(y_ar1_trt ~ period, stepwise=FALSE))

aa_trt_coef <- tidy(aa_trt) |>
  mutate(
  lower_95 = estimate - qnorm(0.975) * std.error,
  upper_95 = estimate + qnorm(0.975) * std.error )|>
  mutate(across(where(is.numeric), \(x) sprintf("%.3f", x)))

aa_trt
```

For the simulated intervention data the order of the fitted model is as follows:

```
# A mable: 1 x 1
                         arima
                       <model>
1 <LM w/ ARIMA(1,0,0) errors>
```

ARIMA has identified the correct order used in the simulation, $ARIMA(1,0,0)$. The estimated coefficients can be extracted using:

```
aa_trt_coef |>
  select(.model, term, estimate, std.error, p.value)
```

The resulting estimates are:

```
# A tibble: 3 × 5
  .model term        estimate std.error p.value
  <chr>  <chr>       <chr>    <chr>     <chr>
1 arima  ar1         0.587    0.073     0.000
2 arima  period_post 7.305    1.652     0.000
3 arima  intercept   101.339  0.880     0.000
```

The intervention impact is measured as 7.30 (period_post) which is statistically significant. The estimated $AR(1)$ coefficient and the intercept terms are also significant with very low p-value. The 95% confidence intervals for the coefficients are extracted as follows:

```
aa_trt_coef |>
  select(.model, term, lower_95, upper_95)
```

The resulting confidence intervals are:

```
# A tibble: 3 × 4
  .model term        lower_95  upper_95
  <chr>  <chr>          <chr>     <chr>
1 arima  ar1            0.444     0.730
2 arima  period_post    4.067    10.542
3 arima  intercept     99.615   103.063
```

The 95% confidence interval (4.07,10.5) from the fitted ARIMAX model, includes the true impact value (i.e., 5). These results are comparable to the estimates obtained from an adjusted linear model that accounted for autocorrelation.

Once you have detected that the intervention has introduced a serial dependency structure in the response data, you should use either a regression model with time-series adjustment or a pure time-series-based model to measure the impact of the intervention. The hypothesis-test-based approach is likely to overestimate or underestimate the true impact. Among the time-series models, ARIMA specifically fits the situation as it can include exogenous regressors.

In the following section, we cover a complex data structure which is a surprisingly common outcome seen from running intervention experiments in business scenarios.

## 15.9 ITS on time series with AR(1) and trends

A/B testing is a well-established tool for measuring the impact of initiatives in business settings. Throughout this chapter, we have focused on types of data where experimental tools can be effectively applied and interpreted. In earlier examples, we primarily examined whether equivalent statistical tests yielded significant results and whether confidence intervals captured the true value of the impact. In this section, however, we explore a scenario where the traditional A/B testing approach falls short: measuring the impact on a slope parameter.

Businesses are often interested not only in changing the level or mean of a metric but also in altering its slope. This scenario arises when the goal is to accelerate growth or slow decline in a particular metric. For example, understanding changes in the sales growth rate is critical for evaluating the effectiveness of marketing campaigns, product launches, or seasonal promotions. A specific use case could be measuring the increase in the slope of sales growth after a new advertising campaign to determine its success. Similarly, monitoring changes in the rate at which customers leave (churn) is essential for assessing customer retention efforts. In such cases, a business might aim to decrease the slope of the churn rate after introducing a new customer loyalty program to gauge improvements in retention.

In the preceding practical examples, it is not possible to "split" the experience of all users into two groups to test the impact of the loyalty program, keeping everything else the same, such as brand inconsistency, demographic attributes, and so on. Hence, a traditional RCT is not feasible. Even when the underlying user base changes as time progresses, the aggregated metric should reflect any change due to intervention. This relaxes the requirements of having control and treatment groups to measure the intervention impact. The impact is measured from the behavior of the group that received the intervention/treatment, while the subjects belong to that group may change. Confounding variables handle situations where the underlying group composition changes significantly, and are easy to pass as exogenous variables in time-series models.

For illustration, we will use data with the $AR(1)$ structure used in the previous example, and add a trend component to it. We will assume the metric showed a $0.5$ growth before the intervention (`trn_coef_pre=0.5`), and the post-intervention growth of $1$ (`trnd_coef_post=1`). We will disregard a fixed impact size because, with non-parallel trends, the interpretation of the coefficient for the time-period dummy differs from the simpler cases discussed earlier. Instead, we will introduce a secondary time index, `tidx2`, to account for the incremental trend in the post-intervention period.

Since observational studies frequently lack control groups, we have excluded them from our simulation to better reflect typical research conditions. The data is generated using the following code:

```
set.seed(3)
trnd_coef_pre <- 0.5
trnd_coef_post <- 1.0

wd_slope <- wd_ar |>
  mutate(tidx2 = c(rep(0,(n_obs - n_trt)), 1:n_trt)) |>
  mutate(
    y_slope_trt = case_when(
      tidx <= (n_obs - n_trt) ~ y_ar1_trt  + trnd_coef_pre*tidx,
      tidx > (n_obs - n_trt) ~ y_ar1_trt + trnd_coef_pre*tidx +
                                 trnd_coef_post*tidx2))
```

The simulated data is shown in *Figure 15.7*:

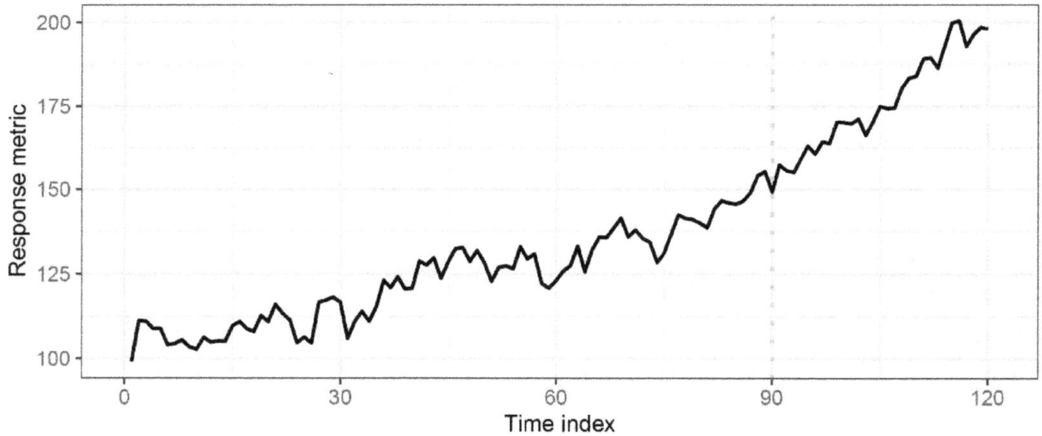


*Figure 15.7: Time plot of a metric with AR(1) and trend*

The plot shows the intervention may cause a shift in the slope of the metric after the intervention period (becoming steeper). Also, notice the slight downward shift in the observed series exactly when the intervention was implemented. We will discuss this further while analyzing this data with ITS. This data will be a major test for an advanced time-series model to evaluate impacts in scenarios where traditional A/B testing is not directly applicable.

### 15.9.1 ARIMAX for measuring the impact from AR(1) data with trends

Traditional statistical tests are not suited for measuring the impact of an intervention that introduces a trend (in other words, a slope parameter), because it violates the crucial assumption of sampled subjects being i.i.d. Previous examples also showed that regression-based models require proper adjustment when post-intervention treatment observations have autocorrelation. These leave the time-series-based approach as an important candidate to properly capture the intervention impact in such a scenario.

To measure the slope change (that is, trend) caused by the intervention, we need to pass the time indicator variable and the period as exogenous regressors in the ARIMA model. We also need to estimate the interaction term between the time indicator and intervention period (`tidx:period`) variables.

Like the previous implementation, the stepwise selection in automated ARIMA should be FALSE to allow searching among all possible model combinations and finding the best one by minimizing the $AIC$, as shown in the following code:

```
aa_ar1slope <- wd_slope |>
  as_tsibble(index = tidx) |>
  model(arima =
        ARIMA(y_slope_trt ~ period + tidx + tidx:period,
              stepwise = FALSE ))

aa_ar1slope_coef <- tidy(aa_ar1slope) |>
  mutate(
  lower_95 = estimate - qnorm(0.975) * std.error,
  upper_95 = estimate + qnorm(0.975) * std.error ) |>
  mutate(across(where(is.numeric), \(x) sprintf("%.3f", x)))

aa_ar1slope
```

The fitted model and the estimated coefficients are presented in the following output:

```
# A mable: 1 x 1
                        arima
                      <model>
1 <LM w/ ARIMA(1,0,0) errors>
```

To examine the estimated coefficients of the model, along with their standard errors and p-values, we can select the relevant columns

```
aa_ar1slope_coef |>
  select(.model, term, estimate, std.error, p.value)
```

Output for the coefficients:

```
# A tibble: 5 × 5
  .model term              estimate std.error p.value
  <chr>  <chr>             <chr>    <chr>     <chr>
1 arima  ar1               0.582    0.074     0.000
2 arima  period_post       -90.321  15.876    0.000
3 arima  tidx              0.508    0.032     0.000
4 arima  period_post:tidx 1.069    0.156     0.000
5 arima  intercept         100.922  1.719     0.000
```

The confidence intervals for each estimated coefficients are as follows:

```
aa_ar1slope_coef |>
 select(.model, term, lower_95, upper_95)
```

Output for the confidence intervals:

```
# A tibble: 5 × 4
  .model term             lower_95   upper_95
  <chr>  <chr>               <chr>      <chr>
1 arima  ar1                 0.437      0.726
2 arima  period_post      -121.438    -59.205
3 arima  tidx                0.445      0.571
4 arima  period_post:tidx    0.763      1.375
5 arima  intercept          97.554    104.291
```

Let's summarize the information provided by the fitted model:

- The model accurately captured the autocorrelation structure in the underlying data by identifying an $ARIMA(1{,}0{,}0)$ process
- It also closely estimated the AR coefficient, `ar=0.582`

- The `period_post=-90.3` represents the immediate change in the level (the step change) of the response variable at the exact moment the intervention occurred. Recall the drop in *Figure 15.7* at $t = 90$.
- `tidx=0.508` measured the baseline trend before the intervention. The true value used in the simulation was $0.5$.
- `period_post:tidx=1.07` estimates the change in slope after the intervention, which also close to the true value $1$.
- `intercept=101` estimated the baseline starting value in the response metric when $t = 0$
- The sum of the original slope (`tidx`) and interaction estimate (`period_post:tidx`) provides the total post-intervention slope. In our case, it is $0.508 + 1.07 = 1.578$, indicating that after the intervention, the metric is increasing at a rate of approximately $1.58$ units per time period.

All the coefficients are statistically significant with very small p-values. This adds credibility to the estimates and demonstrates the reliability of the time series model in capturing the impact of the intervention.

It is important to note that the coefficient for `period_post` (representing the post-intervention period) cannot be interpreted in the same way as in prior examples, despite its conceptual similarity. This coefficient reflects the change in the intercept during the post intervention period. Due to the non-parallel trend in the baseline period, it does not account for an immediate level change caused by the intervention. This nuance highlights the complexity of interpreting coefficients in scenarios involving slope changes and non-parallel trends.

Before concluding this section, it is important to note that while the `auto.arima()` function is designed to identify an appropriate model automatically, further validation often acts as a safeguard. To ensure the integrity of the fit, we can evaluate the model's adequacy by analyzing its residuals plot.

The following code executes the diagnostic plots presented in *Figure 15.8*:

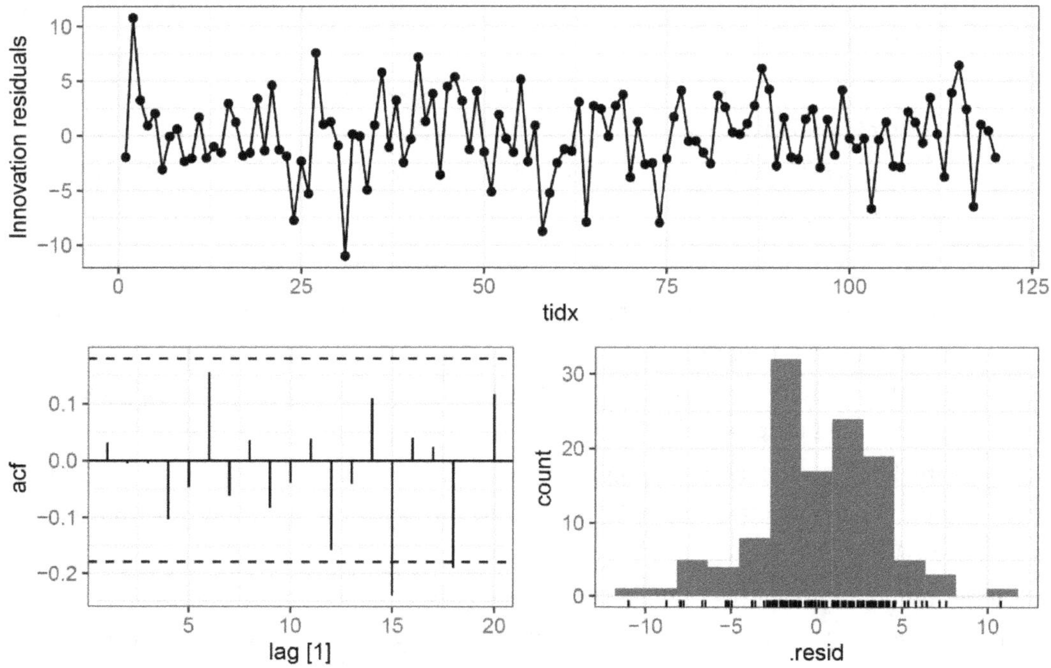


*Figure 15.8: Adequacy check of the estimated ARIMA(1,0,0) on the intervention data with AR(1) and trend*

The residuals from the fitted $ARIMA(1,0,0)$ with errors depicts a white noise pattern with distribution centered around $0$. There are no significant spikes in the ACF, except by a small margin at lag 15. Let's perform the Ljung-Box test to check whether the significant spike is by random chance or not, as shown in the following code:

```
aa_ar1slope_lb <- aa_ar1slope |>
  residuals() |>
  features(.resid, ljung_box)

aa_ar1slope_lb
```

The test result is presented in the following output:

```
# A tibble: 1 × 3
  .model lb_stat lb_pvalue
  <chr>    <dbl>     <dbl>
1 arima    0.123     0.726
```

The large p-value of the Ljung-Box test confirms there is no statistically significant evidence of remaining structure or autocorrelation in the residuals. This indicates that the model fit is adequate and that all predictable time-series patterns have been modeled correctly.

Having established compelling evidence for the use of time-series-based modeling in measuring the impact of business initiatives, we turn our attention to the broader advantages of this framework. The following section summarizes the inherent flexibility this approach can provide.

# 15.10 Flexibilities of ITS-based inference

In order to present a compelling case for the application of time series models for estimating causal impact from time series data, we utilized simple cases, specifically restricting ourselves to $ARIMA(1,0,0)$ in the data simulation process. This choice enabled us to use linear models as alternative tools, given that certain methods allow for valid inference if the residuals from a linear model follow an AR(1) process. Our chosen setting also facilitated the comparison of results from linear models with standard statistical tests used in A/B testing. However, real-world data might originate from a different $ARIMA(p,d,q)$ process. In such a scenario, time-series models become the sole viable method among other alternatives available.

Some key features that highlight the flexibility and suitability of ITS in practice are as follows:

- **Measure impact from only the treatment group**: In contrast to A/B testing, ITS estimates impact exclusively from the treatment group. They do not compare responses from the treatment group against the control group. Rather, the intervention impact is actually measured from the subjects who received the intervention (the treatment group). This approach also allows for the intervention to be applied to a larger cohort, eliminating the need for a control group. By following the same cohort over time, the risk of confounding bias in the impact estimate is reduced. As a regression-based method, it offers the flexibility to adjust for confounders by including them as exogenous regressors in the model. Additionally, it is possible to normalize the outcome variable using other influencing variables, such as the size of the cohort, especially when the cohort is defined by criteria that allow units to be included as time progresses.
- **Allows sample subjects to change over time**: ITS puts no restriction that the subject/items in the treatment group should be exactly the same throughout the intervention period. For example, a major global change has been implemented on a website, such as a major UI redesign or a site-wide navigation update. In this case, the traditional RCT is not feasible because you cannot split the experience of all users without risking technical bugs or brand inconsistency. Crucially, data collected from the sample

subjects are not identical across the timeline; the unique visitors arriving at the site before the intervention are different from those visiting after the change. By measuring the impact of the aggregated metric rather than following individual users' experience, ITS provides a robust measure of impact even when the population is constantly changing. If, in rare scenarios, the underlying sample drastically changes over time, ITS normalizes that effect by adding related confounders as exogenous regressors in the model.

- **Allows seasonality**: Seasonal patterns in the observed metric can be handled by ITS methods by design. The observed data needs to be seasonally adjusted before proceeding with measuring the impact of it. A decomposition method based on non-parametric smoothing, such as LOESS discussed in Chapter 7, can be adopted. This approach ensures that seasonal variations are accounted for, allowing for more accurate estimation of the causal impact. By leveraging these techniques, you can obtain robust and reliable results from time series data, making a strong case for the application of time series models in causal impact estimation.

Based on the discussions covered in this chapter so far, we can summarize the appropriate approaches to measure impact under various intervention data structures in *Table 15.1*:

| Data Characteristics | Groups in Study | Applicable Methods |
|---|---|---|
| Stationary (shifts level only) | Treatment and control | Two-sample t-test, regression model |
| Trend | Treatment and control | Two-sample t-test, regression model (may need adjustments if not adequate enough) |
| Autocorrelated, such as AR(1) | Treatment and control | Regression method with ITS adjustments (Cochrane-Orcutt, Prais-Winsten, Newey-West), time-series model with regressors (ARIMAX) |

| Data Characteristics | Groups in Study | Applicable Methods |
|---|---|---|
| Autocorrelated (AR(1)) with trend | Treatment | Regression-based method with ITS adjustments, time-series model with regressors (ARIMAX) |
| Autocorrelated (AR(1)) with trend and seasonality | Treatment | Time-series model with regressors (ARIMAX) after seasonal adjustments |

*Table 15.1: Applicability of approaches for measuring impact from intervention data*

In intervention analysis, monitoring a guardrail metric is essential to ensure that improvements in the primary business metrics do not come at the expense of long-term business goals or unintended operational trade-offs. In the next section, we discuss how you can incorporate monitoring guardrail metrics in ITS.

# 15.11 Monitoring guardrail metrics in ITS

Traditional A/B testing advocates for checking guardrail metrics. A guardrail metric is a type of secondary metric used in A/B testing to ensure that any changes implemented do not adversely affect important aspects of the user experience or business performance. While primary metrics focus on the main goals of an experiment, such as conversion rates or revenue, guardrail metrics serve as safety checks to monitor potential negative impacts. The importance of checking guardrail metrics in A/B testing cannot be overstated, as they help to identify unintended consequences of experimental changes, ensuring that improvements in primary metrics are not achieved at the expense of other critical factors. For instance, an increase in conversion rates might coincide with a decrease in customer satisfaction, or discounts offered to customers might reduce the overall revenue, and so on.

Several options exist for checking guardrail metrics in an A/B test. By analyzing these metrics alongside primary metrics, researchers can gain a holistic understanding of the experiment's impact. Statistical methods such as t-tests or confidence intervals can be applied to guardrail metrics to assess their significance. Additionally, visualizations such as time series plots or control charts can help identify trends and anomalies. Implementing guardrail metrics in A/B testing ensures that experiments are not only driving desired outcomes but also maintaining or improving overall user experience and operational stability.

When guardrail metrics themselves are non-stationary, t-tests lose their statistical validity. Time-series models can again come to the rescue but have their own caveat. In the realm of time-series analysis, "spurious regression" occurs when two or more non-stationary time series are regressed against each other, leading to a misleadingly high $R^2$ value and significant t-statistics, despite there being no meaningful relationship between the variables (i.e., primary and guardrail metrics). This phenomenon is a consequence of the underlying time-series data being integrated, which means they possess a unit root and exhibit random walks or trends over time. The primary consequence of spurious regression is that it can lead researchers to draw false conclusions about the relationships between variables, potentially informing incorrect business decisions or policy interventions.

To avoid spurious regression, it is crucial to ensure that the time-series data is stationary before performing regression analysis. Stationarity implies that the statistical properties of the series, such as mean and variance, do not change over time. Common techniques to achieve stationarity include differencing the data (detrending). Additionally, performing unit root tests, such as KPSS test, as discussed in *Chapter 8*, can help verify stationarity. By addressing non-stationarity, you can avoid the pitfalls of spurious regression and derive more accurate and reliable insights from the observed time series.

Nevertheless, assessing guardrail metrics in isolation might fail to capture the bi-directional relationship between the primary and guardrail metrics. The time-series-based methods provide a natural solution to this issue. The Vector Autoregression (VAR) modeling discussed in *Chapter 14* offers a robust framework to better understand and analyze these interdependencies. Through VAR modeling, it is possible to account for the dynamic interactions between multiple time series, providing a more comprehensive insight into how changes in the primary metric can affect guardrail metrics over time.

# Summary

In this chapter, you delved into key concepts and practical methods for measuring the impact of initiatives. This content covers the fundamentals of causal inference by explaining the role of experiments, such as A/B testing, in evaluating the impact of initiatives and establishing causality. It also explores statistical significance testing in A/B contexts, focusing on hypothesis-testing tools and their equivalence to linear model–based inference in experimental settings. In addition, it explains how linear models can be used for observational data to measure impact, providing a robust framework for causal analysis. The application of time-series models is discussed as a way to evaluate initiative impact across different scenarios, with particular attention to changes in both level and trend. The material further highlights how the choice of statistical methods can influence conclusions, including differences in level and slope estimates, their standard errors, confidence intervals, and p-values. Finally, it outlines the advantages of time-series modeling, especially its ability to monitor guardrail metrics alongside primary outcomes, ensuring a comprehensive assessment of initiatives.

These insights emphasize the critical importance of selecting appropriate statistical methods to ensure accurate, reliable, and meaningful impact measurement. By understanding the strengths and limitations of different approaches, you will be able to make informed decisions to derive actionable insights to solve your business problems.

# Further reading


- Cohen, J. (2013). *Statistical power analysis for the behavioral sciences. Routledge.*
- Evans, D. (2003). *Hierarchy of evidence: A framework for ranking evidence evaluating healthcare interventions. Journal of Clinical Nursing*, 12(1), 77–84.
- Judge, G. G., Griffiths, W. E., Hill, R. C., Lütkepohl, H., & Lee, T.-C. (1991). *The theory and practice of econometrics* (Vol. 49). *John Wiley & Sons.*
- Kohavi, R., & Thomke, S. (2017). *The surprising power of online experiments. Harvard Business Review*, 95(5), 74–82.
- Montgomery, D. C. (2017). *Design and analysis of experiments. John Wiley & Sons.*
- Turner, S. L., Karahalios, A., Forbes, A. B., Taljaard, M., Grimshaw, J. M., & McKenzie, J. E. (2021). *Comparison of six statistical methods for interrupted time series studies: Empirical evaluation of 190 published series. BMC Medical Research Methodology*, 21(1), 134.

# 16

# Changepoint Detection

In our modern world, data is a continuous, high-speed flow where changes in processes are not just an event, but a constant state. When we deal with continuous signal processing, such as **electrocardiograms (EEGs)** or revenue uplift due to a new product launch, changes in the underlying data-generating process are inevitable. Changepoint detection is a rigorous statistical framework that is used to identify sudden, persistent shifts in the behavior of a system over time. While anomalies in a time series are short-lived spikes that quickly return to the previous baseline, changepoint detection captures persistent shifts, where the underlying data generation process shifts and enters to a "new normal."

Changepoint detection allows us to identify systematic shifts, enabling responses to changes in market trends, equipment failure, patient health, or strategic business decisions. In this chapter, we discuss changepoint detection in the context of a retrospective study. From the methods discussed and the example presented, you will learn about prominent approaches to detecting changepoints in observed data.

The chapter is structured as follows:

- Section 16.1: What is changepoint detection?
- Section 16.2: Breaks for Additive Season and Trend
- Section 16.3: Pruned Exact Linear Time method
- Section 16.4: Bayesian changepoint detection
- Section 16.5: Method selection guidelines

Note

**Your purchase includes a free PDF copy + exclusive extras**

Your purchase includes a DRM-free PDF copy of this book, 7-day trial to the Packt+ library (no credit card required), and additional exclusive extras. See the *Free benefits with your book* section in the *Preface* to unlock them instantly and maximize your learning.

# Technical requirements

Let's load the packages that are required in this chapter using the following code:

```
#install.packages("pacman")
library(pacman)
p_load("conflicted"
       ,"datasets"
       ,"tidyverse"
       ,"feasts"
       ,"patchwork"
       ,"bfast"
       ,"changepoint"
       ,"Rbeast" )

conflicts_prefer(
  strucchangeRcpp::boundary
 ,fabletools::accuracy
 ,fabletools::forecast
 ,fabletools::components
 ,dplyr::filter )
```

The code file is available in the GitHub repo at `https://github.com/PacktPublishing/Modern-Time-Series-Analysis-with-R.git` for you to download and practice.

# 16.1 What is changepoint detection?

Time series often contain structural breaks or abrupt shifts in their underlying patterns. When a persistent shift occurs in fundamental statistical properties, such as mean, variance, or trend, the phenomenon is attributable to a changepoint. A changepoint pinpoints the boundary between two different segments of behavior.

Detection of a structural break is a topic of interest in any discipline with a focus on process monitoring. Some application areas include the following:

- Finding structural breaks in economic time series, such as GDP, unemployment rate, and national debt
- Identifying a regime change in financial time series, such as market volatility, returns, or trading volumes
- Detecting transitions in physiological measurements, such as EEG, ECG, or blood pressure
- Uncovering anomalous behavior or load changes in a telecommunications network
- Detecting shifts in climate patterns, vegetation indices, or hydrological systems in environmental monitoring

The detection of changepoints opens doors to understanding the impact of events in the dataset and accounting for those changepoints to improve subsequent analysis and forecasting. For example, by detecting changepoints, you can reduce forecasting errors and also avoid invalid statistical inferences that could compromise the decision-making process. Changepoint detection is different from measuring intervention impact (covered in *Chapter 15*). In changepoint detection, factors that contribute to change are considered external, and the analysis is retrospective in nature. On the other hand, causal impact measurement assumes that the factors arise from an internal business strategy or decisions, and the business wants to understand the impact of those strategies, whether to claim success or potential refinement of the strategy.

Changepoint detection techniques offer several powerful methodological approaches. The use of a suitable method should match the analytical needs of a problem.

*Table 16.1* provides a brief summary of the three changepoint detection methods that are covered in detail later in the chapter:

| Criteria | Breaks for additive season and trend (BFAST) | Pruned exact linear time (PELT) | Bayesian approach (BEAST) |
|---|---|---|---|
| Overview | A recursive approach that splits data in two parts at a single most significant changepoint, then repeats the process on each sub-segment | An exact search algorithm that minimizes a cost function (mean or variance) across all possible segments, while non-optimal points are discarded | A probabilistic method that updates the likelihood of a changepoint with each new data point, incorporating prior beliefs about the system |
| Application conditions | Works best when segments are relatively long and the number of changepoints is low | Requires a predefined penalty value (e.g., AIC or BIC) to avoid overfitting | Requires selecting appropriate prior distributions |
| Suitability | Suitable for very large datasets due to its computational efficiency | Ideal for identifying the exact number and location of multiple changepoints | Highly suitable for online (real-time) detection, where quantifying uncertainty and immediate response are critical |

| Criteria | Breaks for additive season and trend (BFAST) | Pruned exact linear time (PELT) | Bayesian approach (BEAST) |
|---|---|---|---|
| Common use case | Environmental monitoring, detecting land-cover changes, seasonal shifts in satellite imagery time series | Finance and engineering, identifying regime switches in market volatility, structural breaks in sensor data | Healthcare and cybersecurity, real-time monitoring of heart rates (ECG), detecting sudden spikes in network traffic |

*Table 16.1: Comparison among changepoint detection methods*

All of these methods are applicable to both seasonal and non-seasonal data. In this chapter, we have used the `Nile` data for an example of non-seasonal application and the `AirPassengers` data for a seasonal application. Using the same data for all three methods would also allow you to compare their performance.

Next, we will discuss the BFAST method for changepoint detection.

## 16.2 Breaks for Additive Season and Trend

The Breaks for Additive Season and Trend (BFAST) framework, developed by Verbesselt et al. (2010), is analogous to binary segmentation as it splits the data into two segments in every iteration to check for changepoints. The recursive process terminates when a stopping criterion is met. BFAST is one of the most intuitive and computationally efficient approaches for detecting changepoints.

A central tool used in BFAST is the Moving Sum (MOSUM) test, which allows for detecting an unknown number of changepoints without assuming their locations in advance.

Recall that a time series, $Y = (y_1, y_2, \ldots, y_n)$, in additive form can be written as follows:

$$y_t = T_t + S_t + R_t$$

where,

- $T_t$ is the trend component
- $S_t$ is the seasonal component
- $R_t$ is the remainder (or residual) component

In BFAST implementation, the trend is typically modeled as a piecewise linear regression as shown in the following equation:

$$T_t = \alpha_j + \beta_j t, \quad t \in (\tau_{j-1}, \tau_j]$$

where,

- $0 = \tau_0 < \tau_1 < \cdots < \tau_m < \tau_{m+1} = T$ are unknown changepoints
- $(\beta_{0,j}, \beta_{1,j})$ are regime-specific regression coefficients

A trend break occurs if $\alpha$ or $\beta$ changes at some point in time.

The seasonal component is modeled as a smooth periodic curves, often using harmonic regression:

$$S_t = \sum_{k=1}^{K} \left[ a_k \sin\left(\frac{2\pi kt}{f}\right) + b_k \cos\left(\frac{2\pi kt}{f}\right) \right],$$

where:

- $f$ is the known seasonal frequency (e.g., $f = 12$ for monthly data),
- $K$ is the number of harmonics.

The seasonal component can be written as a combination of repeating waves with known frequency (for example, yearly cycles). The key idea is that this representation captures regular patterns without reacting to short-term noise. Structural breaks in seasonality correspond to changes in the coefficients $\{a_k, b_k\}$. It is worth noting that for simplicity, seasonality can also be modeled using dummy variables. Once the trend and seasonal components are estimated from the time series, they are subtracted from the observed data, $y_t$, to extract remainders, as follows:

$$\hat{R}_t = y_t - \left(\hat{T}_t + \hat{S}_t\right) \qquad (16.1)$$

If the model is stable, these remainders behave like random noise and fluctuate around $0$. If a structural change occurs, the residuals will show a systematic bias, and they will tend to be mostly positive or mostly negative for a period of time. The MOSUM test detects such instability by aggregating remainders over short time windows.

The expression $\hat{R}_t$ is equivalent to the residuals of a regression model. In this sense, the right-hand side in *Equation (16.1)* can be interpreted as the difference between the observed values and the corresponding fitted values. The fitted values are obtained using **ordinary least squares** (**OLS**) estimates of the parameters in the trend and seasonal components of the model. Consequently, the MOSUM statistics employed in BFAST can be viewed as being computed on OLS residuals. This procedure is commonly referred to as OLS-based MOSUM (OLS-MOSUM) in the literature and in the R package we adopted.

In previous chapters, we used STL decomposition to separate (or estimate) remainders from an observed series. Can we use that method in the BFAST scenario? STL decomposition could be an option; however, by expressing trend and seasonality as explicit models, we can test whether their parameters remain stable. The initial definition of "normal" trend and seasonality strongly determines what is detected as change later in the process. Without such a model assumption, we won't be able to perform a test for departure.

The MOSUM test starts with a pre-determined fixed window length, $m$. Then, at each point $t$, the MOSUM is computed as follows:

$$\mathrm{MOSUM}(t) = \frac{1}{\sqrt{m}} \sum_{i=t-m+1}^{t} \hat{R}_i$$

This quantity is simply a local average of scaled residuals, so that values are comparable across window sizes. When there is no change, positive and negative residuals cancel out, and $\mathrm{MOSUM}(t)$ stays close to $0$.

When a change is detected, residuals in the window tend to have the same sign, causing $\mathrm{MOSUM}(t)$ to grow in magnitude. This makes the MOSUM transform from a subtle deviation to a visible signal. To determine whether the observed MOSUM values are unusually large, they are compared to a reference boundary that represents what we would expect under normal random variation. The MOSUM test statistic is defined as follows:

$$M = \max_t |\mathrm{MOSUM}(t)|$$

If $M$ remains below the boundary, the model is considered stable. If $M$ exceeds the boundary, BFAST concludes that a structural change has occurred.

The boundary represents the range of values we would expect the MOSUM to take if the underlying process has not changed. The time at which the boundary is crossed provides an estimate of the location of the break. In the following two sections, you will see how to identify changepoints using BFAST on a non-seasonal and seasonal series, respectively.

Let's see an application of BFAST on non-seasonal data.

## 16.2.1 BFAST implementation on non-seasonal data

In this example, we use the Nile data from the dataset package. The data is in `ts` format and contains a record of historical annual flow from the River Nile between 1871 and 1970. You can apply the BFAST method using the `bfast()` function from the `bfast` package. The function expects input data as `ts` format.

To enforce a non-seasonal additive decomposition on an annual series, the seasonality option is disabled by setting `season = "none"`. We use a minimum segment length of 15% of the total observations (pass in the argument h in the `bfast()` function). There is no strict guidance for choosing the minimum segment length; however, it should not be too small. Apply domain knowledge and context of the observed time series to set a suitable number in your application. We set `max.iter = 10`, which controls the maximum number of iterations allowed for estimation of changepoints in the trend component (and seasonal, where applicable). The following code snippet shows the application and how to extract some information from the model:

```
bfast_nile <- bfast(Nile,
                    h = 0.15,
                    season = "none",
                    max.iter = 10)
## [1] "No seasonal model will be fitted!"
nile_bfast_magnitude <- bfast_nile$Magnitude

bfast_nile
```

The fitted model information is provided in the following output:

```
TREND BREAKPOINTS
    Confidence intervals for breakpoints
    of optimal 2-segment partition:

Call:
confint.breakpointsfull(object = bp.Vt, het.err = FALSE)

Breakpoints at observation number:
  2.5 % breakpoints 97.5 %
1    26          28     31

Corresponding to breakdates:
```

```
   2.5 % breakpoints 97.5 %
1  1896        1898    1901

   SEASONAL BREAKPOINTS:  None
```

The magnitude of change detected at the changepoint by the BFAST method is as follows:

```
nile_bfast_magnitude

# [1] -287.9431
```

A changepoint is detected at observation 28, which happens to be the year 1898, when the Aswan Dam was built. The magnitude of change was $-287.94$, indicating that there was a permanent decline in the average Nile water flow of approximately 288 (unit: $10^8 m^3$Liter).

Let's plot the original data with the detected changepoint using the following code (the `invisible()` function suppresses onscreen information while plotting):

```
invisible(plot(bfast_nile,
          type = c("trend"), main = ""))
```

*Figure 16.1* shows the location of the detected changepoint by a vertical line:

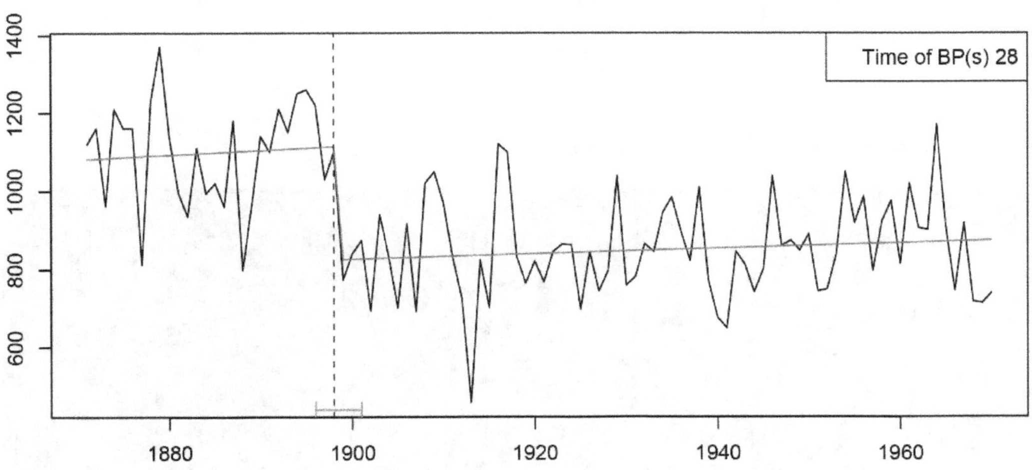


*Figure 16.1: Changepoint detected in the annual Nile flow by BFAST*

BFAST has identified the location of the changepoint at 28, with a 95% confidence interval (26,31) around it. The method has identified the breaks in trend at the changepoint for this non-seasonal series.

In the next section, the BFAST method is applied to seasonal series to detect any changepoint, if it exists.

## 16.2.2 BFAST implementation on a seasonal series

The well-known AirPassengers series that contains monthly totals of international airline passengers from 1949 to 1960, is used to show how you can apply BFAST to detect changepoints in a seasonal series.

Since BFAST decomposes a time series assuming an additive model, a logarithmic transformation is required to ensure the model's appropriateness for this data. For seasonal series, it is better to use a full seasonal period as the minimum segment size. It is computed as the ratio of the frequency in the dataset to the total observations. In this example, it will cover about 8 (12/144) of the total observations.

The `season="harmonic"` argument enforces a harmonic regression for a seasonal component. The maximum iterations are set at 10. The following code implements BFAST on the log of `AirPassengers` data as follows:

```
bfast_ap <- bfast(log(AirPassengers),
                  h = round(frequency(AirPassengers)/
                            length(AirPassengers),3),
                  season = "harmonic",
                  max.iter = 10)

bfast_ap
```

This method produces the following output:

```
  TREND BREAKPOINTS
     Confidence intervals for breakpoints
     of optimal 4-segment partition:

Call:
confint.breakpointsfull(object = bp.Vt, het.err = FALSE)

Breakpoints at observation number:
  2.5 % breakpoints 97.5 %
1    22          23     32
2    52          54     59
3   107         109    114
```

```
Corresponding to breakdates:
  2.5 %    breakpoints 97.5 %
1 1950(10) 1950(11)    1951(8)
2 1953(4)  1953(6)     1953(11)
3 1957(11) 1958(1)     1958(6)

  SEASONAL BREAKPOINTS:  None
```

Three changepoints are detected in the trend; in November 1950, July 1953, and January 1958. There was no seasonal breakpoint, indicating the seasonal distribution stayed the same throughout the observed period.

Let's visualize the location of the detected changepoints presented in *Figure 16.2*:

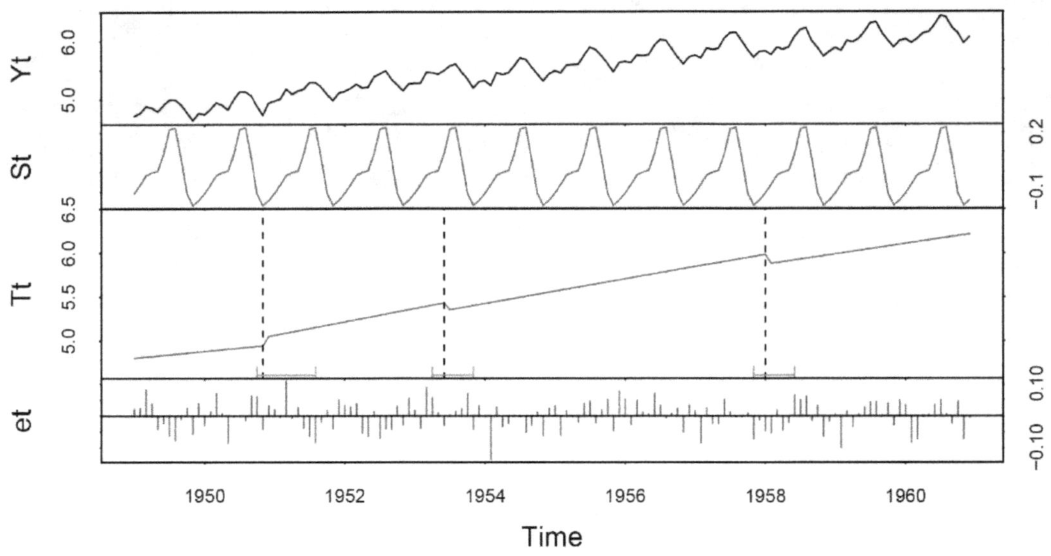


*Figure 16.2: Detected changepoints in the seasonal AirPassengers data by BFAST*

*Figure 16.2* provides visual information of the changepoints in the seasonal component (second panel from top) and in the trend (third panel from top). Combining the model output with the plot, the BFAST method detected three statistically significant changepoints in the trend and none in the seasonal component. Further study identifies the pivotal industry developments around corresponding time points, such as the following:

- **Introduction of jet airliners**: The debut of the De Havilland Comet in 1952, the world's first commercial jet airliner, revolutionized aviation with faster speeds and smoother flights. Despite initial challenges, its legacy enabled advanced successors such as the Boeing 707 (1958) and the Douglas DC-8 (1959). These jets slashed transatlantic flight times from over 14 hours to under 8, combining unprecedented efficiency with passenger comfort.
- **Economy class expansion**: Airlines such as Pan Am pioneered affordable fares, making air travel accessible to a broader demographic. Concurrently, the 1958 U.S. Federal Aviation Act standardized safety regulations, reinforcing public trust in commercial aviation.
- **Pressurized cabin adoption**: Widely implemented in the 1950s, this technology permitted aircraft to cruise above turbulent weather systems, significantly improving ride quality and passenger experience.

It is worth noting that the BFAST algorithm offers the option to avoid overfitting a seasonal model by forcing a non-seasonal model on seasonal data. This can be achieved by setting `season = "none"`. If you would like to consider BFAST for detecting changepoints in trend even for a seasonal series, you could enforce a non-seasonal model by setting `season = "none"` in the `bfast()` function.

Let's look at some advantages and limitations of the BFAST method.

The following are advantages:

- Computationally efficient, making it suitable for long time series
- Intuitive and relatively easy to implement
- Ability to detect multiple changepoints
- Handles both trend and seasonal changes in an integrated framework
- Well-established theoretical properties

The following are limitations:

- The method may not find the global optimum solution
- Performance degrades when changepoints are close to each other
- The method may struggle with heteroscedastic data (changing variance)
- It has a tendency to overestimate the number of changepoints with noisy data
- It requires specification of the minimum segment size parameter

In the next section, we discuss a method that guarantees finding an optimum solution via optimization.

# 16.3 Pruned Exact Linear Time method

The **Pruned Exact Linear Time** (**PELT**) algorithm was introduced by Killick et al. (2012). The method guarantees finding the changepoint in a time series, if it exists, via a global search. This method treats changepoint detection as an optimization problem that finds the segmentation by minimizing a cost function. To avoid overfitting during optimization, the cost function penalizes the model for the number of changepoints.

A general form of cost function minimized by PELT is as follows:

$$\sum_{i=1}^{m+1}\left[C\left(y_{(\tau_{i-1}+1):\tau_i}\right)\right] + \beta m$$

where:

- $m$ is the number of changepoints
- $\tau_1, \tau_2, \ldots, \tau_m$ are the positions of changepoints
- $C(y_{(s+1):t})$ is the cost for the segment $y_{(s+1):t}$
- $\beta$ is a penalty parameter to prevent overfitting

By adopting the general structure, you can use various cost functions based on the context of a problem. Some common cost functions used in practice are as follows:

- Negative log-likelihood based criteria, such as BIC, AIC and so on for parametric distributions
- Residual sum of squares for mean shifts
- Likelihood-based costs for variance changes

The PELT process starts with an initialization step: identify plausible changepoints in an iterative recursion fashion in the second step, then discard non-optimal changepoints in the pruning step, and finally, update candidate changepoints or terminate when the stopping criteria are met. The key component of the PELT algorithm is the pruning step. In this step, the non-optimal changepoints are discarded, which in turn reduces computational complexity.

The detailed steps of the PELT algorithm are as follows:

1. **Initialization step**: The initial cost function $F(0) = -\beta$, and the set of candidate changepoints up to time $t = 1$, $\mathcal{R}_1 = \{0\}$ are set.
   The choice of $F(0) = -\beta$ ensures that when the first segment, $y_{1:t}$, is evaluated, the model offsets the first penalty to ensure the first segment cost, $F(0) + C(y_{1:t}) + \beta = C(y_{1:t})$, is correct. The choice of $\mathcal{R}_1 = \{0\}$ ensures there was no changepoint prior to time $t = 1$.
2. **Iterative recursion step**: For each time step, $t = 1,2,\ldots,n$, the following computation takes place:
   - The optimal cost is computed via $F(t) = \min_{s \in \mathcal{R}_t}[F(s) + C(y_{(s+1):t}) + \beta]$
   - The value of $\tau$ that optimizes the cost function is considered as the most recent changepoint and is stored in set $\mathcal{R}_t$
3. **Pruning step**: From all candidate segments, $\tau$, in the current set, $\mathcal{R}_t$, the optimal one is found by discarding and updating the set as follows:
   - Optimal $\tau_t = \text{argmin}_{s \in \mathcal{R}_t}[F(s) + C(y_{(s+1):t}) + \beta]$
   - Update $\mathcal{R}_{t+1} = \{s \in \mathcal{R}_t \cup \{t\}: F(s) + C(y_{(s+1):t}) + \beta < F(t)\}$
4. **Termination**: Once all the data points $n$ are processed, trace back through $\tau_t$ to retrieve changepoint locations in the optimal partitions.

Now, we will apply PELT to the Nile data to check how many changepoints have been detected by this method.

### 16.3.1 PELT implementation on non-seasonal data

The `changepoint` package offers PELT implementation via the `cpt.meanvar()` function. We will use the **Bayesian Information Criterion** (**BIC**) as a cost function, and 15% of the total observations as the minimum segment length. The `cpt.meanvar()` function can detect the changepoint in both the mean and variance.

You can implement PELT on the non-seasonal ` Nile` data using the following code:

```
pelt_nile <- cpt.meanvar(Nile,
                         method = "PELT",
                         penalty = "BIC",
                         minseglen = length(Nile)*0.15)
```

A summary of the fitted model shows information about the detected changepoints. The following output shows the changepoint detected by PELT:

```
> summary(pelt_nile

Created Using changepoint version 2.2.4
Changepoint type       : Change in mean and variance
Method of analysis     : PELT
Test Statistic  : Normal
Type of penalty        : BIC with value, 13.81551
Minimum Segment Length : 15
Maximum no. of cpts    : Inf
Changepoint Locations : 28
```

According to PELT, there is one changepoint located at the 28the observation, which is the year 1898. The result is the same as that detected by the BFAST method earlier. You can extract the change in mean and variance before and after the changepoint detected from a PELT model object as follows:

```
attributes(pelt_nile)$param.est

# $mean
# [1] 1097.7500  849.9722
```

```
# $variance
# [1] 17573.12 15352.92
```

PELT method estimated the reduction in mean (reduced to 850 from 1098) and variance (reduced to 15,353 from 17,573) in the Nile river flow after the detected changepoint. The reduction in the mean is lower in PELT (−248) compared to what was detected by BFAST (−288).

We can examine the level shift before and after the detected changepoint in 1898 from the plot presented in *Figure 16.3*:

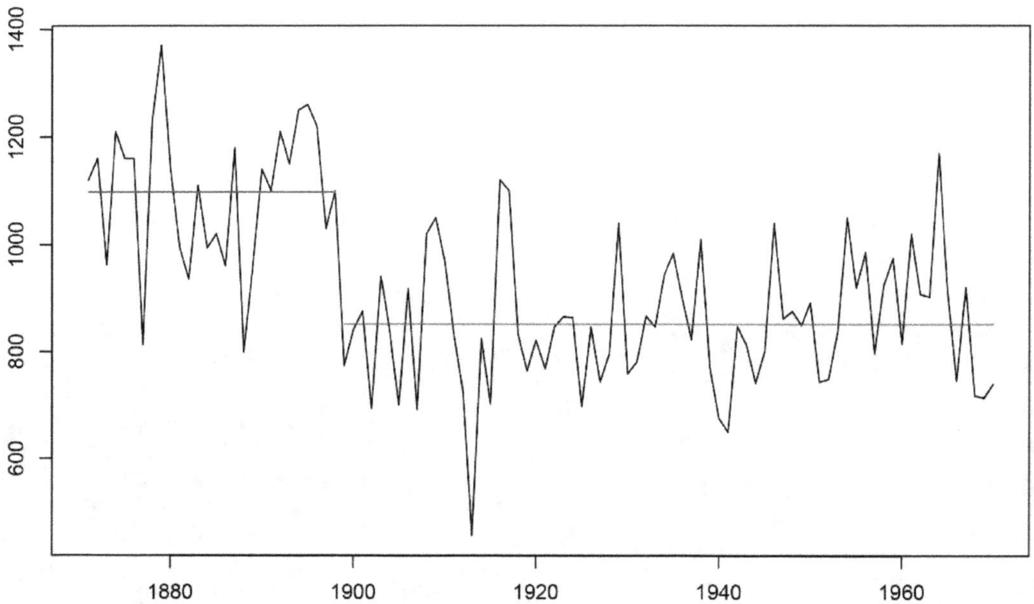


*Figure 16.3: Changepoint detected by PELT method on annual Nile river flow data*

When comparing the plot of PELT with that generated by BFAST in *Figure 16.1*, you will see that the subtle upward trend detected by BFAST before and after the changepoint is not presented in the PELT plot.

Next, PELT is applied to the same seasonal series, and you can compare its performance with that of BFAST.

### 16.3.2 PELT implementation on a seasonal series

PELT works on the fundamental assumption that data between changepoints originates from a stationary distribution; hence, before applying PELT to identify changepoints, an observed series needs to be stationary. The `AirPassengers` data is non-stationary due to the strong presence of trend and seasonality. First, we take the log on the `AirPassengers` series, perform STL decomposition on it, and extract the remainder component. Then check whether the remainder component is stationary by using the KPSS test before applying PELT on it. The following code shows you how to perform the operations:

```
ap_ts <- AirPassengers |>
  as_tsibble() |>
  mutate(log_passengers = log(value))

ap_decomp <- ap_ts |>
  model(stl_decomp = STL(
    log_passengers ~ trend(window = 13) +
      season(window = "periodic"))) |>
  components()

ap_remainder_kpss <- ap_decomp |>
  features(remainder, feasts::unitroot_kpss)

ap_remainder_kpss
```

The result from the KPSS test is shown in the following output:

```
# A tibble: 1 × 3
  .model     kpss_stat kpss_pvalue
  <chr>          <dbl>       <dbl>
1 stl_decomp    0.0264         0.1
```

The KPSS test reports a large p-value ($0.1$), confirming that the remainder series is stationary, and thus we can apply PELT to it. You can also use visual aids, such as an ACF and PACF of the remainder series to examine its stationarity condition.

In the PELT implementation on the remainder series, we impose a minimum segment length of 12 observations (`minseglen=12`) to allow a search within a full seasonal period and use BIC minimization in the optimization process, as presented in the following code:

```
pelt_ap_remainder <- cpt.meanvar(
  ap_decomp$remainder,
  method = "PELT",
  penalty = "BIC",
  minseglen = 12)

summary(pelt_ap_remainder)
```

The following output shows that PELT has detected two optimal changepoints that separate the time series into three segments:

```
Created Using changepoint version 2.2.4
Changepoint type      : Change in mean and variance
Method of analysis    : PELT
Test Statistic  : Normal
Type of penalty       : BIC with value, 14.90944
Minimum Segment Length : 12
Maximum no. of cpts   : Inf
Changepoint Locations : 62 109
```

*Figure 16.4* presents the location of PELT-detected changepoints on the remainder series:

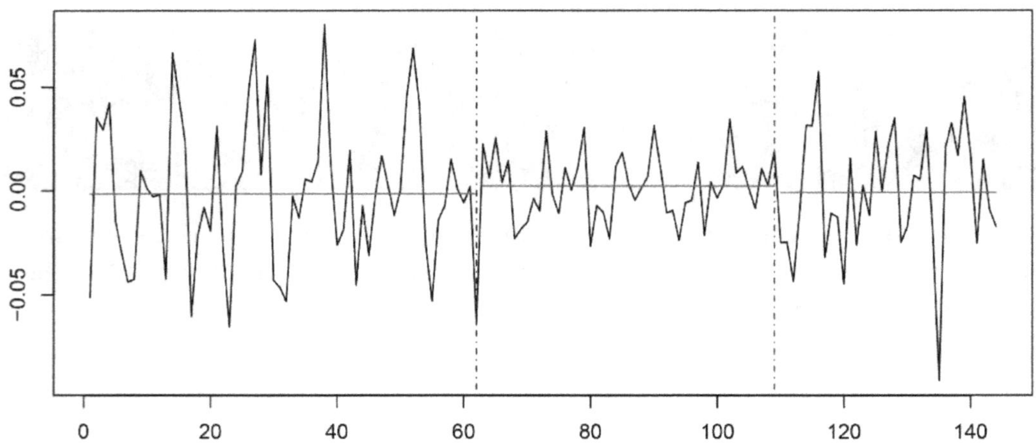


*Figure 16.4: Changepoint detected by PELT on the AirPassengers series*

You can extract the changepoints, as well as the mean and variance, from each segment from the model object as follows:

```
pelt_ap_cp <- pelt_ap_remainder@cpts
pelt_ap_par <-pelt_ap_remainder@param.est
```

The results are given in the following output where 144 being the last time period:

```
> pelt_ap_cp
[1]  62 109 144


> pelt_ap_par
$mean
[1] -0.001516690  0.002190884 -0.000916221

$variance
[1] 0.0012457391 0.0002361172 0.0009013271
```

Unlike BFAST, PELT does not provide the actual year-month reference in the model output; however, you can easily extract it by filtering the data as follows:

```
pelt_ap_ref <- ap_ts |>
  left_join( ap_decomp |> select(index, remainder),
             by = "index") |>
  slice(pelt_ap_cp[1:2])

pelt_ap_ref
```

This provides the actual time points when these changes occurred, as shown in the following output:

```
# A tsibble: 2 x 4 [1M]
     index value log_passengers remainder
     <mth> <dbl>          <dbl>     <dbl>
1 1954 Feb   188           5.24   -0.0640
2 1958 Jan   340           5.83    0.0191
```

These time references are close to two out of the three changepoints detected by the BFAST method. For visual examination, let's plot these two time points on the original AirPassengers series, presented in *Figure 16.5* with vertical broken lines pinpointing the changepoints location:

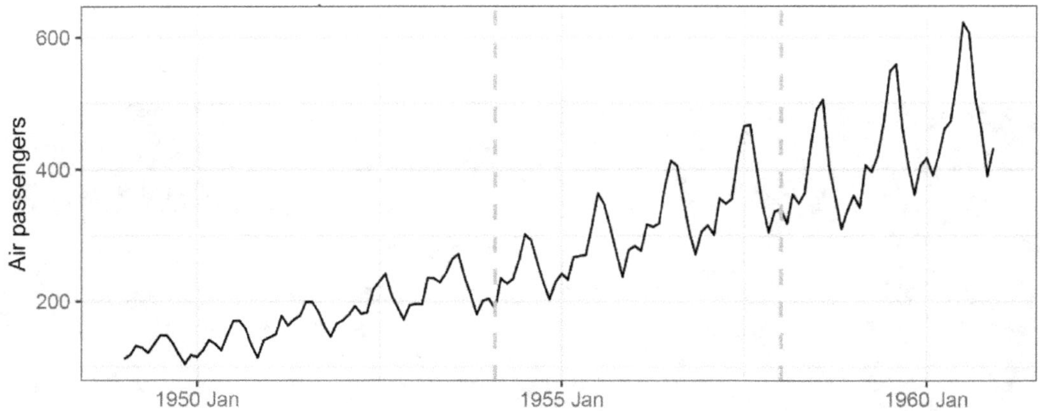


*Figure 16.5: AirPassengers data with changepoints detected by the PELT method from its remainder series*

Instead of finding changepoints in both the mean and variance, you might try different options for assessing results sensitivity in PELT, such as the following:

- Testing mean-only changes using the `cpt.mean()` function
- Examining variance-only changes using the `cpt.var()` function
- Evaluating the impact of changing the `minseglen` argument

Let's summarize the advantages and limitations of PELT method.

The following are advantages:

- It is guaranteed to find the optimal segmentation under the specified cost function
- It has a linear computational complexity for many common cost functions
- It has a flexible framework that can detect changes in mean and variance
- It has a robust theoretical foundation with well-defined statistical properties
- It automatically determines the optimal number of changepoints

The following are limitations:

- It requires specification of a penalty parameter (though information criteria, such as the BIC)
- Its model assumptions must be carefully considered for the chosen cost function
- It is sensitive to model misspecification
- It may struggle with strongly autocorrelated time series

So far, we have covered two frequentist-based methods, BFAST and PELT, that can be applied to both non-seasonal and seasonal series. However, these decomposition-based methods have notable limitations when non-stationarity arises from the complex interaction of trend and seasonality. For instance, seasonal data with complex trends and seasonality might require appropriate differencing to obtain a stationary remainder. Like BFAST, the PELT method is also sensitive to the specification of the minimum segment length (`minseglen`). If we had used `minseglen = 2` in this example, there would have been an additional changepoint detected. This limitation necessitates the exploration of an alternative approach. A Bayesian-based changepoint detection method is discussed in the following section.

# 16.4 Bayesian changepoint detection

The Bayesian approach of changepoint detection begins with specifying prior beliefs about the number and locations of changepoints. This belief is updated using observed data. This process results in the posterior distribution, which quantifies uncertainty about the changepoint configuration.

Unlike frequentist methods, which typically produce point estimates, Bayesian methods provide full posterior distributions, enabling robust uncertainty quantification for all parameters.

Under the Bayesian framework, we want to identify points where the underlying data-generating process changes in an observed time series, $Y = (y_1, \dots, y_n)$. Let, $\tau = (\tau_1, \dots, \tau_m)$ represent a vector of $m$ changepoint locations, where $1 < \tau_1 < \tau_2 < \cdots < \tau_m < n$.

This approach requires specification for the following parameters to initiate the process:

- A prior distribution, $p(m)$, for the number of changepoints
- A prior distribution, $p(\tau|m)$, for the locations of changepoints given their number
- A likelihood model, $p(Y|\tau,\theta)$, describing the data-generating process, with parameters $\theta$
- A prior distribution, $p(\theta|\tau)$, for the model parameters

Bayesian inference then proceeds by computing the posterior distribution:

$$p(m,\tau,\theta \mid Y) \propto p(Y \mid \tau,\theta) \cdot p(\theta \mid \tau) \cdot p(\tau \mid m) \cdot p(m)$$

This posterior encapsulates all available information about the changepoint configuration and model parameters given the observed data.

In Bayesian inference, the posterior distribution can be computed using various methods. When the analytical solution of the posterior distribution is non-trivial, the inference process relies on a numerical parameter estimation algorithm known as Markov Chain Monte Carlo (MCMC) sampling. The algorithm comprises the following steps:

1. Initialize the chain by starting with an initial guess of the parameters $\Theta = \{m,\tau,\theta\}$
2. Iterate through:
    - Propose a new parameter value, $\Theta'$
    - Taking $q(.|.)$ as a proposal distribution, compute acceptance probability $\alpha$
    $$\alpha = \min\left(1, \frac{p(Y \mid \Theta')\, p(\Theta')\, q(\Theta \mid \Theta')}{p(Y \mid \Theta)\, p(\Theta)\, q(\Theta' \mid \Theta)}\right)$$
    - Accept or reject $\Theta'$ by comparing $\alpha$ with a random number, $u$, drawn from a uniform distribution. If $u \leq \alpha$, set $\Theta$ to $\Theta'$.
3. Repeat for many iterations to build a Markov chain whose stationary distribution is the posterior.
4. Estimate parameters by using the chain samples (after discarding burn-in) to estimate quantities, such as the posterior mean, variance, or credible intervals.

Further details on the MCMC can be found in standard textbooks on Bayesian inference.

**Bayesian Estimation of Abrupt changes, Seasonal variation, and Trend (BEAST)** is a Bayesian method-based changepoint detection framework that handles complex time series with multiple components. The method is implemented in the `Rbeast` package (Zhao et al.,

2019). Like BFAST, BEAST uses additive decomposition of time series where a component can undergo structural changes independently, such as the following:

- The trend component, $T_t$, is modeled as a piece-wise linear function with a possible changepoint in both level and slope
- The seasonal component, $S_t$, is modeled as a periodic function that can change in amplitude and phase
- The noise component, $R_t$, is typically assumed to follow a Gaussian distribution with potentially changing variance

The BEAST model allows for an unknown number of changepoints in each component, providing a flexible framework for time-series decomposition and changepoint detection.

The primary steps in the BEAST algorithm are as follows:

1. Initialization: Set initial values for the number and locations of changepoints in trend and seasonal components
2. MCMC sampling: Iteratively update the model configuration through a series of move types. For example, add a new changepoint or remove an existing changepoint.
3. Posterior estimation: Approximate the posterior distributions for changepoint number, locations, and model parameters using the MCMC samples after a burn-in period.

The algorithm automatically balances model complexity and data fit through the Bayesian framework, avoiding the need for explicit penalty terms or cross-validation procedures for controlling overfitting.

Let's see the implementation of the BEAST method in the following sub-sections.

### 16.4.1 BEAST implementation on non-seasonal series

The `beast()` function from the `Rbeast` package performs the algorithm. We use the default settings for model posterior probability computation and MCMC sampling to maintain methodological consistency. The trend component analysis was constrained to test between 0 and 10 potential changepoints using the `tcp.minmax` parameter. For non-seasonal data, the season argument should be set to none.

Apply the BEAST algorithm to the `Nile` data and extract important information using the following code:

```
beast_nile <- beast(Nile, season = "none",
                    tcp.minmax = c(0, 10))

beast_nile_trend_cp <- tibble(
  location = beast_nile$trend$cp,
  prob = beast_nile$trend$cpPr,
  ci_lower = beast_nile$trend$cpCI[,1],
  ci_upper = beast_nile$trend$cpCI[,2],
  magnitude_change = beast_nile$trend$cpAbruptChange ) |>
  na.omit()

beast_nile_trend_cp
```

The summary in the following output presents the year of the changepoint (location), the probability, the confidence intervals, and the magnitude of change:

```
# A tibble: 9 × 5
  location   prob ci_lower ci_upper magnitude_change
     <dbl>  <dbl>    <dbl>    <dbl>            <dbl>
1     1899 0.968     1896.    1901.           -175.
2     1954 0.110     1950.    1957.              5.00
3     1966 0.0733    1962.    1968.             -2.77
4     1946 0.0725    1937.    1951.              2.28
5     1916 0.0644    1907.    1922.              1.61
6     1890 0.0350    1883.    1894.              1.65
7     1929 0.0349    1922.    1937.              0.559
8     1877 0.0345    1874.    1883.             -0.157
9     1961 0.0337    1956.    1964.              0.241
```

The summary is generated for a maximum of 10 changepoints. According to BEAST, there is only one statistically meaningful changepoint detected, in the year 1899 (prob=0.973), with a magnitude change of $-184$. The estimated magnitude change by BEAST is quite different compared to what BFAST and PELT measured for the same data.

An insightful plot generated from the fitted BEAST object is presented in *Figure 16.6*:

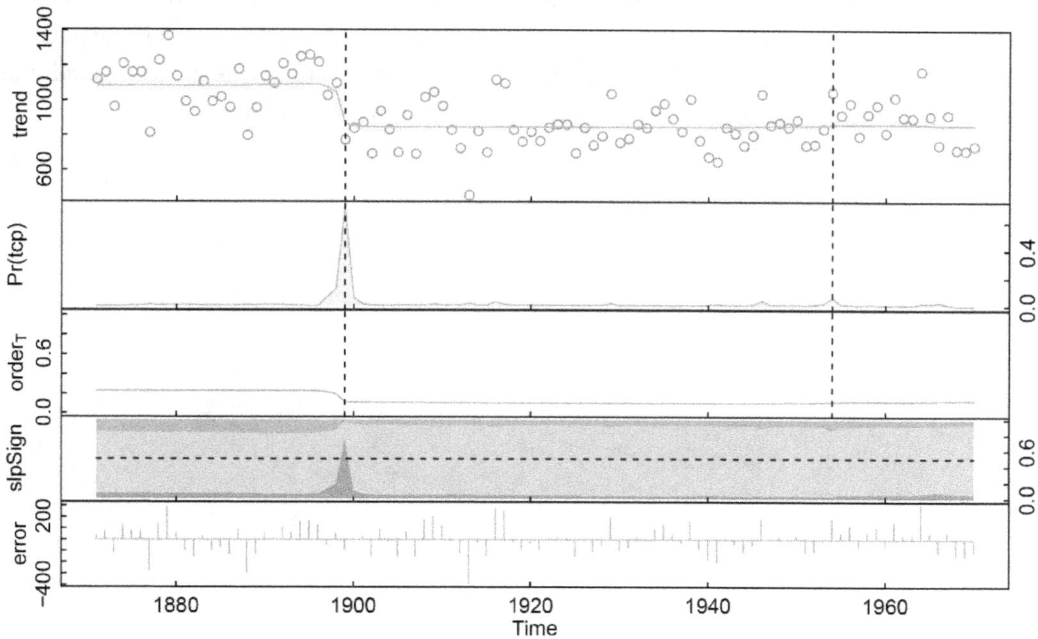


*Figure 16.6: Detected changepoint by BEAST from the Nile data*

The visualization in *Figure 16.6* reveals several interesting findings, such as the following:

- **Trend component**: The green trend line and dashed vertical changepoint markers (overlaid on observed data points) pinpoints 1899 as the significant changepoint year, which is consistent with both BFAST and PELT results. As a Bayesian method, BEAST treats changepoints as model parameters and estimates their posterior probability distribution, rather than identifying a single deterministic changepoint. This approach provides associated uncertainty in changepoint detection. For the identified changepoint in the year 1899, the 95% credible interval spans from June 1896 to April 1901.
- **Changepoint probability in trend**: The `Pr(tcp)` panel shows a distinct probability peak at 1898
- **Slope direction**: The `slpSign` panel indicates slope direction probabilities, with red for positive, blue for negative, and green for neutral (no change in trend).

Now, let's apply the BEAST method to the seasonal AirPassengers data.

## 16.4.2 BEAST implementation on seasonal data

To illustrate BEAST on a seasonal series, the algorithm is applied to the log of AirPassengers data. The logarithmic transformation is applied to transform a multiplicative seasonal pattern to additive. The `season="harmonic"` argument allows the algorithm to detect a changepoint in seasonal patterns. You can apply the method and extract important information from the model object using the following code:

```
beast_ap <- beast(log(AirPassengers),
                  season = "harmonic",
                  tcp.minmax = c(0, 10))

# changepoints in trend
beast_ap_trend_cp <- tibble(
  location = beast_ap$trend$cp,
  prob = beast_ap$trend$cpPr,
  ci_lower = beast_ap$trend$cpCI[,1],
  ci_upper = beast_ap$trend$cpCI[,2],
  magnitude_change = beast_ap$trend$cpAbruptChange
) |>
  na.omit()

# changepoints in seasonality
beast_ap_seas_cp <- tibble(
  location = beast_ap$season$cp,
  prob = beast_ap$season$cpPr,
  ci_lower = beast_ap$season$cpCI[,1],
  ci_upper = beast_ap$season$cpCI[,2],
  magnitude_change = beast_ap$season$cpAbruptChange
) |>
  na.omit()
```

Let's examine the plot presented in *Figure 16.7* generated by the BEAST method:

```
par(mar = c(2, 2, 1, 1), cex=0.8)
plot(beast_ap, main = "")
```

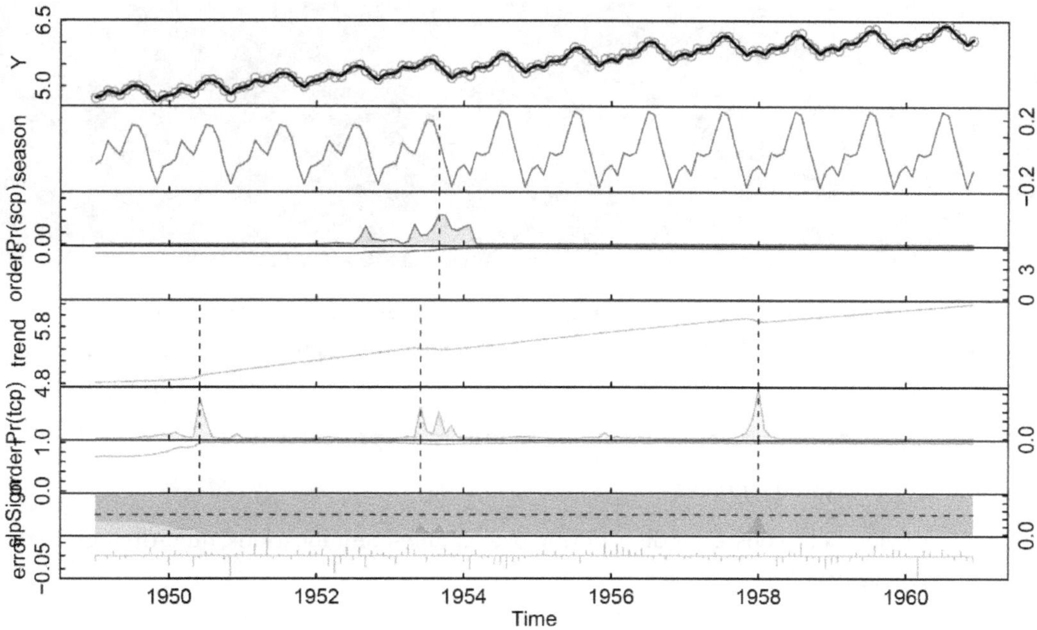


*Figure 16.7: Bayesian estimation of trend and seasonal changepoints in AirPassengers data*

From the plot, you can see that BEAST has identified three changepoints in the trend and one changepoint in the seasonal component. Recall that neither BFAST nor PELT could detect the seasonal changepoint for the same data. You can get the exact location of the trend changepoints from the following output:

```
> beast_ap_trend_cp

# A tibble: 6 × 5
  location   prob ci_lower ci_upper magnitude_change
     <dbl>  <dbl>    <dbl>    <dbl>            <dbl>
1    1958  0.979     1958.    1958.          -0.0412
2    1953. 0.969     1953.    1954.          -0.0353
3    1950. 0.798     1950.    1951.           0.0373
4    1956. 0.127     1956.    1957.           0.0145
5    1955. 0.0881    1954.    1955.           0.0135
6    1952. 0.0297    1952.    1952.           0.0121
```

The changepoint detected in the seasonal pattern is presented in the following output:

```
> beast_ap_seas_cp

# A tibble: 2 × 5
  location  prob ci_lower ci_upper magnitude_change
     <dbl> <dbl>    <dbl>    <dbl>            <dbl>
1    1954. 0.539    1953.    1954.         -0.00756
2    1953. 0.259    1952.    1953.         -0.0886
```

You will notice that more changepoints are detected in the numerical outputs; however, changepoints with a probability higher than 0.5 are considered significant and embedded in the plot.

The advantages and limitations of the BEAST algorithm are presented next.

The following are the advantages:

- It provides a full posterior distribution over changepoint locations, quantifying uncertainty
- It naturally incorporates prior knowledge about changepoint frequency or locations
- It handles multiple types of changes (trend and seasonal) in an integrated framework
- It is less sensitive to tuning parameters compared to frequentist methods

The following are the limitation:

- It is computationally intensive, especially for MCMC-based methods
- It requires specification of prior distributions
- The model complexity increases rapidly for multiple changepoint types
- Its interpretation can be more challenging than simpler methods
- Its implementation may require specialized knowledge of Bayesian statistics

So far, you have learned about three changepoint detection methods with implementation on both non-seasonal and seasonal datasets, along with their advantages and limitations. Are there situations where one method is better suited than the other? The guidelines in the next section will help you identify a suitable method for a business problem.

# 16.5 Method selection guidelines

In real-life applications, when choosing between changepoint detection methods, several factors should be considered beyond their advantages and limitations. These have been summarized and grouped under three major categories to guide you based on your requirements.

The first relates to data characteristics:

- Data size: For very large datasets, binary segmentation may be preferred due to its computational efficiency
- Noise level: Bayesian methods often perform better with noisy data due to their probabilistic nature
- Changepoint frequency: PELT is well-suited for detecting numerous changepoints
- Changepoint magnitude: Smaller shifts may be better detected by PELT or Bayesian approaches

The following are analysis objectives:

- Computational efficiency: Binary segmentation is fastest, followed by PELT and then Bayesian methods
- Theoretical guarantees: PELT provides optimality guarantees under its cost model
- Uncertainty quantification: The Bayesian method excels when uncertainty estimates are needed
- Online detection: Modified Bayesian approaches are particularly suitable for detecting changepoints in streaming data

The final grouping is implementation complexity:

- Technical expertise required: Binary segmentation is simplest, while Bayesian methods typically require advanced statistical and computing skills
- Parameter tuning: The Bayesian method typically requires a wider range of parameter tuning than PELT and BFAST.

# Summary

Changepoint detection represents a critical component of time-series analysis, enabling you to identify structural breaks that may significantly impact subsequent analyses and forecasts. Three methods, BFAST, PELT, and BEAST, were explored in this chapter, along with their unique strengths and limitations.

Binary segmentation (BFAST) provides an efficient and intuitive approach suitable for many applications, particularly when computational resources are limited or when dealing with a very long time series. PELT offers optimal segmentation guarantees while maintaining reasonable computational efficiency, making it an excellent choice for applications requiring high precision. Bayesian changepoint detection excels in quantifying uncertainty and incorporating prior knowledge, providing richer insights at the cost of increased computational complexity.

When applying these methods in practice, careful consideration should be given to the specific characteristics of a dataset, the analytics objectives, and the available computational resources. Often, employing multiple approaches and comparing their results can provide more robust insights than relying on a single method.

As time-series data continues to grow in volume and importance across domains, changepoint detection will remain an essential tool for understanding structural shifts, improving forecasting accuracy, and gaining insights into the underlying dynamics of complex systems.

For time series with pronounced trend and seasonality components, domain expertise becomes essential for the meaningful interpretation of detected changepoints. Blind application of methods without contextual understanding may produce ambiguous or misleading results when confronted with complex temporal patterns.

With this important reminder in mind, let's move on to the last chapter of the book, where we will learn how to detect anomalies in a time series and how to impute missing values, both of which are common situations in business.

# Further reading


- Killick, R., Fearnhead, P., & Eckley, I. A. (2012). *Optimal detection of changepoints with a linear computational cost. Journal of the American Statistical Association*, 107(500), 1590–1598.
- Verbesselt, J., Hyndman, R., Newnham, G., & Culvenor, D. (2010). *Detecting trend and seasonal changes in satellite image time series. Remote Sensing of Environment*, 114(1), 106–115.
- Zhao, K., Wulder, M. A., Hu, T., Bright, R., Wu, Q., Qin, H., Li, Y., Toman, E., Mallick, B., Zhang, X., et al. (2019). *Detecting change-point, trend, and seasonality in satellite time series data to track abrupt changes and nonlinear dynamics: A Bayesian ensemble algorithm. Remote Sensing of Environment*, 232, 111181.

# Join our community on Discord

Join our community's Discord space for discussions with the author and other readers:

`https://discord.gg/G4YmB6DQvf`

# 17

# Anomaly Detection and Imputation

Anomalies or outliers are events in time series that cause sudden deviation from observed systematic patterns for a usually short period. They can arise from various sources, such as equipment failures, extreme weather, or pandemics. Changepoints (discussed in *Chapter 16*) make persistent shifts in the underlying data-generating process. Anomalies, on the other hand, are sudden deviations in a single observation or a sequence, after which the long-term systematic behavior remains unchanged.

In certain situations, replacing anomalies with plausible values is a better approach, particularly when they arise from known, uncontrollable reasons that are unlikely to repeat. Examples include a spike in customer complaints during a one-time database migration or a surge in friendship-band sales during a celebrity music tour.

Completely removing anomalies from the data is generally not advised, as they often contain valuable information about critical events or system failures. Discarding them can make decision-makers blind to genuine disruptions that might signify emerging risks or unique opportunities. Furthermore, simply excluding these observations leaves gaps in the series that make analysis challenging unless filled with prudent methods. Consequently, this chapter explores the dual challenges of detecting anomalies and imputing missing values to preserve the temporal structure of the data.

Anomaly detection in time series presents unique challenges compared to cross-sectional data because "normal" behavior evolves through trend-cycle and seasonal patterns. Effective methods need to distinguish between these expected baseline variations and genuine anomalies. This chapter covers various techniques to address these challenges, highlighting their respective advantages and limitations. Understanding the conceptual background of these methods will allow you to choose the right tools to make informed decisions and solve complex business problems.

The chapter is organized as follows:

- Section 17.1: Anomaly detection in time series
- Section 17.2: STL decomposition-based anomaly detection
- Section 17.3: Isolation forest-based anomaly detection
- Section 17.4: Imputation in time series
- Section 17.5: Guidelines for selecting an anomaly detection method

Note

**Your purchase includes a free PDF copy + exclusive extras**

Your purchase includes a DRM-free PDF copy of this book, 7-day trial to the Packt+ library (no credit card required), and additional exclusive extras. See the *Free benefits with your book* section in the *Preface* to unlock them instantly and maximize your learning.

# Technical requirements

In this chapter, we will need the following packages:

```
#install.packages("pacman")
library(pacman)
p_load("conflicted"
       ,"astsa"
       ,"forecast"
       ,"anomalize"
       ,"patchwork"
       ,"solitude"
       ,"slider"
       ,"tidyverse"
       ,"tsibble"
       ,"tsfeatures"
      ,"wesanderson" )

conflicts_prefer(
  fabletools::accuracy
```

```
    ,fabletools::forecast
    ,dplyr::filter )
```

The complete code files are available in GitHub repo `https://github.com/PacktPublishing/Modern-Time-Series-Analysis-with-R.git`.

Let's start with a summary of time series anomaly detection in the next section.

# 17.1 Anomaly detection in time series

Anomaly detection in time series focuses on identifying deviations from the expected behavior or defined "normality" in the context of the data collection process. The deviation may happen at a single time point or last over a sequence of time points. The identification of anomalies is essential for rigorous analysis. Their presence can distort statistical measures, compromise the extraction and modeling of trend and seasonality, and compromise forecasting performance. By detecting these outliers, we can safeguard the integrity of the analysis, reduce noise, and better understand the underlying patterns. Efficient detection transforms these potential obstacles into strategic advantages by allowing us to control for adverse effects and strengthen business decisions.

Time series anomalies generally fall into two major categories:

- Point anomalies: Single and isolated observations that deviate from the expected pattern
- Collective anomalies: A sequence of observations that together constitute anomalous behavior compared to the observed pattern

Effective time series anomaly detection must account for both temporal dependencies and complex underlying patterns. Defining any anomaly as a deviation from an expected baseline is inherently difficult, as the baseline itself can be distorted by the presence of anomalies. This challenge is magnified in situations where the context for what constitutes "normal" is derived solely from the data rather than established through domain-specific knowledge.

In this chapter, we cover two methods of anomaly detection:

- Seasonal-trend decomposition with LOESS (STL) based detection: A decomposition approach that separates time series components to identify anomalies.
- Isolation forest: An ensemble machine learning method adapted for time series anomaly detection

Each method offers distinct advantages for specific scenarios. You will learn about their foundations and practical considerations to effectively apply anomaly detection in real business problems. Let's start with the STL-based method in the next section.

## 17.2 STL decomposition-based anomaly detection

STL was covered in *Chapter 7*. It provides a robust framework for separating a time series into its components, such as trend, seasonality and remainder. This well-known decomposition technique can be used for time series anomaly detection, too. STL is free from underlying model assumptions, can handle any seasonality type, can capture evolving seasonal patterns over time, and, most importantly, is robust to anomalies. This method can identify anomalies in different components while respecting the underlying patterns extracted from the data.

Recall that the STL decomposition assumes that the components are combined additively in a time series as follows:

$$y_t = T_t + S_t + R_t$$

Let's break this down:

- $y_t$ is the observed series
- $T_t$ is the trend component
- $S_t$ is the seasonal component
- $R_t$ is the remainder (or residual) component

When the components are combined multiplicatively in an observed series, it is transformed into additive form by taking the log before applying the STL decomposition.

The STL-based anomaly detection framework operates on a method designed around quantiles of the distribution of the remainder component, $R_t$. The steps involved in the method are as follows:

1. Decompose the time series using STL and extract the remainder component
2. Establish thresholds for the remainder component: Derive lower and upper limits:

   $$\begin{aligned} \text{LL}_R &= Q_k(R) - \text{SF} \times \left(IQ(R)\right) \\ \text{UL}_R &= Q_{100-k}(R) + \text{SF} \times \left(IQ(R)\right) \end{aligned}$$

   Here, $Q_k(R)$ is the $k$th quantile for $R$.

   $IQ(R) = Q_{100-k} - Q_k$ is the range between quantiles. The typical choice of $k = 25$ makes $IQ$ the interquartile range. SF is a scaling factor, typically chosen between 2 and 3. This determines the width of the threshold and thus the sensitivity of detection.

3. Identify anomalies in each component. Flag observations where the component exceeds its respective threshold. You might like to classify anomalies based on which component contains anomalous values.

Let us apply this anomaly detection technique on the New York Harbor conventional regular gasoline weekly spot price (in cents per gallon) from 2000 to mid-2010 (545 observations) from the `astsa` package. This is seasonal data with a frequency of 52.

In previous chapters, you used the STL implementation from the `fable` package. However, in this chapter, we use the `stats::stl()` function to be consistent with the tidy workflow adopted in the `anomalize` package that we will use to detect anomalies later.

The seasonal component is estimated assuming the same seasonal impact across years by the `s.window = "periodic"` argument. The `robust=TRUE` option ensures that the components are extracted even if there are anomalies in the data. You can use the following code to fit STL decomposition on the gas data and produce a decomposition plot:

```
astsa::gas |>
  stats::stl(s.window = "periodic", robust = TRUE) |>
  autoplot()
```

The plot is presented in *Figure 17.1*:

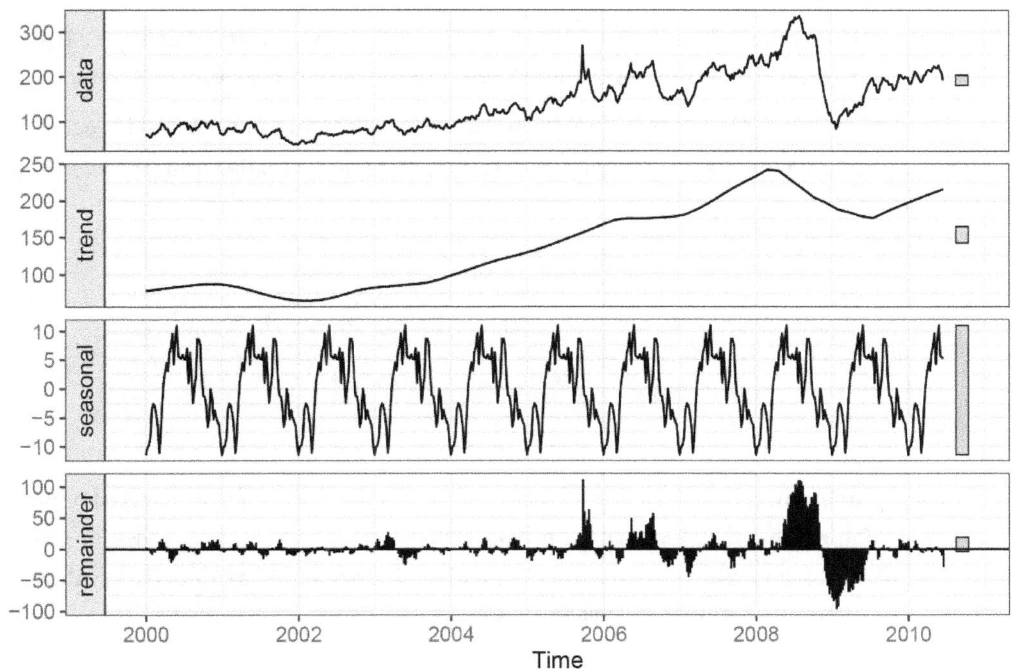


*Figure 17.1: Time plot of gasoline price and its components*

The time plot in the top panel of *Figure 17.1* has an upward trend and a sudden drop in the gas price around 2009. The series is highly seasonal with a consistent weekly pattern, as seen in the third panel. The remainder component had a low variation with noticeable differences occurring from late 2005, which peaked around 2009.

Let's investigate whether there is any anomaly in the gas series. First, we need to convert the data into `tibble` format, as shown in the following code snippet:

```
gas_tbl <- astsa::gas %>%
  as_tsibble() |>
  mutate(date = as.Date(index)) |>
  as_tibble() |>
  dplyr::select(-index, gas_price = value)
```

Next, the anomaly detection process progresses by applying three functions from the `anomalize` package, as follows:

- The `time_decompose()` function, along with the frequency argument, decomposes the original time series.
- Then, the anomaly detection on the remainder term $R_t$ is performed using the IQR-based method via setting the `method="iqr"` argument in the `anomalize()` function.
- Lastly, the `time_recompose()` function creates a lower and upper bound for each observation, which works by separating a threshold of anomalies from the expected observations.

The following code snippet shows how the framework works:

```
gas_anomalies_in_resid <- gas_tbl %>%
  time_decompose(gas_price, frequency = 52) %>%
  anomalize(remainder, method = "iqr", alpha = 0.15/3) |>
  time_recompose()
```

The alpha argument in the `anomalize()` function controls the width of the expected or normal range. The default value of alpha is $0.05 = 0.15/3$, which expands the limits established by expanding the inter-quartile range around the median ($k = 25$ or $25/75$ rule) by a scaling factor (SF) of 3 via the relationship $\text{alpha} = 0.15/\text{SF}$.

It is worth noting that the anomaly detection method can limit the potential maximum anomaly by setting a value for the `max_anoms` argument. This argument is, by default, set to 20%, the maximum percentage of permitted anomalies from a time series.

The combination of the functions produces a tibble with rows equal to the length of the observed series and the following columns:

```
> colnames(gas_anomalies_in_resid )

 [1] "date"           "observed"       "season"
 [4] "trend"          "remainder"      "remainder_l1"
 [7] "remainder_l2"   "anomaly"        "recomposed_l1"
[10] "recomposed_l2"
```

In the `gas_anomalies_in_resid` object, the anomaly column is a binary indicator variable. Once the anomalies are detected, we can visualize them using the `plot_anomalies()` function.

Let's plot the detected anomalies for two different choices of alpha when $SF = 2$ and $3$. This way, the sensitivity of the parameters on the detected anomalies can be easily examined. Any observation that falls outside of the produced limits is deemed an outlier. The following code shows you how to create a plot when $SF = 2$:

```
gas_anomalies_in_resid %>%
  plot_anomalies(time_recomposed = TRUE, alpha_dots = 0.5) +
  theme_minimal() +
  geom_line() +
  labs(title = "Detected anomalies for SF = 2",
       y = "Gasoline spot price",
       x = "Time (week of year)")
```

Let's compare the anomalies detected for two different choices of SF (2 and 3) presented in *Figure 17.2*, in which the grey shaded area represents the thresholds:

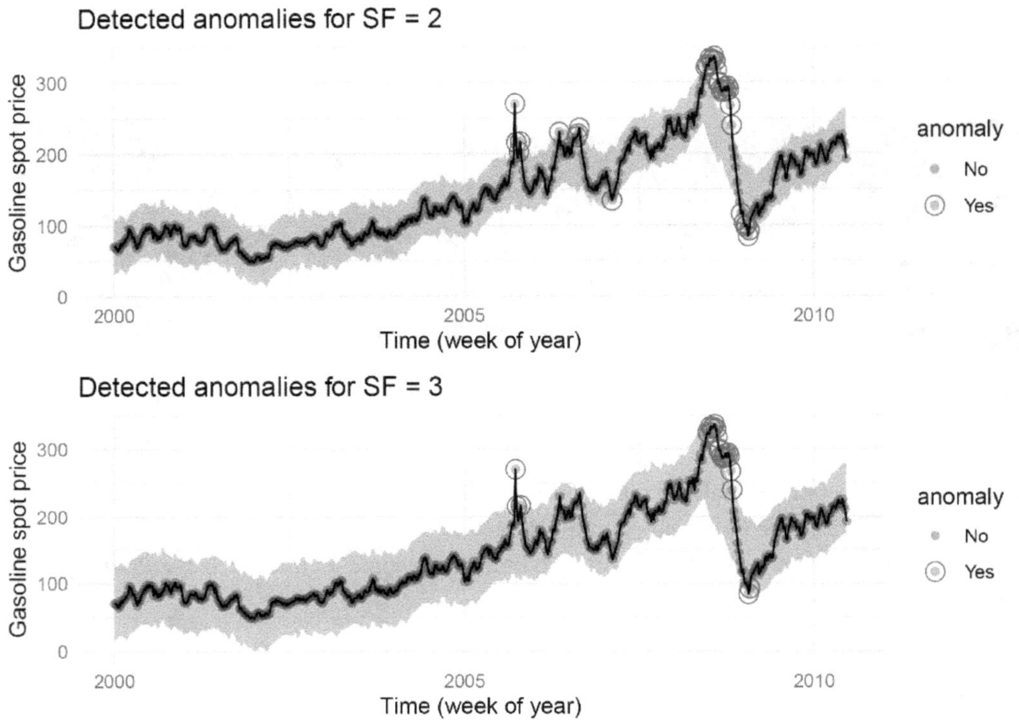


*Figure 17.2: Detected anomalies in the gasoline price for alternative choices of SF*

In the plot, the detected anomalies are highlighted by a red dot with a circle around it. As expected, $\text{SF} = 2$ (`alpha=0.5`) produces a narrow threshold and, consequently, detects more anomalies. On the other hand, the shaded area is wider when $\text{SF} = 3$ (`alpha=0.075` is smaller), resulting in fewer anomalies compared to $\text{SF} = 2$.

The advantages of the STL-based anomaly detection method are as follows:

- It respects the natural structure of time series by separating components
- It can detect anomalies in seasonal, trend, or remainder components separately
- It is robust to evolving seasonal patterns and trends
- It provides interpretable results by indicating which component contains the anomaly
- It works well with strong seasonal patterns where other methods might struggle

However, there are some limitations of the method worth mentioning:

- It requires the specification of the seasonality period in advance
- It may struggle with multiple seasonal patterns at different frequencies
- Its performance depends on appropriate parameter selection for the STL decomposition
- It is less effective for short time series with limited seasonal cycles
- It may not capture complex inter-dependencies between components

Next, you will learn about a machine learning-based approach for anomaly detection.

# 17.3 Isolation forest-based anomaly detection

Isolation forest (Liu et al., 2008) is an ensemble-based anomaly detection algorithm that is based on the principle that anomalies are often easier to isolate than non-anomalous observations. This method was originally designed for high-dimensional cross-sectional data; however, it can be adapted for time series by incorporating temporal features and contextual information in the algorithm.

This intuitively simple but clever algorithm creates a forest of random isolation trees. Each tree partitions the data randomly, and anomalies typically require fewer partitions to be isolated. The anomaly score is derived from the average path length required to isolate each observation.

For a time series $y_t, t = 1,..n$, the algorithm progresses as follows:

1. Feature extraction: Create sequential feature vectors $X_t = [y_t, y_{t-1}, \ldots, y_{t-w+1}]$ to capture temporal context. Such a feature vector includes immediate prior $w$ lagged values of the time series for a chosen width of the window $w$.
   In addition to sequential features, some other contextual features can be added, such as entropy (forecastability), lumpiness (variance of variances), non-linearity and so on.
2. Build isolation trees: For each tree in the ensemble, do the following:
   a. Select a random sub-sample or a bootstrap sample of the data.
   b. Recursively partition the sub-sample by randomly selecting a feature and a split value.
   c. Continue until all points are isolated or a specified depth is reached.
3. Compute anomaly scores: For each time point $t$, do the following:
   a. Pass $X_t$ through each isolation tree

b. Compute the average path length $h(X_t)$ across all trees

c. Calculate the anomaly score using the following:

$$s(X_t, n) = 2^{-\frac{E[h(X_t)]}{c(n)}}$$

Here, $c(n)$ is the average path length of an unsuccessful search in a binary search tree:

$$c(n) = 2H(n-1) - \frac{2(n-1)}{n}$$

$H(i)$ is the harmonic number, approximated by $\ln(i) + 0.5772$. This number helps normalize the anomaly score.

4. Anomaly detection: Finally, flag observations with anomaly scores above a threshold (typically $0.5$ to $0.6$) as anomalies.

In order to use an isolation forest on time series, we need to extract information from the time series and arrange it in a way that mimics a high-dimensional data scenario. The extracted information from the time series should encapsulate temporal and contextual information efficiently. In machine learning, such a data curation process is commonly referred to as feature engineering. It is important to understand that the distinction between a useful versus useless machine learning model can often be determined by the set of features curated in the feature engineering phase. Finding the right set of features is a common challenge in the machine learning process.

Let's implement an isolation forest on the New York gasoline spot price data. The feature set includes lagged values along with the following features (detailed discussion available in *Chapter 8*):

- Crossing points: defined as the number of times a time series crosses the median line
- Entropy: measures the "forecastability" of a time series, where low values indicate a high signal-to-noise ratio
- `Holt_parameters`: estimated smoothing parameter for level ($\hat{\alpha}$) and the smoothing parameter for trend ($\hat{\beta}$) of Holt's linear trend method
- Lumpiness: the variance of the variances for tiled (non-overlapping) windows of a width of 10 for this data
- Nonlinearity: takes large values when the series is nonlinear, and values around $0$ when the series is linear
- Stability is the variance of the means for tiled (non-overlapping) windows of width $10$ for this data

We create a custom function to compute relevant features within a sliding window. The following code shows the custom `create_features()` function that takes observed series, `ts_data`, and the width of the sliding window, `win_wid` as arguments:

```
# Create features for time-series context
create_features <- function(ts_data, win_wid) {
  n <- length(ts_data)
  X <- matrix(NA, nrow = n - win_wid + 1,
                  ncol = win_wid + 7)
  for (i in win_wid:n) {
    win_data <- ts_data[(i - win_wid + 1):i]
    row_idx <- i - win_wid + 1
    # Window values
    X[row_idx, 1:win_wid] <- win_data
    holtPars <- holt_parameters(win_data)
    # Additional cross Mean, SD, Max and Min respectively
    X[row_idx, win_wid + 1] <- crossing_points(win_data)
    X[row_idx, win_wid + 2] <- entropy(win_data)
    X[row_idx, win_wid + 3] <- holtPars["alpha"]
    X[row_idx, win_wid + 4] <- holtPars["beta"]
    X[row_idx, win_wid + 5] <- lumpiness(win_data)
    X[row_idx, win_wid + 6] <- nonlinearity(win_data)
    X[row_idx, win_wid + 7] <- tsfeatures::stability(win_data)
  }
  # Assign column names to X
  colnames(X) <- c(
    paste0("lag_", (win_wid - 1):0),
    "xpoints", "entropy", "holt_1", "holt_2",
    "lumpiness",  "nonlinearity", "stability")
  return(as.data.frame(X))
}
```

Now, apply the `create_features()` function on the gas data using window size same as the total number of weeks in a year, by running the following code:

```
win_wid <- ceiling(52.14)
gas_features <- create_features(astsa::gas, win_wid)
```

The `gas_features` object is a matrix of 493 rows and 60 columns. Computing features this way aligns with how this object needs to feed into the `isolation forest` function in the next step. We use the isolation forest algorithm implemented in the `solitude` package and set the arguments, number of trees (`num_trees=100`) and maximum depth (`max_depth=9`). The latter adopts the same formula used by the algorithm itself.

The `isolationForest()` function specifies the model, the `fit()` function trains the model, and the `predict()` function computes the anomaly score for each data point. The complete process is provided in the following code snippet:

```
# Train Isolation Forest model
iso_forest <- solitude::isolationForest$new(
  sample_size = nrow(gas_features),
  replace = TRUE,
  num_trees = 100,
  seed = 101,
  max_depth = ceiling(log2(nrow(gas_features)))
)

# fit
isof_fit <- iso_forest$fit(gas_features)

# predict
isof_pred <- iso_forest$predict(gas_features)

isof_pred
```

A snapshot of the prediction from the trained model is presented in the following output:

```
        id average_depth anomaly_score
     <int>         <num>         <num>
  1:     1          8.71     0.5930571
  2:     2          8.73     0.5923460
  3:     3          8.78     0.5905721
  4:     4          8.78     0.5905721
  5:     5          8.68     0.5941253
 ---
489:   489          8.83     0.5888035
490:   490          8.81     0.5895103
491:   491          8.87     0.5873924
492:   492          8.91     0.5859847
493:   493          8.84     0.5884504
```

The definition of anomaly depends on an arbitrary cut-off value of the anomaly score from the prediction. We allow a maximum of 10% of the data to meet the threshold for anomaly. Hence, the threshold score in this case is the 90-th percentile of the anomaly score derived from the isolation forest predictions.

The next step is to apply this threshold and identify possible anomaly observations. Using the maximum anomaly proportion (`max_anom_prop=10`), calculate the 90-th percentile of the score derived from the prediction (`anomaly_threshold`), and create a binary indicator column anomaly to pinpoint the time index when an anomaly has occurred under the given threshold. Executing the following code will organize the results:

```
# Subjective choice: Maximum 10% data are permitted to be anomaly
max_anom_prop <- 0.1

gas_isof_anomaly <- gas_tbl |>
  mutate(
    isof_anomscore = c(
      rep(NA, win_wid-1),
      isof_pred$anomaly_score
    )
  ) |>
  mutate(
    anomaly_threshold = quantile(
      isof_pred$anomaly_score,
      probs = 1 - max_anom_prop
    )
  ) |>
  mutate(
    is_anomaly = case_when(
      is.na(isof_anomscore)  ~  NA,
      isof_anomscore >
        anomaly_threshold & !is.na(isof_anomscore)  ~ "Yes",
      .default = "No"
    )
  )
```

For the gas data, the threshold for being anomaly (90-th quantile of the anomaly score) identified as:

```
gas_isof_anomaly$anomaly_threshold |>
  unique()
```

```
# [1] 0.614421
```

This threshold depends on the curated features set used, parameter choices in the isolation forest model, and the maximum anomaly proportion cut-off. Different combinations of these factors could detect less or more anomalies in the same data.

```
gas_isof_anomaly |>
  filter(is_anomaly == "Yes") |>
  select(gas_price, date, isof_anomscore)
```

In this example, 49 observations are detected as anomalous by the isolation forest as seen the following output:

```
# A tibble: 49 × 3
   gas_price date       isof_anomscore
       <dbl> <date>              <dbl>
 1      56.5 2002-01-15          0.622
 2      53.5 2002-02-12          0.618
 3      54.5 2002-02-26          0.621
 4      54.5 2002-03-12          0.619
 5      55.7 2002-03-19          0.622
 6      64.2 2002-03-26          0.615
 7     216.  2005-10-01          0.617
 8     194.  2005-10-08          0.632
 9     220.  2006-08-06          0.616
10     180.  2006-09-24          0.624
# i 39 more rows
```

We can visualize the detected anomalies using the following code:

```
gas_isof_anomaly |>
  ggplot() +
  geom_line(aes(x = date, y = gas_price)) +
  geom_point(aes(x = date, y = gas_price, col = is_anomaly)) +
  scale_color_manual(values = wesanderson::wes_palette("Royal1")) +
  labs(y = "Gasoline spot price cents/gallon",
       x = "Time (week of year)") +
  theme(legend.position = "top")
```

The plot is presented in *Figure 17.3* in which the red colored points are the detected anomalies:

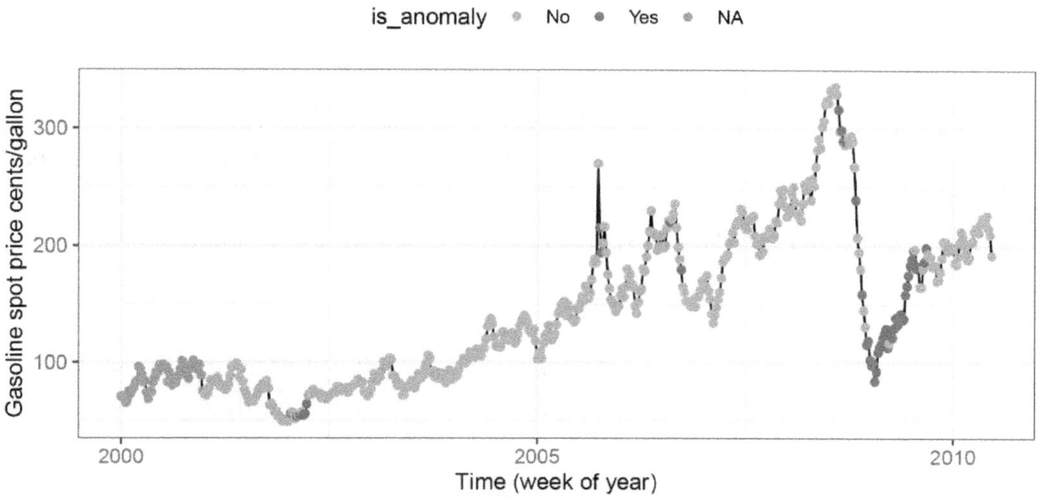


*Figure 17.3: Detected anomalies in the gasoline price using isolation forest*

In the plot, the legend "NA" is for the first window where feature engineering could not be performed, and consequently, the isolation forest could not detect any anomalies in that section of the data. The isolation forest-based method has identified most observations during 2009 as anomalies.

When compared, the isolation forest and the STL-based anomaly detection method highlight some overlaps as well as notable differences. In particular, a large spike after 2005 in the gas price was picked as an outlier by the STL method but not by the isolation forest. Such differences are not unexpected as both methods depend on some subjective decision variables. In addition, the isolation forest uses feature engineering, which has potential for improvement with deep domain knowledge.

The advantages of the isolation forest methods are as follows:

- It handles high-dimensional feature spaces effectively
- It does not make strong assumptions about data distribution
- It is computationally efficient
- It can detect both global and local anomalies
- It performs well with limited training data
- It can capture complex patterns when properly adapted for time series

The following limitations need to be considered when we apply the isolation forest method for detecting anomalies in practice:

- It requires careful feature engineering to capture temporal dependencies
- It may miss contextual anomalies if temporal features are poorly designed
- Its performance is affected by the selection of hyperparameters (window size, number of trees, etc.)
- It is not directly suited for detecting collective anomalies spanning multiple time points
- It may struggle with extremely noisy time series or highly complex seasonal patterns

Isolation forest for time series anomaly detection is particularly valuable in scenarios where many features can be generated and collected to explain the underlying data-generating process.

Once detected, a way to handle an anomaly is to exclude it. However, that leaves a gap in the observed time series. Missing values are also common when you work with real business data. It is important to handle those missing values with a proper method instead of just excluding them from further analysis and modeling time series. This leads to the topic of imputing missing values in time series in the next section.

# 17.4 Imputation in time series

A critical question practitioners often face after detecting anomalies is how to address the unwanted influence of anomalous data points in subsequent analysis. Instead of removing anomalies, it is advisable to use robust methods to reduce the influence of anomalous observations on the analysis. Logarithmic transformation is one of the simplest but most effective ways, as it reduces the impact of anomalous observations. However, the problem might get complicated in a collaborative environment in which domain knowledge about data cleaning is not adequately shared with the analytics team. In fact, this is a common setup in modern business-to-business (B2B) analytics environments, especially where data is provided as a service. In this workflow, when anomalies are detected and removed before sharing for analysis, the important information about the anomalies might get missed, too. The subsequent analysis requires a rigorous scientific approach to impute these missing values for maintaining data integrity and performing accurate analysis.

With knowledge of forecasting, one might be inclined to use a forecasting model to fill those gaps (imputation), using the forecasting model for forecasting or backcasting (applying a forecast on a reversed time series). However, it is important to recognize that the time series imputation method needs to be model-agnostic, meaning imputation should not be done under any model assumption. If a model-based imputation is performed, subsequent analysis and models built on the imputed data will favor the model used for imputation. Keeping the caveat in mind, we adopt the linear interpolation method for imputation using a tidy workflow. The following subsections cover methodology and implementation.

## 17.4.1 Linear interpolation-based imputation of anomalies

Linear interpolation is a method of predicting plausible values for anomalous data points using neighboring non-anomalous data points of a time series. The linear interpolation method can be used directly on non-seasonal series. For seasonal series, a robust STL decomposition is first computed. Then a linear interpolation is applied to the seasonally adjusted data, and the seasonal component is added back. The linear interpolation method is applicable to any non-linear series that can be transformed into a linear series.

The linear interpolation process (Hyndman, 2021) is a two-step process involving (i) detection of anomalies and (ii) linear imputation. The process begins by decomposing the time series into trend, seasonal, and remainder components using an additive model $y_t = T_t + S_t + R_t$. The seasonal component may contain several seasonal patterns corresponding to the seasonal periods in the data. For example, daily data might have a seasonality of $7$ and $365.25$ to allow for weekly and daily seasonality, which can be estimated iteratively using the `feasts::STL()` function.

From the resulting decomposition, the strength of seasonality is measured by calculating the quantity:

$$F_s = 1 - \frac{\mathrm{Var}(y_t - \hat{T}_t - \hat{S}_t)}{\mathrm{Var}(y_t - \hat{T}_t)}$$

If the strength of seasonality exceeds the threshold of $0.6$, a seasonally adjusted series is computed as $y_t^* = y_t - \hat{S}_t$; otherwise, set $y_t^* = y_t$.

The benefit of using a strength of seasonality threshold is to avoid over-fitting and not to forcefully estimate seasonality from noisy data when the underlying seasonality is too weak (or non-existent). Forcefully estimated seasonal components may mask anomalies by absorbing them.

In the next step, the trend component is re-estimated from $y_t^*$ values. For non-seasonal time series (e.g., annual), this is a necessity, whereas for series with weak or non-existent seasonality, this is immensely useful, as trend estimation was not performed accurately in the initial decomposition due to uncertainty around the existence of a seasonal component. This re-estimation of the trend $\hat{T}_t$ is performed by applying a super smoother (Friedman, 1984) to the $y_t^*$ data, a non-parametric regression method designed to adaptively smooth data while minimizing bias and variance trade-offs.

The anomalies are identified from the estimated remainder series $\hat{R}_t = y_t^* - \hat{T}_t$, using the quantile-based technique described in the STL decomposition-based anomaly detection section for the choice $SF = 3$. If the distribution of $\hat{R}_t \sim N(0, \sigma_R^2)$, the probability of an observation being identified as an anomaly is approximately $2.34 \times 10^{-6}$, i.e, $\approx 1$ in 427,000.

Anomalies identified by this method can be replaced by the linear interpolation method and repeated for identifying and imputing additional anomalies that might exist in the time series.

## 17.4.2 Tidy workflow for simultaneous anomaly detection and imputation

The `forecast` package in R provides a streamlined functionality of anomaly detection using STL decomposition and its imputation using linear interpolation in time series data. Anomaly detection and its imputation simultaneously can be viewed as cleaning of time series, and its automatic implementation is provided by the `forecast::tsclean()` function. If the `replace.missing` argument set to TRUE, the function removes anomalies and impute those using linear interpolation. The function expects input data in a `ts` format.

We will perform the anomaly detection and linear imputation on the New York weekly gasoline spot price dataset. First, we compute the cleaned (imputed any anomalous observations) gas price and then create a binary indicator to pinpoint its time index (location) by comparing the imputed series with the original series.

The following code shows how to implement the linear interpolation to detect anomalies and replace it with imputation:

```
gas_tsclean <- gas_tbl |>
  mutate(
    clean_gas_price = tsclean(
      astsa::gas, replace.missing = TRUE)
  ) |>
  mutate(is_anomaly = case_when(
    gas_price != clean_gas_price ~ "Yes",
    .default = "No"))
```

We make a time plot of the cleaned data to check whether missing data due to the exclusion of anomalies was successfully replaced by the imputation process presented in *Figure 17.4*:

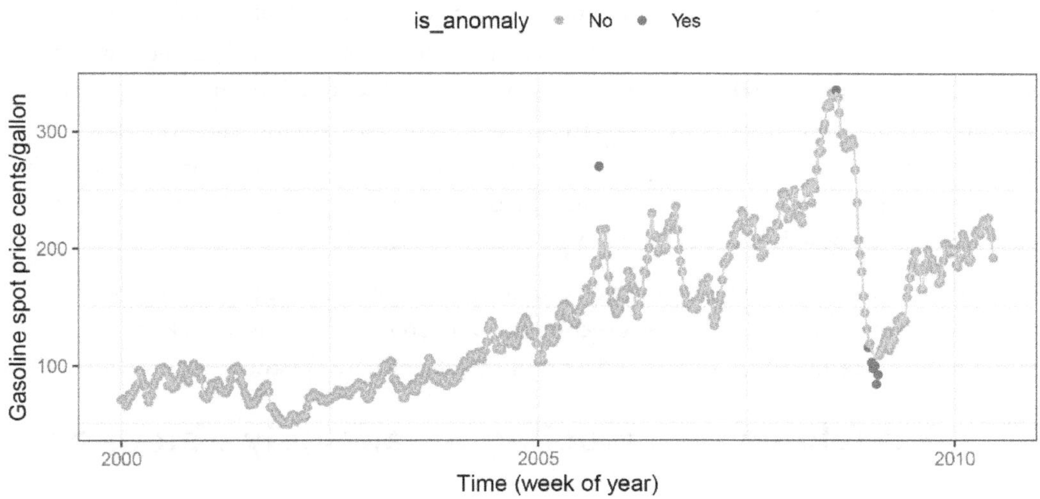


*Figure 17.4: Detected anomalies and imputed values by the linear imputation-based method on the gasoline price*

The plot shows that the spike after 2005 was successfully picked by the anomaly detection and imputed accordingly. Similarly, imputation were performed on the anomalous observations detected during 2009.

The total number of observations detected as anomalies and imputed by the linear interpolation method are small compared to the previous methods. The following output shows that for gas price data, linear interpolation flagged only 8 observations as anomalous as follows:

```
> gas_tsclean |> filter(is_anomaly == "Yes")

# A tibble: 8 × 4
  gas_price date       clean_gas_price is_anomaly
      <dbl> <date>               <dbl> <chr>
1     270.  2005-09-24            200. Yes
2     336.  2008-07-30            334. Yes
3     116.  2008-12-17            125. Yes
4     103.  2009-01-01            113. Yes
5      97.9 2009-01-08            111. Yes
6     100.  2009-01-15            108. Yes
```

```
7        84.2 2009-01-22             106. Yes
8        92.4 2009-01-29             110. Yes
```

It is worth noting that the number of anomalies detected by the `tsclean()` function is different from what we have seen in the STL decomposition-based anomaly detection section (*Section 17.1*). This is due to the difference in the process `tsclean()` adopts in the anomaly detection phase. These differences are relative strengths of the methods. The STL-based method solely focuses on detecting point and contextual anomalies, which can be tuned to be considered as anomalies that are consistent with domain knowledge. On the other hand, the linear imputation based method is a purely data-driven technique.

From the discussion of the various methods, which one should you use in your business problem to detect anomalies? The next section provides some useful tips to make an informative choice given the context of the problem and the data at hand.

# 17.5 Guidelines for selecting an anomaly detection method

In this chapter, we explored two distinct anomaly detection methods, STL-based and isolation forest. The linear imputation method, which simultaneously detect and replace anomalies, also use an STL-decomposition and assessing strength of trend and seasonal components for anomaly identification, and once detected they are replaced via a linear imputation technique.

Applying these techniques to the same dataset revealed noticeable differences in performance, highlighting how method selection can significantly impact results. The implementation of each approach depends on the careful selection of certain parameters, which often involves subjective judgment. However, these parameters alone do not encompass all the factors necessary for choosing a method in practice. When selecting an appropriate anomaly detection method, the following considerations should also be taken into account:

- Data characteristics
    - Seasonality: STL-based methods excel for time series with strong seasonal patterns, while isolation forest may require additional pre-processing for seasonal data
    - Trend: All methods can handle trends, but STL explicitly models the trend component, making it more interpretable

- Stationarity: Both STL and the adapted isolation forest can handle non-stationary data
- Data volume: Isolation forest scales well to large datasets, while STL decomposition may become computationally intensive for very long time series with complex seasonal patterns.

- Anomaly types
  - Point anomalies: All methods are effective at detecting individual outliers, with STL generally being the simplest option
  - Collective anomalies: Isolation forest with appropriate window features can detect patterns of anomalies, while standard STL methods may struggle with this task
- Implementation considerations
  - Computational resources: STL-based methods are typically the least computationally intensive, followed by isolation forest
  - Parameter tuning: All methods require parameter selection, with isolation forest often requiring the most domain expertise for effective calibration
  - Interpretability: STL provides the most interpretable results by separating anomalies by component, while isolation forest offers the least transparent decision mechanism
  - Online detection: Neither STL nor standard isolation forest can be used for online detection, but they are good for batch processing

In addition, the summarized criteria presented in *Table 17.1* will help you decide on a method.

| **Criteria** | **Selected method for treating anomalies in time series data** |
|---|---|
| The time series is relatively well-behaved, and computational efficiency is the prime concern | STL-based method, especially when domain expertise is not available for parameter tuning in an isolation forest |
| The time series exhibits strong seasonal patterns | STL-based detection to separate and analyze anomalies in different components |
| The time series is complex with multiple interacting features or robust performance is expected with minimal assumptions | Isolation forest with carefully engineered temporal features |

| Criteria | Selected method for treating anomalies in time series data |
|---|---|
| Multiple types of anomalies need to be detected, or uncertainty quantification is important | Consider combining methods or using ensemble approaches that leverage the strengths of each method |

*Table 17.1: Assessment criteria to choose between anomaly and imputation methods*

# Summary

In this chapter, you learned about anomaly detection and missing value imputation, essential tools for navigating today's dynamic business landscape. You looked at two anomaly detection methods: the STL-based anomaly detection technique and isolation forest. Both methods were presented and implemented in the same series to compare their performance. Following that, we discussed the linear imputation method to handle missing values in the observed series. Finally, we presented a method that operates iteratively to detect anomalies and impute using linear interpolation. A tidy framework showed how to implement this method.

# Further reading


- Friedman, J. H. (1984). *A variable span scatterplot smoother.* SLAC-PUB-3477, STAN-LCS 005.
- Hyndman, R. J. (2021). *Detecting time series outliers.* `https://robjhyndman.com/hyndsight/tsoutliers/`
- Liu, F. T., Ting, K. M., & Zhou, Z.-H. (2008). *Isolation forest. 2008 Eighth IEEE International Conference on Data Mining,* 413–422.


# Get this book's PDF version and more

Scan the QR code (or go to `packtpub.com/unlock`). Search for this book by name, confirm the edition, and then follow the steps on the page.

UNLOCK NOW

*Note: Keep your invoice handy. Purchases made directly from Packt don't require an invoice.*

# 18

# Unlock Your Exclusive Benefits

Your copy of this book includes the following exclusive benefits:

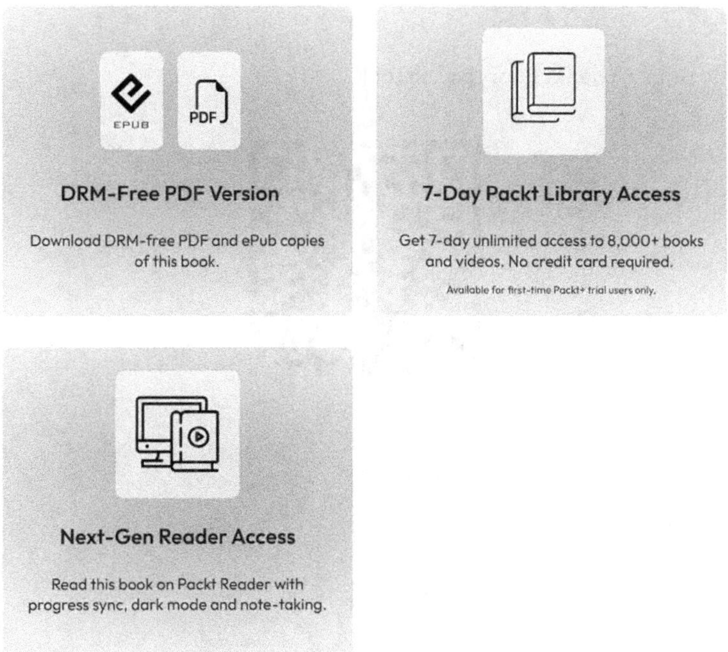


Follow the guide below to unlock them. The process takes only a few minutes and needs to be completed once.

# Unlock this Book's Free Benefits in 3 Easy Steps

## Step 1

Keep your purchase invoice ready for *Step 3*. If you have a physical copy, scan it using your phone and save it as a PDF, JPG, or PNG.

For more help on finding your invoice, visit `https://www.packtpub.com/en-us/unlock?step=1`.

> Note
>
> **Note**: If you bought this book directly from Packt, no invoice is required. After *Step 2*, you can access your exclusive content right away.

## Step 2

Scan the QR code or go to `packtpub.com/unlock`.

On the page that opens (similar to *Figure 18.1* on desktop), search for this book by name and select the correct edition.

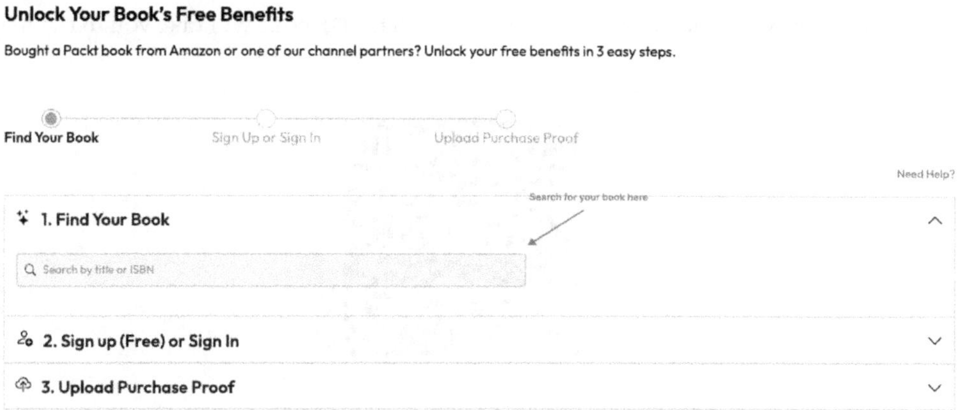


*Figure 18.1: Packt unlock landing page on desktop*

## Step 3

After selecting your book, sign in to your Packt account or create one for free. Then upload your invoice (PDF, PNG, or JPG, up to 10 MB). Follow the on-screen instructions to finish the process.

## Need Help

If you get stuck and need help, visit `https://www.packtpub.com/unlock-benefits/help` for a detailed FAQ on how to find your invoices and more. This QR code will take you to the help page.

> Note
>
> **Note**: If you are still facing issues, reach out to `customercare@packt.com`.

# Other Books You May Enjoy

If you enjoyed this book, you may be interested in these other books by Packt:

**Time Series Analysis with Spark**

Yoni Ramaswami

ISBN: 9781803232256

- Understand the core concepts and architectures of Apache Spark
- Clean and organize time series data
- Choose the most suitable modeling approach for your use case
- Gain expertise in building and training a variety of time series models
- Explore ways to leverage Apache Spark and Databricks to scale your models
- Deploy time series models in production
- Integrate your time series solutions with big data tools for enhanced analytics
- Leverage GenAI to enhance predictions and uncover patterns

**Time Series Analysis with Python Cookbook**

Tarek A. Atwan

ISBN: 9781805124283

- Understand what makes time series data different from other data
- Apply imputation and interpolation strategies to handle missing data
- Implement an array of models for univariate and multivariate time series
- Plot interactive time series visualizations using hvPlot
- Explore state-space models and the unobserved components model (UCM)
- Detect anomalies using statistical and machine learning methods
- Forecast complex time series with multiple seasonal patterns
- Use conformal prediction for constructing prediction intervals for time series

## Packt is searching for authors like you

If you're interested in becoming an author for Packt, please visit `authors.packtpub.com` and apply today. We have worked with thousands of developers and tech professionals, just like you, to help them share their insight with the global tech community. You can make a general application, apply for a specific hot topic that we are recruiting an author for, or submit your own idea.

## Share your thoughts

Now you've finished *Modern Time Series Analysis with R*, we'd love to hear your thoughts! Scan the QR code below to go straight to the Amazon review page for this book and share your feedback or leave a review on the site that you purchased it from.

`https://packt.link/r/1805124013`

Your review is important to us and the tech community and will help us make sure we're delivering excellent quality content.

# Index

## A



## B



## C

## H



## I



## J



## K



## L



## M

## Q



## R



## S

# U



# V



# W

www.ingramcontent.com/pod-product-compliance
Lightning Source LLC
LaVergne TN
LVHW060011210926
848481LV00013B/3688